ROUTLEDGE LIBRARY EDITIONS:
POLITICAL GEOGRAPHY

Volume 12

POLITICAL FRONTIERS AND BOUNDARIES

POLITICAL FRONTIERS AND BOUNDARIES

J. R. V. PRESCOTT

Routledge
Taylor & Francis Group
LONDON AND NEW YORK

First published in 1987

This edition first published in 2015
by Routledge
2 Park Square, Milton Park, Abingdon, Oxon, OX14 4RN

and by Routledge
711 Third Avenue, New York, NY 10017

Routledge is an imprint of the Taylor & Francis Group, an informa business

British Library Cataloguing in Publication Data
A catalogue record for this book is available from the British Library

ISBN: 978-1-138-80830-0 (Set)
eISBN: 978-1-315-74725-5 (Set)
ISBN: 978-1-138-81419-6 (Volume 12)
eISBN: 978-1-315-74765-1 (Volume 12)
Pb ISBN: 978-1-138-81420-2 (Volume 12)

Publisher's Note
The publisher has gone to great lengths to ensure the quality of this reprint but points out that some imperfections in the original copies may be apparent.

Disclaimer
The publisher has made every effort to trace copyright holders and would welcome correspondence from those they have been unable to trace.

POLITICAL FRONTIERS AND BOUNDARIES

J. R. V. Prescott
Department of Geography
University of Melbourne

London
UNWIN HYMAN
Boston Sydney Wellington

Published by the Academic Division of
Unwin Hyman Ltd
15/17 Broadwick Street, London W1V 1FP, UK

Unwin Hyman Inc.
955 Massachusetts Avenue, Cambridge, MA 02139, USA

Allen & Unwin (Australia) Ltd
8 Napier Street, North Sydney, NSW 2060, Australia

Allen & Unwin (New Zealand) Ltd
in association with the Port Nicholson Press Ltd
Compusales Building, 75 Ghuznee Street, Wellington 1, New Zealand

First published in 1987
First published in paperback 1990

British Library Cataloguing in Publication Data

Prescott, J. R. V.
Political frontiers and boundaries.
1. Boundaries
I. Title II. Prescott, J. R. V. Boundaries and frontiers
320.1'2 JC323
ISBN 0–04–445948–3

Library of Congress Cataloging in Publication Data

Prescott, J. R. V. (John Robert Victor)
Political frontiers and boundaries.

Rev. and expanded ed. of: Boundaries and frontiers. 1978.
Includes bibliography and index.
1. Boundaries. I. Prescott, J. R. V. (John Robert Victor). Boundaries and frontiers. II. Title.
JC323.P723 1987 320.1'2 86–32092
ISBN 0–04–445948–3 (alk. paper)

Typeset in 10 on 12 point Bembo
by Columns, Caversham, Reading
and printed in Great Britain by Mackays of Chatham

This book is dedicated to
Greg, Margaret, Matthew,
and Elise

Preface

This book consists of two parts. The first six chapters treat frontiers and boundaries in a systematic fashion. These chapters are based on the work originally published as *The geography of frontiers and boundaries* by Hutchinson in 1965 and revised in *Boundaries and frontiers* published by Croom Helm in 1978.

All the chapters have been rewritten to take account of new ideas and new publications in this field. The Introduction has been completely rewritten. The chronological review of the works in this field has been changed to an examination of the major themes developed in the literature. Two additional sections have been written in the same style dealing with maritime boundaries and boundaries in the air. The chapter on maritime boundaries has been completely rewritten to deal with the new régime for the oceans provided in the 1982 Convention on the Law of the Sea.

The remaining seven chapters treat frontiers and boundaries in a regional manner. Each chapter deals with the principal characteristics of the evolution of boundaries in one major region of the world and provides an account of the remaining serious boundary disputes.

Throughout the text, wherever possible I have used place names as they appear in the comprehensive edition of *The Times atlas of the world* published in 1980. When names do not appear in this atlas I have used the form employed by the country concerned.

I willingly acknowledge my debt to Professors Harm de Blij and Jean Gottmann for many helpful suggestions on the content and structure of the book. For helping me in the preparation of this study I thank Rob Bartlett for drawing the maps so skilfully and Wendy Nicol for photographing so well the finished maps and the maps needed for research. I am also grateful to Patrick Singleton of the Baillieu Library of the University of Melbourne for his considerable assistance.

VICTOR PRESCOTT
University of Melbourne

Contents

List of figures

1 *Introduction*

Land boundaries

Political frontiers and boundaries separate areas subject to different political control or sovereignty. Frontiers are *zones* of varying widths which were common features of the political landscape centuries ago. By the beginning of the 20th century most remaining frontiers had disappeared and had been replaced by boundaries which are *lines*. The divisive nature of frontiers and boundaries has not prevented them from forming the focus of interdisciplinary studies by lawyers, political scientists, historians, economists, and geographers. Scholars from these fields have produced a rich literature dealing with frontiers and boundaries. Any survey of this extensive literature will reveal that the following themes have attracted the most attention.

National histories and international diplomacy

When the histories of countries are unravelled it is plain that most of them did not emerge at one time within a single set of international limits which have remained unchanged. That is certainly the case with countries in Europe, North and South America, and Asia. Although it is true that many African countries, such as Somalia and Mozambique, became independent within a set of boundaries which have not been altered, research into their colonial antecedents reveals a variety of colonial boundaries. A significant part of the history of several countries concerns the struggle for territory, and the identification of previous national boundaries on a single map provides a shorthand account of stages in the progress to the present pattern of states. Often the events which established new boundaries were sufficiently important to mark the division between important periods in the political history of countries or in the diplomatic and military history of continents.

This point can be illustrated by Figure 1.1, which shows the boundaries of Greece since 1832, the year when the modern state of Greece emerged from nearly three centuries of Turkish rule. The rebellion began in the Peloponnisos in April 1821, and this area was quickly cleared of Turks. The Greek sailors also enjoyed success against the Turkish navy in the Aegean Sea, but the tide of rebellion had been halted by June 1827 and other foreign powers decided it was time to intervene to avoid continued instability in this region. The destruction of the allied Turkish and Egyptian fleets by naval squadrons from England, France, and Russia in

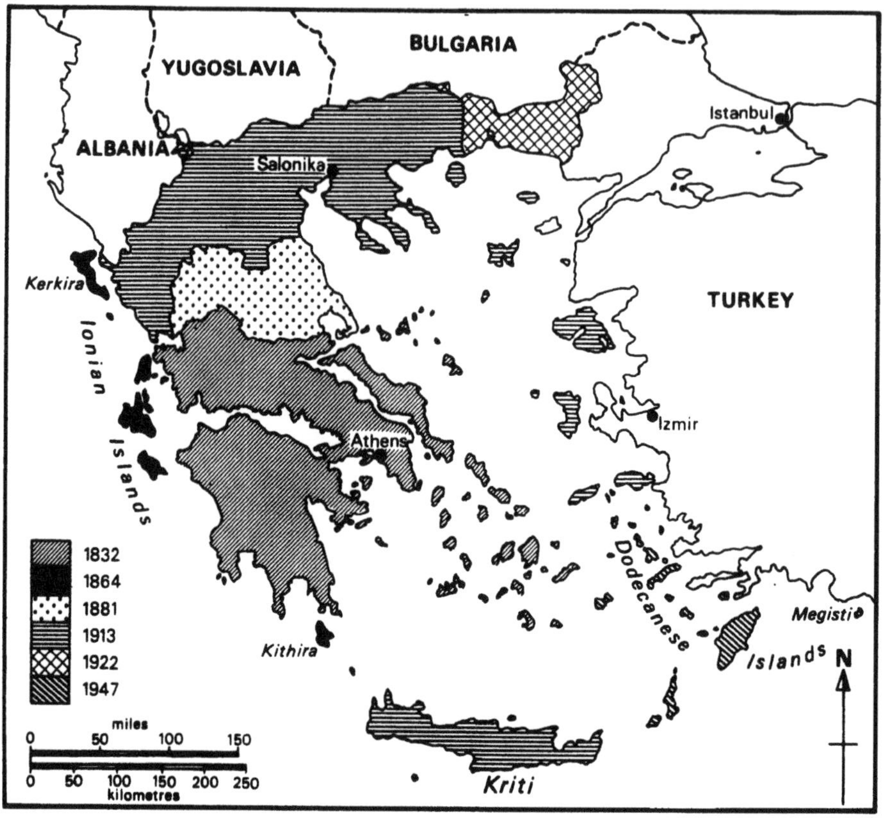

Figure 1.1 The evolution of modern Greece.

Navarino Bay on 20 October 1827 paved the way for an enforced settlement on the terms of these countries. In 1830 it was proposed that the northern boundary of Greece should run southwestwards from the vicinity of Lamia to Mesolongion, but Palmerston then persuaded his allies that the line should run from Pagasitikos Kolpos to Amvrakikos Kolpos, as shown in Figure 1.1. However, Palmerston defeated proposals to give to Greece the islands of Samos and Kriti on the grounds that the former was too close to the Turkish coast and the latter was too valuable and had a large Turkish population which should not be subject to Greece.

On 29 March 1864, six months after Greece had installed a Danish King, Britain ceded the Ionian Islands to Greece. These islands stretch down the west coast of Ipiros and Peloponnisos from Kerkira (Corfu) in the north to Kithira in the south. These islands had been formed into the United States of the Ionian Islands in 1815 at the Treaty of Versailles and placed under the protection of Britain. Their union with Greece followed a unanimous vote of support by the Legislative Assembly in the islands.

The war between Russia and Turkey in 1877 presented Greece with an opportunity to claim the Greek provinces in Turkish Europe, but the decision to take action was delayed so long that the Greek army had no chance to march before the war was ended a month later. The peace treaty signed by Russia and Turkey at San Stefano on 3 March 1878 contained no territorial gains for Greece. Fortunately for that country the other major powers in Europe were dissatisfied with the territorial arrangements which Russia had forced on Turkey, and the treaty was revised at the Congress of Berlin attended by all the major powers in June and July 1878. Greece was not represented at this Congress, but Britain persuaded the other powers to require Turkey to make concessions to Greece along their common boundary. At first it was proposed that the new boundary should run from Stoupi in the east to the mouth of the Thiamis River in the west, opposite Kerkira. This would have given Greece the region of Thessalia and most of Ipiros. Turkey could not resist the cession of the former region with its pronounced Hellenic character, but it was able to retain the areas of Janina and Preveza, which then constituted most of modern Ipiros. The arguments in favour of Turkish control of this area centred on their large Moslem minorities. The final treaty in this period was signed on 24 May 1881, although Greece was not a party to these arrangements.

The Greek authorities overplayed their hand in 1897 when they tried to force further concessions from Turkey in Kriti and on the continent. The European powers had to intervene to prevent a Turkish victory, and Greece was forced to cede 11 small areas which had particular strategic significance along its northern boundary to Turkey. The scale of Figure 1.1 does not allow these areas to be shown.

The next major territorial advance for Greece came in 1912 and 1913 in wars first with Turkey in alliance with Bulgaria and Serbia and then with Bulgaria in alliance with Serbia. These campaigns enabled Greece to move northwards to its present boundary and eastwards along the Macedonian coast as far as the Nestos River. In the Aegean Sea, Greece gained many Turkish islands, apart from Gokceada and Bozcaada, which Turkey retained, and the Dodecanese Islands, which had been occupied by Italy in 1912. Italy seized these islands from Turkey in April and May 1912 as part of its campaign against Turkey over control of Tripoli and Cyrenaica.

The rôle of Greece in World War I was rewarded with Bulgaria's coastlands on the northern shore of the Aegean Sea in western Thraki (Thrace) and the Turkish area of eastern Thraki as far east as Catalca within 20 miles of Istanbul. A large area of the Turkish mainland around Izmir (Smyrna) was placed under Greek control, and it was arranged that a plebiscite in five years could opt for union with Greece. The Treaty of Sevres, which conferred these gains on Greece, was never ratified, and the

Treaty of Lausanne on 24 July 1924 left Greece only with western Thraki. This dramatic change had been produced by the revival of a nationalist government in Turkey, agreements between that government and the Soviet Union and France, and by Greek military defeats at the hands of Turkish forces. Turkey was also helped by the unpopularity of King Constantine with the western powers because of his pro-German policy of 1915. The Treaty of Lausanne marked a retreat from the line marking the largest territorial extent of modern Greece and Turkey's boundary of 1914 along the Meric River was restored.

The final extension of Greece occurred in 1947 when Italy ceded the Dodecanese Islands and the island called Megisti, which Italy had obtained from France, in 1920.

In tracing the evolution of national boundaries there is no substitute for the treaties, protocols, agreements, and conventions which specify the formal arrangements between states. Sometimes these can be difficult to find, but there is now a fairly comprehensive range of publications dealing with this subject. For the period from 1648, which is 'classically regarded as the date of the foundation of the modern system of States' (Parry 1969, Vol. 1, p. 3), until 1919 Parry has edited a collection of treaties filling 231 volumes. From 1920 to the present many treaties are published in the *League of Nations Treaty Series* and its successor the *United Nations Treaty Series*. These collections contain only treaties lodged with these international organizations and therefore they are not comprehensive. One serious problem in interpreting treaties relates to the need to consult maps used by the negotiators and often published with the treaty. The collection by Parry does not include maps; the other two treaty series sometimes do. The importance of using maps which were contemporary at the time the treaty was signed turns on the facts that modern place names might be completely different and sometimes the maps were inaccurate and those inaccuracies were carried into the written description of the boundary. For example, the boundary between British Rhodesia and Portuguese Mozambique south of parallel 18°30′ south was defined in the following terms on 11 June 1891:

> . . . thence it follows the upper part of the eastern slope of the Manica plateau southwards to the centre of the main channel of the Sabi . . . (Brownlie 1979, p. 1119)

The demarcation team sent by both countries to fix this line could not agree on the location of the upper part of the eastern slope of the Manica plateau and the matter was referred to the arbitration of Senor Vigliani, an Italian senator. He handed down his decision on 30 January 1897 (The Geographer 1971, p. 2). The original negotiators can be excused for producing such an uncertain definition in view of the maps in the archives in London and Harare. It is quite clear that the maps and accompanying cross-sections sent by officers in the field show the edge of the plateau as a prominent and apparently unmistakable feature.

Fortunately there are also regional guides to the literature and treaties dealing with the evolution of national boundaries. Nicholson (1954) and Paullin (1932) have provided detailed accounts of international boundaries in North America, and the atlas edited by Paullin contains several useful, large-scale coloured maps. Ireland (1938) has provided a comprehensive account of the stages by which boundaries in South America evolved before World War II. Brownlie (1979) has published an encyclopedia which records the boundary agreements determining the boundaries of Africa and includes clear maps of the present period. Brownlie does not provide the detailed background to boundary evolution which Ireland makes available and which is also included in the study of the evolution of boundaries in Asia by Prescott (1975). Documents connected with national claims in Antarctica are published in the collection prepared by Bush (1982) and the background to the various agreements and proclamations is found in the analysis by Prescott (1984).

There is no equivalent work for Europe or for Central America. Hertslet (1875–91) has edited a useful collection of boundary treaties for Europe, including several indispensable maps. Fortunately most of the gaps in the literature have been filled by individual studies produced by The Geographer of the State Department of the United States of America. Under the general title of *International boundary study* this office has now issued nearly 200 separate analyses.

Contemporary international conflict and co-operation

Boundaries represent the line of physical contact between states and afford opportunities for co-operation and discord. Lord Curzon of Kedleston, who was a Viceroy of India, summed up this situation in words that have often been quoted:

> Frontiers are indeed the razor's edge on which hang suspended the modern issues of war and peace. (Curzon 1907, p. 7)

Most commentators use these words to introduce discussion of boundary conflicts, but Curzon's reference to peace should not be forgotten. Regrettably, newspapers and radio and television bulletins seem to regard boundary disputes as more worthy of attention than co-operation to solve boundary issues. In fact both conflict and co-operation over boundaries are important subjects for study. Indeed a survey of the volumes in the *United Nations Treaty Series* (United Nations 1945–) reveals dozens of treaties dealing with co-operation between states along common boundaries, as the following examples show.

On 23 October 1950 Belgium and the Netherlands defined an underground mining boundary in the vicinity of the Meuse which was independent of their boundary on the surface. This was done to reduce to a minimum the amount of exploitable coal which had to be left in the ground (United Nations 1964, Vol. 507, pp. 207–2). Austria and Yugoslavia agreed on 19 March 1953 to create frontier strips on each side of their common boundary. These strips consisted of 195 Austrian communes and 41 political communes in Yugoslavia. Citizens living within these zones were entitled to cross the boundary without the usual formalities if they owned property which straddled the boundary or if they were concerned with herding livestock or forestry on the opposite side of the line. This agreement specified 34 crossing points which could be used for this purpose (United Nations 1963, Vol. 467, pp. 380–427).

On 17 May 1963 Burma and Thailand set up a joint committee at ministerial level to confer and agree on measures to strengthen border security, to solve specific boundary problems, and to devise measures to promote economic and cultural co-operation (United Nations 1963, Vol. 468, pp. 320–9). Later, on 30 November 1963, Romania and Yugoslavia signed an agreement dealing with power generation and improved navigation on the River Danube between Sip and Gura Vaii in the vicinity of the Iron Gate. This defile in the Danube valley was made navigable in the 1890s, and the new agreement provided for the construction of two locks, two power plants, a spillway dam, and roads and railways (United Nations 1964, Vol. 512, pp. 42–66).

On 5 November 1977 India and Bangladesh signed an agreement which allocated water in the Ganges to each country (*Keesing's Contemporary Archives* 1978, p. 28763). This agreement settled a dispute which had arisen when India built the Farakka Barrage to divert water into the Hooghly River so that sediments could be flushed into the Bay of Bengal. The agreement defined a period of low flow in the Ganges River from January to May. This period was then divided into intervals of ten days and the amount which each side could draw in each interval was fixed. For part of this dry period the amount discharged into the Hooghly River would be less than 40 000 cubic feet per second which is considered the minimum flow necessary to prevent silt accumulation in the port of Calcutta.

In his study of claims to territory and their importance in international affairs, Hill (1976, p. 3) noted that boundary disputes have been conspicuous amongst the causes of war. Recent border wars in Asia can be traced back many years to boundary agreements which later proved unsatisfactory. The 1962 war between China and India centred on different interpretations of the McMahon Line in the eastern Himalayas. This line was settled in an agreement reached between Britain and Tibet in 1914 and was named after the British chief delegate. The war that began in September 1980 between Iraq and Iran was caused by Iraq's desire to regain control of the Shatt al Arab. This river had been awarded to the

Ottoman Empire by Persia in the second Treaty of Erzurum, dated 31 May 1847. Iraq had ceded half the river to Iran in various treaties dated 1914, 1937, and 1975.

Boundary disputes which last for a long time sometimes serve as a barometer of the condition of relations between countries. Periods of discord are often marked by bellicose statements about the dispute and sometimes the mobilization of reserves or the conduct of manoeuvres in the borderland. Contrasting periods of harmony are characterized by an apparent disinterest in the dispute or by political assurances that the dispute will not be allowed to stand in the way of improved relations. Relations between Argentina and Chile have passed through such phases on a number of occasions since 23 July 1881 when the two countries signed a treaty which defined their boundary along the Andes to the tip of the continent. The dispute appeared to be finally solved on 2 May 1985 when a treaty was ratified at the Vatican by the two countries. This treaty solved the final problem, which concerned islands in the Beagle Channel. The third article of the 1881 agreement included the following phrase:

> . . . and to Chile shall belong all the islands to the south of the Beagle Channel up to Cape Horn. (Lovering & Prescott 1979, p. 179)

This definition created problems because there were some large islands *in* the Beagle Channel which Argentina claimed were not covered by the phrase 'south of the Beagle Channel'. The three main islands are called Picton, Lennox, and Nueva. By the 1985 agreement Chile retained control of these islands, but they were discounted when offshore claims by each country were defined.

Soon after Lesotho gained independence in October 1966 it began to demand certain areas of the Orange Free State in South Africa. The claim was based chiefly on the ground that this land had once belonged to the Basuto Kingdom before White settlers moved into the area in the 1820s. There was sporadic conflict between settlers and tribesmen and the intervention of British authority settled a boundary which included areas in Basutoland which had previously lain beyond the writ of King Moshweshwe. After the creation of the Orange Free State in 1854 the new Republic tried to negotiate with the Basuto king to settle the matter permanently. These efforts failed and in three wars, in 1858, 1865–6, and 1867, the matter was brought to a head. Britain had to intervene in the last war to prevent the complete subjection of the Basuto Kingdom, and the war was formally ended when the second Treaty of Aliwal North was signed on 12 February 1869. This treaty fixed the boundary in its present location. The present Lesotho claim is based on the argument that the 1869 treaty was forced on the Basuto nation in an unequal war after connivance between the Boers and the British (Africa Research 1979, p. 5220).

Boundary disputes are not only associated with national boundaries that have existed for a long time; some modern lines have also created difficulties. For example, the Indian Independence Act of 18 July 1947 awarded Sind Province to Pakistan. This meant that the southern boundary of that province became the international boundary between India and Pakistan. Unfortunately these two countries disagreed over the course of that line through the Great Rann of Kutch, which consists of an extensive salt marsh. Fighting between these two countries in April 1965 led to a cease-fire and an agreement to submit the question to arbitration. The appointed tribunal handed down its decision on 19 February 1968 and defined a boundary which gave each side part of what it had claimed (Prescott *et al.* 1977, p. 40).

On Christmas Day 1985 fighting between forces belonging to Burkina Faso and Mali caused 300 casualties. This was another battle in a dispute which began in September 1960 when the two countries became independent. They were separated by a French colonial boundary which was elevated to international status. Unfortunately both sides claimed the Agacher Strip, which measures about 100 miles by 15 miles and is reputed to contain valuable mineral deposits (Africa Research 1986, p. 7888).

The factors that encourage co-operation or conflict along international boundaries and the consequences that follow from policies connected with these two activities involve many aspects of national life. These aspects include strategy, administration, economics, politics, and culture. No single discipline deals exclusively with this field of scholarship and the most comprehensive understanding of any particular situation will usually be found by reading and comparing separate studies by lawyers, political scientists, economists, geographers, and historians.

Generalizations about international boundaries

Attempts to produce a set of reliable theories about international boundaries have failed. Attempts to devise a set of procedures by which boundaries can be studied have been successful.

The German geographer Ratzel made a determined effort to produce a set of laws which would enable the behaviour of states in respect of boundaries to be predicted. Ratzel believed that each state had an idea of the possible limits of its territorial dominion and he called this idea 'space conception'. This notion appears to be similar to the concept of *les limites naturelles* (natural boundaries) which Pounds (1951, 1954) explored in respect of France. Pounds established that for much of France's history after the 16th century successive rulers regarded France's desirable boundaries as coinciding with the sea, the Alpine watershed, the Pyrenees, and the Rhine.

Ratzel's view on the space conception of states followed logically from his belief that the state was like a living organism which grew and decayed. The boundary and the adjacent territory, which is called the border, formed the epidermis of this organism and provided protection and allowed exchanges to occur. So for Ratzel the border was a dynamic feature and when it was fixed in position we were witnessing a temporary halt in political expansion. He enunciated two laws of territorial growth:

> The law of the evolution of boundaries can be defined as a striving towards simplification and in this simplification is contained a shortening of borders. (Ratzel 1897, p. 555)

> In accordance with the general law of growth of historical spatial phenomena the borders of the larger areas embrace the borders of the smaller one. (Ratzel 1897, p. 557)

Evidence for devising these rules can be found in the evolution of Germany by an amalgamation of small marches, kingdoms, and principalities. Many examples involving other countries can also be found in the collection of treaties prepared by Hertslet (1875–91). For example, by a treaty dated 14 April 1816 Austria and Bavaria exchanged territories in their borderlands. Bavaria ceded parts of Hausruckviertel and Innviertel to Austria and in return received parts of the Department of Mount Tonnerre and the city and fortress of Landau (Hertslet 1875, pp. 434–43). Ratzel would also have found support for his laws in the procedures by which Britain, Belgium, Germany, and France were acquiring colonies in Africa.

Ratzel also made strong assertions about the nature of borders.

> Political balance [between countries] is to a large extent dependent on the [characteristics of] borders between them. (Ratzel 1897, p. 584)

> We have seen how growth and decline of a region not only find expression in the areal form and protective measures of the border but also in a way prepare and foreshadow themselves therein. (Ratzel 1897, p. 605)

This view that the border was the area within which the growth and decline of the state was organized and became evident was taken up by students of *Geopolitik* thirty years later. *Geopolitik* was the name given to the school of political geography established by Major-General Haushofer who emphasized the rôle of geography in creating policies which would make the Germany of the 1920s strong again (Gyorgy 1944). Haushofer proposed that a cultural boundary should be established around a population which had a high degree of ethnic homogeneity and that beyond this cultural boundary there should be a military boundary which would prevent a surprise attack on the cultural homeland. When Haushofer (1927) classified boundaries he did so into categories labelled 'attack', 'defence', 'growth', and 'decay'.

Spykman, an American political scientist, also gave geography a very important rôle in the development of national policies and the conduct of international relations. In a paper dealing with geographical objectives and foreign policy the following assertion was made:

> Boundary changes will be indications of a shift in the balance of forces, caused either by an increase in driving force on one side of the frontier [boundary] or by a decrease in resistance on the other. (Spykman & Rollins 1939, p. 392)

This view of boundaries as temporary lines where opposed power of neighbouring states is neutralized also found favour with Ancel. In writing of boundaries as lines of power equilibrium this French political geographer referred to international boundaries as political isobars (Ancel 1938). This is not a useful analogy since an isobar links points experiencing identical air pressure; on one side the air pressure will be higher and on the other side it will be lower. However, a more serious criticism of Ratzel's original concept and its modern embellishments is that since 1945 most changes in the balance of power between adjoining states have not been accompanied by any changes in the position of international boundaries. For example, the domination of Vietnam over Laos and Cambodia in the 1980s has not been accompanied by any boundary changes. The internal problems of Iraq and Iran and their destructive war against each other has not been accompanied by any territorial losses to their neighbours such as the Soviet Union, Turkey, or Syria.

It is also possible to demonstrate that several boundary changes in the 20th century resulted from causes other than changes in the relative strengths of countries. For example, Britain made territorial concessions to the Italian colonies of Libya and Somalia after World War I in return for Italian participation in the war. On 16 February 1963 Mali and Mauritania agreed to modifications to the colonial line they had inherited from France three years earlier in order to make it correspond more closely to the geographical realities of this desert region. At the end of the last century many colonial boundaries were defined as straight lines when they were first drawn because the area they traversed was imperfectly known. Later explorations revealed situations where the two parties could obtain mutual benefits by exchanging the straight line for one which followed an irregular course related to the political and geographical realities of the region.

Of course there will always be cases where some powerful states are able to seize and hold the territory of a weaker neighbour. Thus Indonesia incorporated the territory of Portuguese Timor in 1975 and for the past decade Libya has occupied the Aouzou strip of northern Chad. Fortunately, such situations are rare compared with the multitude of cases where international boundaries have been fixed in location for a very long time, even though they separate states with wide differences in military and economic strength.

Although scholars have failed in their efforts to produce laws or even guidelines by which national behaviour in respect of boundaries can be predicted with confidence, they have succeeded in identifying reliable procedures by which the evolution of boundaries and boundary disputes can be studied.

A French lawyer called Lapradelle and an American geographer called Jones defined a series of stages through which ideal boundaries would pass in their history. In each case these major studies, which were published 17 years apart, do not bear the impress of the period in which they were written. This characteristic sets them apart from studies by Ratzel, Haushofer, Spykman, and Ancel.

Lapradelle identified three stages in the evolution of a boundary which he called 'preparation', 'decision', and 'execution'. The equivalent stages described by Jones (1945) are called 'allocation', 'delimitation', and 'demarcation'.

> The processes of preparation precede true delimitation. The problem of the boundary's location is debated first at the political level then on the technical level. The question is, in general, of determining, without complete territorial debate, the principal alignment which the boundary will follow. . . The decision involves the description of the boundary or delimitation . . . The execution consists of marking on the ground the boundary which has been described and adopted, an operation which carries the name demarcation. (Lapradelle 1928, p. 73)

In adopting and embellishing this pioneer proposal Jones added the final stage of administration which describes the maintenance of the boundary monuments, cleared lines, and fences.

Neither of these authors insisted that all boundaries would pass through each stage. In some cases the first boundary which allocates territory might be a straight line and this line is then demarcated either because there is no more suitable line or because agreement cannot be reached on modifications. In other cases the terrain is so well known it is possible to proceed directly to delimitation as the first stage. This situation was encountered often in the territorial arrangements which followed World Wars I and II. On many occasions boundaries which were delimited were not demarcated or the original demarcation was not maintained, and the location of the boundary became blurred as forests grew over cleared lines and peasants raided boundary monuments for metal pipes and building blocks.

In a series of perceptive comments Jones pointed to the reasons why the search for laws about boundaries was unproductive:

> Each boundary is almost unique and therefore many generalizations are of doubtful validity. (Jones 1945, p. vi)

> The process of boundary-making is smoothed by considering each boundary as a special case with individuality more pronounced than resemblance to a theoretical type. (Jones, 1945, p. 11)

It is probably this fact which has prevented those devoted to the quantification of political and social data from identifying the laws or consistencies which eluded Ratzel and others. For example, Boggs (1940) tried to devise an index of the interruptive quality of boundaries by calculating the continental ratios of boundary length and continental area. He then accepted that pressure against the boundary was a direct function of national population and obtained another index by multiplying the interruptive index by the value for the population density of the continent. Hinks (1940) demonstrated the unreliability of these calculations as a basis for serious continental comparisons. Dorion (1963) had no more success than Boggs in trying to devise an index of sinuosity in his study of the boundaries of Quebec and Newfoundland.

The host of case studies about boundary disputes by lawyers such as Johnson (1966), who wrote about the Columbia River between Canada and the United States, by historians such as Lamb (1966), who analysed the McMahon Line, by political scientists such as Widstrand (1969), who considered African boundary problems, and by geographers such as Glassner (1970), who studied the dispute over the River Lauca between Bolivia and Chile, have enabled the essential elements of boundary disputes to be identified. Analyses of such matters should focus on the cause of the dispute, the reason for its development at a particular time, the aims of the countries concerned, the arguments used to justify particular actions, and the results which follow from the settlement or continuation of the dispute.

Borders

One of the special contributions which geographers have made is to study the boundary in its territorial context. It was noted earlier that the adjacent areas which fringe the boundary are called the 'border'. This is an old concept which Ratzel expressed neatly in the following sentence:

> The border fringe is the reality and the border line is the abstraction thereof. (Ratzel 1897, p. 538)

Fawcett (1918) was mainly concerned with the zonal character of frontiers and borders and stressed their transitional nature.

This emphasis provided part of the basis for Lapradelle's work. He called the border *le voisinage* and he distinguished three parts of this region on the basis of law. The central area immediately adjacent to the boundary was called *le territoire limitrophe*, which is the zone where international law may apply. The word *limitrophe* is difficult to translate because it is a technical term for land set aside to support troops in the border. The phrase *le territoire limitrophe* may be translated as 'neighbouring territory' or 'adjacent territory'. The other two parts flanked this central region and are subject only to the internal laws of the states. Lapradelle called such zones *les frontières* (the frontiers). In his conclusion he stressed the need to study the legal realities of the border and to avoid the belief that the entire state is subject to uniform internal boundaries right up to the boundary.

It is not surprising that political geographers should have been at the forefront of border studies. Such a perspective is traditional as shown by the way geomorphologists study rivers in relation to the run-off provided by the catchment and to the structure of the basin drained by the river whereas economic geographers examine the traffic generated in the region served by a railway.

Ritter & Hajdu (1982) have provided an account of the influence of the boundary between East and West Germany on the surrounding region. After tracing the successive stages in the demarcation of this boundary after World War II, a statistical analysis shows the decline in economic and human links between the two new states. Districts which had previously been located near the heart of Germany between the wars now had a peripheral location which added to the disadvantage of interruption in previous connections and patterns of circulation:

> Major cities such as Lubeck lost most of their traditional hinterland and saw their commercial functions seriously impaired. The rural district of Luchow-Dannenberg lost its link with its higher order centre, and the city of Hof found that the vertical integration of its textile industry with plants in Saxony ceased to exist. A loss of administrative functions, employment opportunities and population became a general feature of these peripheral districts. (Ritter & Hajdu 1982, p. 257)

The terminology of land boundaries

> Over many years some words have developed a specific connotation in respect of frontiers and land boundaries.

Boundary refers to a line, but *frontier* refers to a zone. The terms *allocation*, *delimitation*, and *demarcation* are used in the sense outlined by Jones (1945). Allocation means the initial political division of territory between two states. Delimitation means the selection of a boundary site and its definition. Demarcation refers to the construction of the boundary in the landscape. *Borderland* refers to the transition zone within which the boundary lies; it corresponds to Lapradelle's *voisinage*. It has not been found necessary for geographical purposes to distinguish *le territoire limitrophe* from the borderland. Lastly, there are the sequential terms proposed 50 years ago by Hartshorne (1936). These terms describe the relationship between the boundary and the landscape through which it was drawn. An *antecedent* boundary was drawn before the development of most of the features of the cultural landscape, and if a boundary was drawn through an uninhabited area it was called a *pioneer* boundary. *Subsequent* boundaries were drawn after the development of the cultural landscape. If the boundary coincided with some physical or cultural divide, it was described as *consequent*. If, however, the boundary was not drawn within such a feature, it was described as *superimposed*, for which the synonym *discordant* was occasionally used. A *relict* boundary is one which has been abandoned but is still marked by differences in the landscape which developed during its lifetime.

Maritime boundaries

So long as there have been states with access to the sea, governments have had a proper interest in the adjacent coastal waters. In tracing the evolution of maritime claims, it becomes evident that there were three basic problems to be solved. The first concerned the establishment of a legal concept empowering states to exercise jurisdiction over parts of the sea. The second involved deciding how much of the seas and seabed would be legally appropriated by states. The third concerned accurately defining the claimed zones.

The legal concept of maritime jurisdiction

Fenn (1926), who identified the four main contributors to the establishment of the concept of maritime jurisdiction as a doctrine of international law, has provided the most detailed analysis of this question. A careful reading of his arguments and sources does not reveal that geography played an important rôle, although the chief contributions were made by lawyers living around the Mediterranean or the North Seas.

The first important step in the evolution of this concept was taken by Roman glossators who determined that the emperor had the right to punish wrongdoers at sea in precisely the same way that he punished them on land. The application of right was primarily concerned with the suppression of piracy. Azo, who lived at the beginning of the 13th century, advanced the concept by establishing that a private right could be granted by the sovereign either as a privilege or as a result of long and uninterrupted use. These rights included sole use of shallow fishing grounds in bays, control over salt deposits in tidal marshes, and exemption from port dues.

Bartolus, an Italian jurist who lived in the first half of the 14th century, developed the concept by asserting successfully that countries own islands within 100 nautical miles of the coast. Gentilis introduced the final, essential change in the 16th century when he persuasively argued that coastal waters are a continuation of the territory of the state whose shores they adjoin. It therefore followed that the territorial rights which the sovereign possessed on land extended over the coastal waters.

Fenn has no doubt that 'after Gentilis it is literally correct to speak of territorial waters in international law' (Fenn 1926, p. 478). He also noted that by solving the first problem Gentilis had cleared the way for a resolution of the remaining problems:

> There remains a problem of placing a limit on these [territorial] waters. The theory, however, is complete with Gentilis. The delimitation of the territorial waters is a mere matter of detail and becomes a problem for statecraft and not for lawyers to settle. (Fenn 1926, p. 481)

The detail to which Fenn referred has proved harder to settle than the theory on which it is based, as the following sections show.

The extent of national maritime claims

There did seem a time in the 19th century when the question of how much of the sea fell under national jurisdiction had been answered. All the major naval powers claimed territorial seas 3 nautical miles wide and they refused to recognize wider claims by other countries such as Spain. When Russia proclaimed waters 12 nautical miles wide in 1909 there was no expectation that such claims would become general, but that is what happened in the period after 1945.

The process by which claims to zones 3 nautical miles wide became general in the 19th century has been unravelled by Baty (1928), Walker (1945), and Kent (1954). The twin characteristics of claims to a continuous belt of territorial waters 3 nautical miles wide had different origins. The concept of a continuous belt of territorial waters was born in Scandinavia; the specific distance stemmed from the rule related to the range of cannon, which was common in the Mediterranean Sea and the southern reaches of the North Sea.

Denmark played an important part in establishing the concept of a continuous zone of territorial waters. Following the rediscovery of Greenland in 1585, Denmark, which had acquired Iceland and Norway in 1381, completed its control over the shores of the northern Atlantic Ocean. This encouraged the Danish authorities to assert a claim to sovereignty over all this oceanic region, and it willingly gave licences to aliens to navigate and fish in these waters. However, these licences did not apply to a zone of coastal waters adjacent to the coast. For example, in 1598 a Danish decree reserved a belt of waters 2 leagues wide around Iceland for the exclusive use of Danish fishermen. A Danish league measures 4 nautical miles.

These unilateral claims had to be adjusted according to the strengths of countries which contested them and Denmark was forced by Holland, Britain, France, and Russia to reduce this claim at various times. In 1743 Russia managed to force a reduction in the claim to 1 league. The governor of Finnmark was fearful that Russia might close its land boundary with Norway and prevent trade and the transhumance movement of the reindeer herds of Lapps; to placate the Russian authorities he granted Russian fishermen access to waters more than 1 league from the coast (Meyer 1937, p. 500). The Danes charged a small fee, so that it could be argued that the wider claim had not been abandoned, but in 1745 Denmark announced a general reduction of its claim to 1 league. Kent (1954, p. 545) has suggested that this action was taken on economic grounds because the narrower zone offered less sanctuary to alien traders, who stayed away, and at the same time allowed the more frequent capture of prizes which were sold in Danish ports to the benefit of national revenue.

Fulton (1911, p. 156) has traced the earliest mention of the cannon shot rule to 1610 during a fishing dispute between Britain and the Netherlands. Bynkershoek is generally credited with introducing the concept of territorial waters extending as far as a cannon's range into the literature of international law in 1703. However, Walker (1945, pp. 211–18) has shown conclusively that the rule had been applied by various countries, including France and the Netherlands, throughout most of the 17th century. Bynkershoek judged that states could possess the coastal seas which they could command from their shores:

> Therefore it evidently seems more just that the power of the land [over the sea] be extended to that point where missiles are exploded . . . the power of the land [over the sea] is bounded where the strength of arms is bounded; for this as we have said guards possession. (Cited in Balch 1912, p. 414)

Bynkershoek formulated this rule because he regarded claims to the sea within sight of land as being too imprecise. Unfortunately his regulation is also open to conflicting interpretations. The cannon shot rule could apply only to the field of fire of actual pieces mounted on the shore and the claim would vary with the type of cannon, its height above the sea, the charge used, and the weight of ball. Under this interpretation if there was no cannon mounted on the shore there would be no claim to territorial waters and prizes could be taken up to shoal water. On the other hand, the cannon shot rule could be interpreted to apply to the waters commanded by a line of imaginary guns of known range mounted along the entire coast of the country. There is evidence in the writings of Walker (1945) and Kent (1954) that both these views were held. The specific interpretation operated in favour of those countries which were strong and well armed and which had a high density of ports along their coast; Britain, France, and the Netherlands favoured this application of the rule. The general interpretation made all states equal whatever the condition and number of their fortresses and whatever the nature of their coastlines. It was probably this conclusion which led Galiani in 1782 to propose that instead of waiting to see what guns a neutral state might mount along its coast a belt of 3 nautical miles be fixed as the zone of territorial waters (Galiani 1782).

It is through Galiani that the concept of a continuous zone used in Scandinavia was finally united with the cannon-shot rule of the Mediterranean region. His practical suggestion was quickly accepted by diplomats. When Britain and France engaged in war in 1793, President Washington of the United States declared that his country would insist on neutrality within 3 nautical miles of its coasts. This position was justified to the British authorities in the following terms by Secretary of State Jefferson:

> The greatest distance to which any respectable assent amongst nations has at any time been given is the extent of human sight, estimated at upwards of twenty miles, and the smallest distance I believe, claimed by any nation whatever, is the utmost range of a cannon ball, usually stated at one sea league. (Crocker 1919, p. 636)

The French also accepted Galiani's proposal quickly, for on 25 May 1795 a treaty was signed with Tunis which stipulated that the neutral zone was measured as the cannon shot whether the guns were actually in position or not. It is impossible to disagree with the American Ambassador in Paris who informed the French government in 1864 that 'no other rule than the three mile rule was known or recognized as a principle of international law' (Crocker 1919, pp. 659–60).

Most countries before 1945 claimed only a single maritime zone. Additional fishing zones were claimed off ten territories. Five of them were the independent states of Argentina, Brazil, Colombia, Ecuador, and Uruguay and the others were the colonial territories of Cambodia, Lebanon, Morocco, Palestine, and Vietnam. None of these claims exceeded 12 nautical miles. In some cases conservation zones were declared to preserve seal populations. For example, in 1893 regulations were proclaimed to stop British and American citizens from capturing or pursuing fur seals within 60 nautical miles of the Pribilov Islands in the Bering Sea. Unfortunately the rules did not apply to Japanese fishermen who hunted the seals all year round to within 3 nautical miles of the islands (Fulton 1911, pp. 694–6).

Although there were no claims to the whole of the continental shelf by countries in the period before 1945, there were claims to some of the resources associated with that feature. As early as 1758 Vattel had asked the rhetorical question, 'Who can doubt that the pearl fisheries of Bahrain and Ceylon may be the lawful objects of ownership?' (Vattel 1758, p. 107). By the Colonial Act of 1811 authorities on Ceylon were permitted to seize boats which drifted or anchored over banks where pearl oysters were located. Some of these banks lay 21 nautical miles from the coast. After 1904 the Tunisian authorities claimed the right to regulate sponge fishing outside the territorial waters to the 50 metre isobath.

Claims to minerals on or under the seabed became common only after 1942, following the agreement between Britain and Venezuela to divide the seabed in the Golfo de Paria. Minerals had been mined from under the sea long before this. Wenk (1969) refers to galleries driven under the sea from collieries in western Scotland in 1620, and Gidel (1932, p. 510) has provided a list of mines which penetrated beneath the seabed off the shores of Cornwall, Cumberland, Nova Scotia, Vancouver, Australia, Chile, and Japan. In 1858 a British act proclaimed that all minerals won from beneath the low-water mark under the open sea were the property of the British sovereign. The first petroleum was mined from beneath the sea in 1899 by using pipes driven into the seabed from piers.

Even though the Institute of International Law (1928, p. 517) had observed in 1894 that it was wrong to try to create a single zone where authorities could both control fishing and preserve the security of the state, the practice of claiming only territorial waters remained general until the 1940s.

Since 1945 claims to the continental shelf and fishing or economic zones have become common. There are also other claimed waters such as the contiguous zone, archipelagic waters, and safety zones. The 1958 Convention on the Territorial Sea and the Contiguous Zone and the 1982 Convention on the Law of the Sea have defined the various zones which may be claimed. They have also sought to answer the third question by providing rules which govern the delimitation of these zones, but the general, sometimes vague, language of these rules leaves plenty of scope for continued disagreement between countries.

The delimitation of maritime zones

Until 1945, when claims to the continental shelf became increasingly common, there were two basic problems in delimiting maritime boundaries. First, it was necessary to select a baseline from which the width of the territorial sea was measured. This is a comparatively easy task along coasts free from indentations and lacking offshore islands and low-tide elevations. The English coast south of Bridlington, the Dutch coast north of 'sGravenhage, and the French coast north of Bayonne provide excellent examples of such simple coasts. In such cases the maritime zone can be measured seawards from a low-water mark which can be identified by observations over a long period. It is usual for countries to select a low-water mark because this will push the claim as far seawards as possible. Of course, many coasts are deeply indented and fringed with islands and low-tide elevations, and such features make the selection of a baseline difficult. Although coastal geomorphologists can classify coasts according to many criteria such as process, stage of development, plan, profile, and the arrangement of islands, no two sections of coast are identical; this makes it necessary to devise rules which would ensure that any two surveyors working independently would draw the same baseline for the same section of coast.

The second problem was to divide waters between two countries whose territories lay close together. This problem always existed for adjacent countries which had to decide how a land boundary would continue seawards from its terminus on the shore. Since most countries only claimed territorial seas 3 nautical miles wide, it was only necessary to consider drawing boundaries between opposite states when there was less than 6 nautical miles of water between them. Because the delimitation of boundaries between two states is a matter solely for them, there is no need to have general rules on this subject; they can select any boundary which is mutually satisfactory. However, if countries cannot agree on a boundary, it would obviously be useful to have one or more general rules which could be applied by impartial tribunals which might be called upon to give an opinion.

Considerable progress was made in forming rules regarding baselines before the 1940s. This progress occurred in two unequal periods. Prior to 1930 there was a long period when states made unilateral assertions about their baselines, when challenges to these lines produced compromises, and when regional groups of states agreed on general rules. In 1930 there was an international conference at 'sGravenhage attended by 47 countries; although none of the proposed precise rules was adopted by a formal vote, there is no doubt that the useful discussions at the conference paved the way for rapid progress at the United Nations Conference on the Law of the Sea in 1958. The baselines rules defined in the 1958 Convention on the Territorial Sea and the Contiguous Zone were adopted with only slight changes for the 1982 Convention on the Law of the Sea.

A survey of practices before 1930 reveals three main points. First, bays were considered as a special case from earliest times. The granting of exclusive fishing rights in bays by Roman emperors has already been mentioned. From the 13th century, Venice and Genoa claimed control over their bays and gulfs. During the reign of King Edward II (1307–27) the English authorities claimed as internal waters all bays where the limiting headlands could be seen from each other on a clear day. In the 17th century Britain's neutral waters were bounded by straight lines linking 27 headlands around the English coast from Holy Island in the north-east to the Isle of Man in the west (Fig. 1.2). In 1804 President Jefferson of the United States invoked the same practice as the advisers of Edward II five centuries before and pointed the way for the present rule of closing bays by straight lines and measuring the territorial seas from those lines:

> The rule of common law is that wherever you can see from land to land all the waters within the line of sight is in the body of the adjacent country and within the common law jurisdiction . . . The 3 miles of maritime jurisdiction is always to be counted from this line of sight. (Crocker 1919, p. 641).

At this stage there was no general agreement on the maximum size of bays which constituted special circumstances.

The second main point which the survey reveals is that the countries which possessed offshore islands measured their maritime claims from them. On 18 June 1745 Denmark defined its territorial waters in the following terms:

> . . . no foreign privateers shall be permitted to capture any ship and vessel within one league of our coasts and the shoals and rocky islets which are situated there are also included in that term. (Kent 1954, p. 545)

A declaration by the same country in 1812 measured the territorial sea from those outermost islands or islets which are not submerged by the sea. Similar claims were made by Sweden in 1899 and Russia in 1893.

The third main conclusion is that the low-water mark came to be accepted as the baseline from which maritime claims should be measured in those cases where straight lines did not close bays, In 1911 the Russian government claimed that its territorial waters were measured from 'the lowest ebb-tide or from the extremity of coastal standing ice' (Jessup 1927, p. 28). The 1930 conference at 'sGravenhage tried to make these three general rules sufficiently precise for use as part of international law.

A subcommittee of the conference considered the technical question of boundary delimitation and its report was made under 13 headings which can be grouped into five subjects. The first subject dealt with the baseline to be used on uncomplicated coasts. There was general agreement that a low-water mark should be used, and the line which was specifically recommended was the mean low-water spring tide. The height of this tide is obtained by averaging the readings for two successive low waters

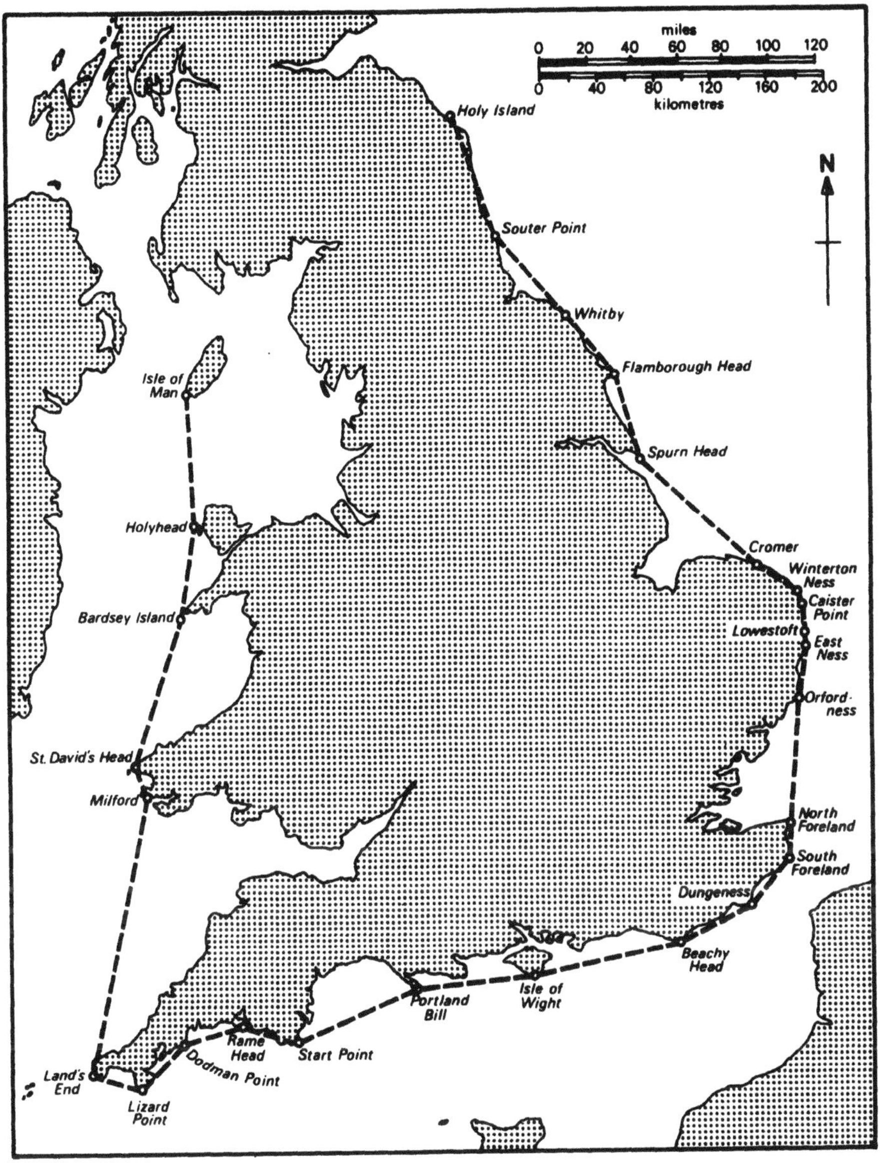

Figure 1.2 The King's Chambers.

during the periods of 24 hours throughout the year when the maximum declination of the moon is 23°30′.

The second subject concerned the delimitation of a baseline around ports, roadsteads, and river mouths. There was complete agreement that the outermost permanent harbour works should be considered as part of the land for the purposes of drawing baselines. It was also recommended that where rivers flowed directly into the sea their mouths should be closed by a straight line. Where the river entered the sea through an estuary the rules for closing bays should apply.

Offshore islands and low-tide elevations formed the third subject. There was no dissent from the views that islands should possess their own territorial waters and that low-tide elevations that fell within the territorial sea should be used for measuring an additional zone of territorial waters.

The fourth subject was narrow straits and the recommendations distinguished three situations. Where the straits linked the high seas and internal waters the rules for bays should be applied. Where the strait linked two areas of high seas and the opposite shores were owned by a single country, then the territorial seas should be drawn in the normal way. If the opposite shores in such straits were owned by different countries, it was recommended that a median line should be drawn between the two shores. The median line is sometimes called a line of equidistance and it is a line which at every point is equidistant from the nearest points of the shores which it separates. In short, a median line guarantees that each country secures the waters which are closer to its shore than to the shore of any other state. It is an old concept which was expressed by Lord Chief Justice Hale in 1636 when he wrote that the British king had jurisdiction over 'so much at lest as adjoines to the British coast nearer than to any forren coast' (Fulton 1911, p. 543).

The fifth and most difficult subject related to the status of bays. The subcommittee tried to define bays by geometric formulae but was unable to choose between those put forward by the delegations of France and the United States. The American formula would have resulted in many fewer bays being closed than the French method. Nor could the subcommittee make any firm recommendations about claims to historic bays or bays which were shared by states, and those particular problems were not solved by the Conventions of 1958 or 1982.

After the 1930 conference ended, Boggs produced two excellent technical papers on maritime boundaries. He was then The Geographer of the United States' State Department and he attended the 1930 conference. His first paper (Boggs 1930) argued in favour of eliminating small pockets of high seas which are sometimes enclosed when territorial seas are delimited around a group of islands. Even though the inconvenience of such pockets is well known, the subsequent conventions have not dealt with this problem. Boggs (1937) also suggested that when land boundaries are continued seawards they should follow median lines. Prior to this suggestion it had been usual for countries to argue that the boundary through the waters should be projected along the azimuth of the last section of the land boundary (Lapradelle 1928, pp. 215–16). An alternative line was offered by a boundary drawn perpendicular to the general alignment of the coast. Either of these lines could be unfair to a particular country, depending upon the configuration of the coast near the boundary's terminus and upon the presence of offshore islands belonging to one or both states.

The 1930 conference was more important than its lack of formal results suggests. It successfully distilled state practices for previous decades and simplified the search for agreed solutions which were reached in the Conventions of 1958 and 1982. It is clear, as the history of the international conferences are traced from 1930 to 1982, that the rules for drawing baselines have not changed very much. Unfortunately when contemporary baselines are considered in Chapter 5 it will become clear that some of the greedy national claims in recent years contrast sharply with modest claims in the 1940s and 1950s. The explanation is not to be found in changes in the rules but in different interpretations of those rules and sometimes in the flouting of them.

The conference in 1930 failed to produce any general principle which impartial tribunals might apply in resolving disputes brought before them by neighbouring states. That is not a matter for criticism since the Conventions in 1958 and 1982 similarly failed to establish any reliable principles for the settlement of such disputes. Further, bilateral negotiations to delimit maritime boundaries were comparatively rare in the period before 1930. It is only since the 1940s that such negotiations have become a frequent occurrence.

The terminology of maritime boundaries

There is a large technical vocabulary associated with modern maritime boundaries and these terms will be encountered and explained in Chapter 5. The point which must be made in this section is that there are no precise terms which have special meanings for the process of creating maritime boundaries. At sea there are no comparable stages such as allocation, delimitation, and demarcation which can be identified for international boundaries on land. Maritime boundaries suddenly appear in final form after a period of usually secret preparation. If the boundary is a domestic limit, such as a baseline, the preparation is undertaken by government departments responsible for matters of survey and law. If it is an international boundary, the preparation involves various government departments in each country and then national delegations. An inspection of the 1982 Convention of the Law of the Sea reveals no consistent use of terms to describe the fixing of different kinds of maritime boundaries. For example, the term 'delimitation' is used in Articles 15, 16, 74, 75, 83, and 84 and in Article 9 of Annex II to refer to the fixing of bilateral boundaries. It is also used in Article 50 to describe the fixing of internal waters of archipelagic states and in Articles 60(8), 147(2e), and 259 to describe the unilateral fixing of the territorial sea, the exclusive economic zone, and the continental shelf about artificial islands. Although the term 'delineation' is only used to refer to the fixing of the outer edge of the continental margin, in Article 76(4 & 7) the terms 'establish', 'determined', and 'drawn' are also used to refer to the same process in the same article.

A comparison of boundaries on land and sea

There are a few similarities between international boundaries on land and sea. During the second half of the 19th century the European colonial powers were very active in drawing international boundaries in Africa and Asia, and this was also the period when many boundaries were settled in South and North America. This historic surge in drawing boundaries on land has been matched since the 1940s by what Eckert (1979) has called 'the enclosure of the oceans', as countries have asserted unilateral claims to adjacent seas and divided semi-enclosed waters with their neighbours.

The motives for drawing boundaries on land and sea are identical. States wish to secure a clear unchallenged title to parts of the earth which are valued either for the human or material resources they contain or the strategic advantages which they confer. They also wish to create lasting arrangements with neighbours which will at least minimize the risk of friction either between governments or between citizens.

The procedures by which states conduct bilateral negotiations to produce boundaries on land and sea have much in common. Each country carefully prepares its case by marshalling all the technical, historic, economic, cultural, and legal arguments which might be useful. Delegations are then appointed, given precise instructions, and charged with the responsibility of obtaining the best possible result. Sometimes when a satisfactory compromise cannot be achieved the contending parties either take the dispute to arbitration or establish a neutral zone or a provisional boundary.

An important difference which the negotiators have to face when seeking to fix maritime boundaries is that there are some imprecise rules in the 1958 and 1982 conventions which might bear upon the question. In addition, it is most likely that arguments about the location of a maritime boundary will be mainly of a technical nature concerned with the characteristics of coastlines and the seabed and the proper consideration which should be given to very small islands. Issues of history and culture which seem to be prominent in any negotiations about boundaries on land are much less important in the fixing of maritime boundaries.

Although countries usually fix a single international boundary on land, they often draw a number of lines at sea to delimit zones where they possess various rights and responsibilities. There are already cases where one country owns the fishing rights over the seabed and another owns the mining rights and the rights to harvest sedentary species of fish.

Maritime boundaries in some situations are much more permeable than boundaries on land. Unless special arrangements have been made for the benefit of those citizens who live close to a boundary on land, there are usually some formalities which must be observed when anyone crosses an international boundary. But there are no comparable formalities if a master decides to sail his vessel through the territorial waters of another country without calling at a port, providing his passage is innocent. Nor is there any prohibition against aliens catching fish in the waters over the continental shelf of another country, providing they stay outside the exclusive economic or fishing zone.

Maritime boundaries are rarely demarcated except where two countries share a comparatively narrow, navigable waterway, whereas it is fairly common practice to demarcate boundaries on land. In addition, it is noticeable that almost all maritime boundaries are defined by straight lines or arcs of circles, whereas boundaries on land may follow irregular courses.

One encouraging difference between boundaries on land and sea is that the latter have not yet been the cause of war between countries, although there have been occasional incidents. In contrast, nations have sometimes used war as a means of seeking a satisfactory solution to some boundary dispute on land as recent wars between China and India, India and Pakistan, and Iraq and Iran testify.

Boundaries in the air

Although national claims to land, lakes, rivers, and adjoining seas have been common since Roman times, national claims to the air above these features were delayed until this century. Matte (1981, pp. 53–5) has described individual legal rights to airspace from Roman times to the 19th century. The legal principle was summed up in the gloss prepared by Accursius in the Middle Ages:

> Cujus est solum ejus debet esse usque ad coelum. [He who possesses land possesses that which is above it to the sky.] (Matte 1981, p. 54)

This meant that if water draining from one roof fell onto and damaged the home of a neighbour then that neighbour could claim compensation. The first flight by balloon occurred on 21 November 1783, and 32 years later a British judge ruled that the passage of a balloon over private property did not involve any infraction of ownership rights (Matte 1981, p. 55).

National claims to airspace started when aeroplanes began to cross national boundaries and when they were used in warfare. The earliest use of balloons in war recorded by Matte occurred in 1793. French Republican forces used the tethered balloon *Entreprenant* as an observation platform during the battle of Fleurus (Matte 1981, p. 93). The first use of dirigibles and aeroplanes in a war occurred in a conflict between Italy and Turkey over Tripolitania in 1911 and 1912. More than 40 years earlier Chancellor Bismarck had advised the French government that aeronauts who crossed enemy lines would be treated as spies (Kroell 1934, p. 2).

National claims were firmly established in the Paris Convention which was signed on 13 October 1919 and which entered into force on 11 July 1922. Article 1 of this Convention includes the following statement:

> . . . the High Contracting Parties recognise that every Power has complete and exclusive sovereignty over the airspace above its territory. (Martin *et al.* 1985, Vol. 1, IV/3)

The 1944 Chicago Convention on International Civil Aviation included the identical principle in its first Article:

> The contracting States recognize that every State has complete and exclusive sovereignty over the airspace above its territory. (Matte 1981, p. 605)

In both Conventions territory was considered to include all land and territorial waters of the state and any colonies or mandates which it might possess. Martin *et al.* (1985, IV/5) believe that this rule would also apply to territory leased by one state to another.

It is now necessary to consider the limits of the airspace within which countries exercise sovereignty.

The horizontal extent of airspace

The Paris and Chicago Conventions defined territory as all land and territorial waters. This means that on land the horizontal extent of airspace is defined by international boundaries whereas over the sea it is defined by the limits of the territorial sea.

There are three possible combinations of these limits. The horizontal extent of airspace over *landlocked countries* will be bounded entirely by international boundaries agreed with neighbours. Of course, if there is some dispute over the location of those boundaries and conflict over which state owns a particular piece of territory, that dispute will also extend to the airspace above the contested zone. However, the debate over the ownership of territory never seems to involve arguments related to the airspace above it.

The airspace of *coastal states* will be bounded by their international boundaries with neighbours on land and through the territorial waters and by the outer limit of their territorial seas measured from a specified baseline. Because many neighbours have not formally extended their land boundaries through the territorial waters, it is likely that it will be assumed that the land boundary is continued seawards from its terminus on the shore by a median line. Such a line is always equidistant from the nearest point of opposite shores.

When the two Conventions were ratified and entered into force the usual width of territorial seas was 3 nautical miles. At the beginning of 1986 only 19 countries still claimed territorial seas 3 nautical miles wide. Another 93 states claimed territorial waters up to 12 nautical miles wide and 27 countries claimed waters more than 12 nautical miles wide. The 1982 Convention on the Law of the Sea stipulates that territorial seas should not exceed 12 nautical miles (United Nations 1983, p. 3). It is therefore uncertain whether states which claim seas in excess of 12 nautical miles can properly claim the airspace beyond 12 nautical miles from the baseline.

There is another problem which applies to some countries which claim only 12 nautical miles. The outer limit of territorial waters can be pushed seawards by the improper use of straight baselines (Prescott 1986). When this is done the outer limit of territorial seas off the coasts of countries such as Bangladesh, Malaysia, and Vietnam are much more than 12 nautical miles from the nearest land. At one point in the Strait of Malacca between the tiny islands called Perak and Jarak a section of Malaysia's territorial waters is 59 nautical miles from the nearest land. It will not be surprising if some countries object to such unreasonable claims and refuse to recognize national sovereignty over the airspace above these extended territorial seas.

The horizontal extent of airspace over states composed entirely of *archipelagos* will be the limit of territorial waters whether measured from a baseline or agreed with a near neighbour.

The 1982 Convention on the Law of the Sea also contains some rules about the airspace over maritime zones, and these must now be examined. Freedom of overflight is guaranteed over waters in an international strait by Article 38 (United Nations 1983, p. 12). An international strait is used for international navigation from one part of the high seas or an exclusive economic zone to another part of the high seas or an exclusive economic zone and does not possess a continuous route through the high seas or an exclusive economic zone. At the narrowest point of such a strait the territorial seas claimed from each shore overlap. For discussion, it can be assumed that such a strait will measure less than 24 nautical miles at its narrowest point since 12 nautical miles is expected to be the maximum claim to territorial seas. A selection of these straits is shown in Table 1.1.

Flight through these straits is described as *transit passage* which should be continuous and expeditious. Those involved in the flight should not offer any threat or use of force against the sovereignty, territorial integrity, or political independence of states bordering the strait, and they should not engage in any activities other than those incident to their normal mode of continuous and expeditious passage. Dispensation from this second

Table 1.1 A selection of international straits less that 24 nautical miles wide.

Strait	Width in nautical miles
North America	
Juan de Fuca Strait	9
Bering Strait	19
South and Central America	
Dominica Channel	16
Martinique Passage	22
St Lucia Channel	17
St Vincent Passage	23
Estrecho de Magallanes	2
Europe	
Karadeniz Bogazi (Bosporus)	1
Canakkale Bogazi (Dardanelles)	1
Stretto di Messina	2
Karpathos Strait	23
Strait of Bonifacio	6
Strait of Gilbraltar	8
Strait of Dover	18
North Channel	11
Oresund	2
Alands Hav	17
Asia	
La Perouse Strait	23
Korea Strait	23
Mindoro Strait	20
San Bernadino Strait	8
Balabac Strait	27
Selat Lombok	11
Selat Sunda	12
Strait of Malacca	8
Palk Strait	3
Strait of Hormuz	21
Africa	
Bab el Mandeb	14
Oceania	
Torres Strait	2
Bass Strait	14
Bougainville Strait	15
Dampier Strait	13
Cook Strait	12

Source: The Geographer.

requirement is given in the case of *force majeure* or distress. Aircraft in transit passage must also observe rules established by the International Civil Aviation Organization and continually monitor the radio frequency assigned by air traffic control authorities (United Nations 1983, p. 13). There is also an obligation on states bordering the international strait to publicize any known dangers to overflight.

The right of transit passage through an international strait does not apply in one special case. This occurs when the strait is formed by an island owned by the state owning the mainland coast and when there is an alternative route seawards of the island of similar convenience with regard to navigational and hydrographical characteristics.

Archipelagic states may draw straight baselines around the outer edge of their outer islands if they can satisfy certain rules, which are examined in Chapter 5. The waters contained within such baselines are called 'archipelagic waters' and sections of them may be much more than 12 nautical miles from land. For example, the Banda Sea lies entirely within the archipelagic baselines proclaimed by Indonesia on 18 February 1960, and some parts of that sea are 80 nautical miles from the nearest land. In the southern part of the archipelagic waters claimed by Fiji on 1 December 1981 some sections are more than 60 nautical miles from the nearest islands. According to Article 49 of the Convention on the Law of the Sea, sovereignty extends to the airspace over the archipelagic waters as well as to the seabed (United Nations 1983, p. 16). Aircraft of other countries may be flown through archipelagic airspace according to the right of archipelagic sea lanes passage. This right seems to be very similar to transit passage through international straits.

Archipelagic states may designate air routes suitable for continuous and expeditious passage by a series of continuous axes from the point of entry to the point of exit. Such routes must include all the normal routes used for international navigation, although duplication of routes of similar convenience between the same entry and exit points is not required. Planes flying along these routes may not deviate by more than 25 nautical miles from the defined axis. However, the plane may not pass closer to land than 10 per cent of the distance between the nearest points of land bordering the sea lane. This means that the full deviation of 25 nautical miles will only be possible when the sea lane is at least 63 nautical miles wide.

If the state does not designate archipelagic sea lanes, planes may fly along the routes normally used for international navigation (United Nations 1983, p. 17).

The Law of the Sea Convention also places an obligation on states to prevent, reduce, and control pollution of the marine environment from atmospheric sources. Article 212 entitles the country to adopt laws and regulations dealing with this subject which shall apply to its airspace and to planes registered by that country (United Nations 1983, p. 76).

The vertical extent of airspace

The vertical extent of national sovereignty was not defined in either the Chicago or Paris Conventions. Nor is there any reference to the vertical limit of airspace in the Law of the Sea Convention. Escalada (1979, pp. 51–3) discusses the lack of a precise vertical limit and notes that most of the writers on the laws relating to airspace believe that a definite vertical limit should be set.

When a precise limit is agreed it will presumably separate national airspace from international outer space. The treaty dealing with outer space which was opened for signature on 27 January 1967 stresses the international nature of outer space in Article 2:

> Outer space . . . is not subject to national appropriation by claims of sovereignty, by means of use or occupation or by any other means. (Martin *et al.* 1985, IV/1)

Various proposals have been made about the vertical definition of airspace and they fall into three groups. First, there are those related to the physical properties of the atmosphere. Such limits could be related to specified values for air pressure or air density. Unfortunately the surfaces marking equal air pressures or air densities would resemble the surface of the ocean with very large waves. In short, the elevation of any specified value would not be the same for various places at the same time, nor the same for one place at different times. Such a variable limit would simply create unrestricted opportunities for legal quibbling in the future.

Secondly, the vertical limit could also be related to the performance of aircraft or satellites. For example, it has been suggested that the limit should be set at the level where flight based on aerodynamic lift becomes impossible. This level is known in the literature as the von Karman Line (Escalada 1979, p. 53, Martin *et al.* 1985, IV/1). Although it is agreed that this line will occur at about 80 kilometres, it will vary above and below that height, depending upon the density of the air at any time.

At a meeting of the International Law Association in 1968 at Buenos Aires it was agreed that the zone of outer space should begin at the perigee of the orbit of a satellite closest to the Earth's surface.

Thirdly, there is a limit fixed at some arbitrary height above the Earth's surface. This type of definition would correspond to the selection of 12 nautical miles as the proper width for territorial waters or 200 nautical miles as the width of the exclusive economic zone. It would be necessary to fix a height above the maximum elevation that planes could reach in aerodynamic flight. If the von Karman Line is valid, the arbitrary limit would be in the order of 90 or 100 kilometres above sea level.

Although there would be locations in the world's oceans and seabeds where drawing the outer limit of the exclusive zone at 210 nautical miles rather than 200 nautical miles would confer economic benefits to the coastal state, it seems unlikely that any country would make significant gains of any kind if the vertical limit of airspace was set at 95 rather than 90 kilometres. At that elevation the main activities are likely to be travel and scientific research, and such activities would be freely available in outer space just above the limit of national sovereignty. Escalada (1979, p. 53) may be right in asserting that the drawing of a line between airspace and outer space is imperative, but there is no evidence that the majority of countries in the world give this matter a high priority!

Disputes associated with national sovereignty over airspace

In recent years the most dramatic dispute concerned with national control of airspace occurred in August 1983 when a South Korean civil airliner was shot down by military planes of the Soviet Union in that country's airspace. Following this event the International Civil Aviation Organization adopted an amendment to the Chicago Convention. This amendment laid down the principle that states must refrain from resorting to the use of weapons against civil aircraft in flight (Martin *et al.* 1985, IV/4). The amendment also permits the sovereign state to intercept a plane flying illegally in its airspace though ensuring that the lives of persons on board and the safety of the aircraft are not endangered. Intercepted planes may be required to land at a designated airport. Further, the amendment provides for each contracting state to establish laws which punish severely the crew of any registered plane which refuses a proper order to land.

Another type of dispute is associated with pollution carried in the atmosphere from one country to another. In January 1986 William Davis, a former Premier of Ontario Province, and Drew Lewis, a former Secretary of the United States Transportation Department, issued a joint report on acid rain in the borderland between their two countries. Acid rain is produced when large quantities of sulphur dioxide and nitrogen oxides are discharged into the atmosphere and reactions within the atmosphere produce dilute sulphuric, nitric, and, in some cases, hydrochloric acid (Brown *et al.* 1981, pp. 17–21, and Ostmann 1982, pp. 10–14). These dilute acids are present in rain and snow in some areas and they can cause damage to crops, the soil, buildings, fish, animals, and people.

It is considered by many scientists that the level of acidity in precipitation is directly related to the burning of coal with a high sulphur content and discharging the smoke from very tall chimneys, which are, ironically, designed to avoid pollution in the immediate vicinity of the factory (Ostmann 1982, p. 35).

The Davis–Lewis report recommended that Canadian and American industries should share the cost of developing a technology to ensure that coal is burned cleanly to avoid the discharge of pollutants. The report calculates the cost at US$ 5 billion. It is considered by authorities in the United States that this would be a cheaper solution than installing pollutant scrubbers or requiring factories to switch to cleaner fuels.

The issue of atmospheric pollution dominated headlines in Europe in late April and early May 1986 after the nuclear accident at Chernobyl in the Soviet Union on 26 April 1986. A high-pressure system which lasted for some days after this date created prevailing southerly winds which carried a plume of radioactive gases northwards across Scandinavia. In early June a report gave some indication of the incidence of unusually high levels of radioactivity in Scandinavia. The agricultural district around Gavle recorded average radioactive readings of 137 000 becquerels per square metre after heavy rains on the 28 and 29 April 1986. Gavle, which is on the east coast of Sweden, had a highest specific reading of 200 000 becquerels. Average figures declined westwards to 13 000 becquerels in Trondheim and 2500 becquerels in Scotland. These radioactive levels were produced by caesium 137, iodine 131, plutonium 239, and strontium 90 (*The Observer* 1 June 1986, p. 14).

A number of governments in Europe took precautions to prevent the importation of contaminated food from the Soviet Union and Eastern Europe. They have also monitored levels of radioactivity since the end of April 1986, and it is possible that claims for compensation will be made against the Soviet Union.

It is possible to consider the atmosphere as a common resource which straddles international boundaries. Just as the water of the river Nile flows successively from Ethiopia to the Sudan to Egypt, so weather systems cross international boundaries sometimes in one direction and at other times in the opposite direction. Just as downstream states demand that the river should not reach them as a polluted or reduced flow, so states have the right to demand that neighbours will not allow the pollution of the circulating atmosphere. Steps have already been taken to try to reduce the incidence of acid rain, and it is likely that international concern will encourage measures to reduce the risks of radioactive material being emitted into the atmosphere.

References

Africa Research 1979. *Africa Research Bulletin*. Political, social and cultural series Vol. 16. Exeter: Africa Research.

Africa Research 1986. *Africa Research Bulletin*. Political, social and cultural series Vol. 23. Exeter: Africa Research.

Ancel, J. 1938. *Les frontières*. Paris: Armand Colin.

Balch, T. W. 1912. Is Hudson Bay closed or open sea? *American Journal of International Law* **6**, 409–59.

Baty, T. 1928. The three mile limit. *American Journal of International Law* **22**, 503–37.

Boggs, S. W. 1930. Delimitation of the territorial sea. *American Journal of International Law* **24**, 541–55.

Boggs, S. W. 1937. Problems of water boundary definition: median lines and international boundaries through territorial waters. *Geographical Review* **27**, 445–56.

Boggs, S. W. 1940. *International boundaries: a study of boundary functions and problems.* New York: Columbia University Press.

Brown, C., B. Jones, G. B. Sperling & T. White 1981. *Rain of death: acid rain in western Canada.* Edmonton: NeWest Press.

Brownlie, I. 1979. *African boundaries: a legal and diplomatic encyclopaedia.* London: Hurst.

Bush, W. M. 1982. *Antarctica and international law* (3 volumes). London: Oceana.

Crocker, H. G. 1919. *The extent of marginal seas.* Washington, DC: Government Printing Office.

Curzon of Kedleston, Lord 1907. *Frontiers.* The Romanes Lecture. Oxford: Oxford University Press.

Dorion, H. 1963. *La frontière Quebec–Terreneuve* (The boundary between Quebec and Newfoundland). Quebec: Centre d'Études Nordiques.

Eckert, R. D. 1979. *The enclosure of ocean resources.* Stanford, Calif.: Hoover Press.

Escalada, V. 1979. *Aeronautical law.* Alphen aan den Rijn: Sijthoff & Noordhoff.

Fawcett, C. B. 1918. *Frontiers, a study in political geography.* Oxford: Oxford University Press.

Fenn, P. T. 1926. Origins of the theory of territorial waters. *American Journal of International Law* **20**, 465–82.

Fulton, T. W. 1911. *The sovereignty of the sea: an historical account of the claims of England to the dominion of the British seas, and of the territorial waters; with special reference to the rights of fishing and the navy salute.* London: W. Blackwood.

Galiani, P. 1782. *Dei Doveri dei Principi Neutrali Versi i Principi Guerregianti, e di Questi Verso i Neutrali* (On the obligations of neutral princes towards belligerent princes and the latter towards neutral princes). Milan.

Gidel, G. C. 1932. *Le droit international public de la mer* (International public law of the sea). Chateauroux: Mellottée.

Glassner, M. I. 1970. The Rio Lauca: dispute over an international river. *Geographical Review* **60**, 192–207.

Gyorgy, A. 1944. *Geopolitics: the new German science.* University of California Publications in International Relations, Vol. 3, 141–304.

Hartshorne, R. 1936. Suggestions on the terminology of political boundaries. *Annals, Association of American Geographers* **26**, 56–7.

Haushofer, K. 1927. *Grenzen im ihrer Geographischen und Politischen Bedeutung* (The geographical and political significance of boundaries). Berlin: K. Vowinckel.

Hertslet, E. 1875–91. *The map of Europe by treaty* (4 volumes with continuous pagination) London: Butterworth.

Hill, N. L. 1976. *Claims to territory in international law*. New York: Greenwood Press.
Hinks, A. R. 1940. Review. *Geographical Journal* **96**, 286–9.

Institute of International Law 1928. *Annuaire de l'Institut de Droit International* (Yearbook of the Institute of International Law), new abridged edition, Vol. 3. Paris: A. Pedone.
Ireland, G. 1938. *The possessions and conflicts of South America*. Cambridge, Mass.: Harvard University Press.

Jessup, P. C. 1927. *The law of territorial waters and maritime jurisdiction*. New York: G. A. Jennings.
Johnson, R. W. 1966. The Canada–United States controversy over the Columbia River. *University of Washington Law Review* **41**, 676–763.
Jones, S. B. 1945. *Boundary-making, a handbook for statesmen, treaty editors and boundary commissioners*. Washington, DC: Carnegie Endowment for International Peace.

Keesing's Contemporary Archives 1978. Volume 24. Bath: Keesing's Publications.
Kent, H. S. K. 1954. The historical origins of the three mile limit. *American Journal of International Law* **48**, 537–53.
Kroell, J. 1934. *Traité de droit international public aerien* (Treatise of international law dealing with airspace). Paris: Les Éditions Internationales.

Lamb, A. 1966. *The McMahon Line* (2 volumes). London: Routledge & Kegan Paul.
Lapradelle, P. de 1928. *La frontière: etude de droit international* (The boundary: a study of international law). Paris: Les Éditions Internationales.
League of Nations 1920–45. *Treaty series*. Geneva: League of Nations.
Lovering, J. F. & J. R. V. Prescott 1979. *Last of lands – Antarctica*. Melbourne: Melbourne University Press.

Martin, P., J. D. McClean, E. de M. Martin & R. D. Margo 1985. *Shawcross and Beaumont: air law*, Vol. 1. London: Butterworth.
Matte, N. M. 1981. *Treatise on air–aeronautical law*. Toronto: Carswell.
Meyer, C. B. V. 1937. *The extent of jurisdiction in coastal waters*. Leiden: A. W. Sijthoff.

Nicholson, N. L. 1954. *The boundaries of Canada its Provinces and Territories*. Memoir 2, Department of Mines and Technical Surveys, Geographical Branch, Ottawa.

Ostmann, R. 1982. *Acid rain: a plague upon the waters*. Minneapolis: Dillon.

Parry, C. 1969–79. *Consolidated treaty series* (231 volumes). New York: Oceana.
Paullin, C. O. 1932. *Atlas of the historical geography of the United States*. New York: Carnegie Institution and the American Geographical Society.
Pounds, N. J. G. 1951. The origin of the idea of natural frontiers in France. *Annals, Association of American Geographers* **41**, 146–57.
Pounds, N. J. G. 1951. France and 'les limites naturelles' from the seventeenth to twentieth centuries. *Annals, Association of American Geographers* **44**, 51–62.
Prescott, J. R. V. 1975. *Map of mainland Asia by treaty*. Melbourne: Melbourne University Press.

Prescott, J. R. V. 1984. Boundaries in Antarctica. In *Australia's Antarctic policy options*, S. Harris (ed.), 83–112. Canberra: Centre for Resource and Environment Studies.
Prescott, J. R. V. 1986. Delimitation of marine boundaries by baselines. *Marine Policy Reports* **8** (3), 1–6.
Prescott, J. R. V., H. J. Collier & D. F. Prescott, 1977. *Frontiers of Asia and Southeast Asia*. Melbourne: Melbourne University Press.

Ratzel, F. 1897. *Politische geographie* (Political geography). Berlin: Oldenbourg.
Ritter, G. & J. Hajdu 1982. *Die Deutsch–Deutsche Grenze* (The boundary between East and West Germany). Koln: Geostudien.

Spykman, N. J. & A. A. Rollins, 1939. Geographic objectives in foreign policy. *American Political Science Review* **32**, 28–50, 213–36.

The Geographer 1971. *Mozambique–Southern Rhodesia boundary*. International Boundary Study No. 118, Department of State, Washington DC.
The Geographer n.d., *Widths of selected straits and channels*. Department of State: Washington DC.
The Observer 1 June 1986. London.

United Nations 1945– . *Treaty Series*, Vols 1–701. New York: United Nations.
United Nations 1983. *The law of the sea: United Nations Convention on the Law of the Sea*. New York: United Nations Secretariat.

Vattel, E. de 1758. *The law of nations or the principles of natural law*, Vol. 3. Washington DC: published by the Carnegie Institution of Washington in 1916, after translation by G. G. Fenwick.

Walker, W. L. 1945. Territorial waters: the cannon shot rule. *British Yearbook of International Law* **22**, 210–31.
Wenk, E. 1969. The physical resources of the ocean. *Scientific American* September, 167–76.
Widstrand, C. G. 1969. *African boundary problems*. Uppsala: Nordiska Afrikainstitutet.

2 *Frontiers*

Political geographers use the term 'frontier' in two senses; it can either refer to the political division between two countries or the division between the settled and uninhabited parts of one country. In each sense the frontier is considered to be a zone. There is no excuse for geographers who use the terms 'frontier' and 'boundary' as synonyms.

This chapter considers those aspects of frontiers which are of interest to political geographers. It is divided into three parts. First, there is a consideration of settlement frontiers which exist within a single country. Secondly, political frontiers which used to separate neighbouring countries are examined. Thirdly, an account is given of frontiers in the area occupied today by Nigeria.

Settlement frontiers

Two kinds of settlement frontiers are recognized. Primary settlement frontiers exist when a state is taking possession of its territory for the first time. The classical example is the westward expansion of American sovereignty through its territory in North America. Secondary settlement frontiers are found in many countries today and mark zones which separate settled and uninhabited regions of the state. The two types have different characteristics.

Primary settlement frontiers are historic features, whereas secondary settlement frontiers are currently found in many countries where an adverse physical environment or inadequate techniques hinder further advances of land use and settlement. The primary settlement frontier marked the actual limit of the state's political authority, whereas the political authority of modern states extends beyond the secondary settlement frontiers and can be exerted when necessary. Any country, such as Australia, which includes a large area of desert provides special services which can operate in those uninhabited areas. The range of potential economic activities in a primary frontier is generally greater than in the secondary frontiers. Fur-trapping, timber-felling, semi-subsistence cultivation, grazing, mining, and manufacturing and service industries were all found at some point on the American frontier, or developed as it passed. On the other hand, the advance of secondary settlement frontiers is likely to be accompanied by the extension of irrigated farming as in Mali on the Samanko scheme near Bamako; by extensive ranching as in Matabeleland in Zimbabwe; and by the use of new mineral or fuel deposits as in the Northern Territory of Australia. Becker (1980) has described the rôle of mining settlements in extending the frontier in the Yukon Territory and Alaska. The discovery of rich placer deposits of gold provided the impetus for centres such as Dawson, Forty Mile, and Grand Forks. His analysis shows that once the prospectors had discovered the deposits the main influence on the growth of the settlement was provided by traders. They were searching for the best locations for commerce and the chance of profitable speculation in real estate. Governments played a minor rôle, which was mainly concerned with surveying and laying out a rectangular town site often after the first buildings had been erected.

Secondary frontiers normally reflect the limited range of economic activities by a population of low density, but on primary settlement frontiers densities may be moderate to heavy. There are exceptions to this generalization in some secondary frontiers in Pakistan and Bangladesh. The United States Census Bureau's definition of the frontier as areas having a population density of two to six persons per square mile would have excluded many of the early frontiers in Georgia. The development of secondary settlement frontiers is usually carefully planned and is based on a satisfactory communications network, in contrast to the haphazard development of primary frontiers, which were also characterized by 'rudimentary socio-political relations marked by rebelliousness, lawlessness and/or absence of laws' (Kristof 1959). Lastly, the primary settlement frontiers often advanced rapidly. In 1783, 4 million acres of the Cumberland Valley were sold in seven months, and in 1795, during only two months, 26 000 migrants crossed the Cumberland River in search of cheap land to the west (Billington 1960). The advance of secondary settlement frontiers usually involved small areas and comparatively few people.

Primary settlement frontiers

Much has been written about the primary settlement frontier. American historians are largely responsible for the thorough documentation of the American primary frontier in their efforts to support or refute Turner's frontier hypothesis that the 'existence of an area of free land, its continuous recession and the advance of American settlement westward explain American development' (Turner 1953). The small number of contributions by geographers suggests that historians have preempted this field. Whittlesey (1956), writing on the expansion and consolidation of the United States, makes only passing reference to the frontier and no reference to the detailed historical studies. Despite this situation, geographers can make a useful contribution beyond the mapping of frontier phenomena, which has been done for the American and Canadian frontiers (Paullin 1932, Adams 1943, and Kerr 1961).

The position of the frontier, which represents the actual zonal limits of political authority, and its width are of prime interest to geographers. In order to determine the frontier's extent some criteria must be developed to distinguish it from non-frontier areas. A simple basis of population density is unsatisfactory, and a more satisfactory measure is likely to be found in the degree of political and economic organization. This is a task calling for training in historical and political geography. In considering South Africa's frontiers, Christopher (1982) has drawn attention to the inadequate statistical information available for the 19th century, and he has proposed a careful examination of population distribution, land alienation, cultivation, and stock numbers to identify the position of the frontiers and their characteristics.

Information about the position of the frontier at any time will give some indication of the factors which have influenced its rate of advance. Any advance of the frontier probably resulted from a combination of factors, which can be principally divided into forces of attraction based on the nature of the environment and forces of pressure from the frontier's hinterland, often of a social, economic, or political nature. The rôle of unusually favourable soil groups, such as are found in the Blue Grass country of Kentucky and the cotton lands of the Gulf plains, in promoting the rapid advance of the American frontier are well known. In a similar fashion, discoveries of precious mineral deposits have caused spectacular frontier advance as in the case of the Transvaal.

Pressures within the frontier hinterland take many forms. Turner (1953) and Billington (1960) have shown how the frontiersmen were seeking to avoid high land prices, heavy taxation, and political and religious disabilities imposed either by the first, well established settlers or by the governments of the country of origin. Further, the experience gained on one frontier, in respect of land legislation, mining laws, and Indian treaties, was applied at subsequent frontiers and often allowed speedier settlement of these problems. Periods of standstill or retreat along the frontier resulted either from the unfavourable nature of the environment or the inadequacy of techniques for utilizing it – the armed resistance of indigenous groups, or the preoccupation of the government with more urgent considerations:

> In these successive frontiers we find natural boundary lines which have served to mark and affect the characteristics of the frontiers, namely: the 'fall-line'; the Allegheny Mountains; the Mississippi; the Missouri where its direction approximates north–south; the line of the arid plains, approximately the ninety-ninth meridian; and the Rocky Mountains. The fall line marked the frontier of the seventeenth century; and Alleghenies that of the eighteenth; the Mississippi that of the first quarter of the nineteenth; the Missouri that of the middle of this century

(omitting the Californian movement); and the belt of the Rocky Mountains and arid tract, the present frontier. Each was won by a series of Indian wars. (Turner 1953, p. 9)

The maps of the Indian battles (Paullin 1932) show that the fiercest resistance by indigenes often coincided with the most difficult terrain, which offered superior strategic opportunities for defence. It was in the scrub country of Queensland that Aborigines offered the greatest resistance to the extension of pastoral activities. In Cameroun, during the early years of this century, the Germans faced their greatest problems in pacifying the Chamba and other pagan groups, who were located in the heavily dissected borderland with Nigeria. The stagnation of the American frontier during the twenty years before 1795 resulted from the preoccupation of the colonies with securing independence, establishing a federal constitution, and defeating the Indians in the area already settled. Porter (1979) has noted that the delayed settlement of the western areas of Maryland can only be understood after consideration of the contemporary physical, financial, and political conditions.

In some cases attempts to halt frontier advance were made by the government or by interested trading organizations. Some of the earliest coastal states in North America tried to restrict the frontier advance in order to retain political power and to avoid a further drain on their population. In the hinterland of New York and Pennsylvania, the Iroquois Confederation blocked for a century the route through the Catskill and Berkshire ranges, by the Mohawk and Hudson valleys, in order that the Indians who supplied the fur trade should not be driven away (Billington 1960). Konrad (1981) has described how the desire of Iroquois to continue their profitable trade in furs, especially beaver, resulted in an advance of their frontier from the southern shore to the northern shore of Lake Ontario. In the second half of the 17th century this area was a devastated no-man's-land as far as the British and French authorities were concerned. However, the Iroquois established their settlements to provide bases from which winter expeditions could be conducted north and west of Lake Ontario and as staging posts for fur brigades on their way south to trade with the English.

Turner maintains that 'each frontier leaves its trace behind it, and when it becomes a settled area the region still partakes of the frontier characteristics' (Turner 1953, p. 4). This suggests a fruitful field for geographical research. Is it possible to attribute any elements of the cultural landscape to the period when the area was a primary frontier? There is probably a connection between present property boundaries and the original policies of land allocation and appropriation. It seems unlikely that the present economy will reveal many features which can be traced to frontier times, since the earliest economic activities of hunting and grazing will survive only if the land is unsuitable for cultivation and lacks the resources on which can be built towns with secondary and service industries. It has been noted by Clarke (1959) that when a period of standstill allowed an accumulation of population, as in Georgia and Tennessee, eventual advance was more orderly and complete. Rapid advance, without resistance, resulted in scattered and discontinuous settlement patterns. It would be interesting to know whether settlement analysis reflects this process of development.

Mitchell (1972) has published a good geographical analysis of the American primary settlement frontier in the Shenandoah valley in the 18th century. His analysis proceeds along three main themes. First, he considers the movement of migrants into and through the area, for such an investigation provides valuable information regarding the political organization of the frontier and the distribution of groups with different cultural characteristics. Further, a knowledge of migration routes allows the scholar to understand how the frontier moved and developed differentially. Secondly, he is concerned with the developments which occurred while the frontier occupied various positions. The processes of land acquisition from the Indians in the first case and subsequently land subdivision and redistribution by the landowners in the second case attract Mitchell's attention. The nature of pioneer and post-pioneer economies receives detailed treatment. Thirdly, the changing location of the frontier relative to the settled areas and the uninhabited areas is assessed at intervals throughout the frontier's history. Mitchell concluded that during the frontier phase the Shenandoah valley was socially more complex and economically more sophisticated than is generally acknowledged. His paper forms a valuable example of useful research which geographers can undertake.

Secondary settlement frontiers

Secondary settlement frontiers are found in all countries which include areas of unfavourable environment, such as tropical or temperate desert, heavily dissected uplands, or thick tropical rain forest, and areas which require the use of advanced and often expensive techniques if they are to be used for purposes other than mineral extraction. These are the areas which are often bypassed by the primary frontier, when the population is concerned with rapid exploitation and profit. These inferior regions will be attacked later if circumstances require it and if new techniques or discoveries make it possible to revalue the environment. Burt has recorded the following interesting observation about the Canadian frontier in the 19th century: 'The expansion of Canadian settlement ran up against the rocky Pre-Cambrian Shield, with the result that the Canadian frontier movement crossed the [American] border, where it became merged in the greater movement to the northern Middle Western States' (Burt 1957, p. 71). Only when the availability of land in the American West was reduced did the frontier cross the boundary, again promoting the development of western Canada. Straight (1978) has shown that there was also a northward movement by American settlers across the Canadian boundary in New England. This movement started after the 17th century and continued without much apparent interference.

In many European states only short secondary settlement frontiers surround small sectors of unfavourable environment. In Australia, on the other hand, a long secondary settlement frontier surrounds the central desert. Attempts to thrust forward secondary frontiers usually depend on some incentive, such as shortage of land, shortage of food in times of war, strategic needs, or the discovery of new mineral deposits. In the 1950s shortage of land in African Reserves in southern Matabeleland led to the cultivation of areas where there is a high risk of drought or rainfall deficiency. In Java, population pressure on available land has caused the cultivation of slopes with a high erosion hazard. In many cases efforts to advance secondary settlement frontiers are guided and controlled by government, because of the need for large amounts of capital which are unlikely to yield satisfactory returns. Anderson (1981) has described the efforts of the Brazilian government to establish agricultural settlements in the Bragantina region, which lies between Belem and Braganca. Nearly 30 settlements were established between 1889 and 1920; they produced a densely populated area which satisfied the short-term goals of the authorities. However, the long-term goals were not achieved, and Anderson attributes this failure to limiting factors in the humid tropics which were not perceived by the planners.

Examples of the part played by improved techniques can be found in many countries. The waterless areas of the Kalahari sandveld in South Africa were not settled until after 1903, when the well-drill made it possible to tap underground water reserves, which could be brought to the surface by wind pumps. In Australia the development of heavy machinery and the stump-jump plough allowed more intensive use of the Mallee areas. These areas were covered by species of eucalypt which form a tabular mass of hard wood at, or just below, the surface. The foliage was unpalatable to stock, sheep could not be mustered because the vegetation was so dense, and watering points were scarce. In Canada, the development of new strains of wheat has allowed the use of areas with short growing seasons. The spectacular extension of Soviet agriculture in central Asia was stimulated during the 1940s by Germany's capture of areas in eastern Europe.

Nearly a century ago, the threat of Russia's advance into Hokkaido encouraged the Japanese government to foster the rapid colonization of that island. The government distinguished between those immigrants who travelled independently and those who travelled with the aid of a government subsidy. Independent farmers received implements, seed, and 10 yen for every quarter acre cleared. Subsidized farmers received a rice ration for three years in addition to seed and implements. Their bonus for clearing land was 2 yen per quarter acre. Independent artisans and merchants received a gift of 150 yen for three years, which eventually had to be repaid (Harrison 1953). In the decade which followed 1869 nearly 65 000 Japanese migrants entered Hokkaido.

The discovery of new mineral deposits or the change in world trading conditions, which make the mining of known mineral deposits a profitable undertaking, have often caused the advance of secondary settlement frontiers into the Arctic and tropical deserts. Lawless (1974) has described the establishment of a new mining and industrial centre at Arlit in Niger. The French Atomic Energy Commission discovered a major uranium deposit 250 kilometres north-west of Agades, the nearest town. A town capable of housing 5000 people has been built and roads leading to the coast have been improved. Production from this site reached 1500 tonnes in 1975, when Niger was the fifth largest producer of uranium in the world.

Sometimes for strategic reasons countries such as China, Canada, and the Soviet Union have established settlements in inhospitable deserts and along the shores of the Arctic Ocean.

Stone (1979) has provided a useful summary of the general location and characteristics of the world's settlement frontiers. He classifies the land into three categories: continuous settlement, discontinuous settlement, and uninhabited. Then he proceeds to divide the areas of discontinuous settlement into the inner fringe zone, the middle fringe zone, the outer fringe zone, and the outermost fringe zone. The inner and middle fringe zones are deemed to constitute settlement frontiers that might have a reasonable potential for new settlement, and where these areas occupy at least 7000 square miles they are mapped at the world scale. Stone notes that there will be some areas measuring less than this and they should be identified by the use of maps with a scale larger than 1 : 125 000, which were the base maps he used. The frontiers he identifies are described according to their population densities and their recent changes in numbers of people.

Stone's work builds on the global view of secondary settlements set out by Bowman (1931) nearly 50 years before. He called the secondary settlement frontiers 'the pioneer fringe'.

> Without wading too deeply into the sea of technology we may define the remaining pioneer areas of the world as regions of potential settlement in which man may have a reasonably safe and prosperous life; but regions in most of which he is required to make certain special

> adaptations. What these adaptations are the settler may discover by painful experiment, as in the past, or by less painful experience if government and science step in to help him. The rainfall of the undeveloped pioneer lands is on the whole not so favourable as in most of the settled areas of the earth, but it is sufficient if the settler uses it rightly. The temperature may be too hot or too cold to suit him, but it is tolerable. (Bowman 1931, p. 51)

Although the world maps produced by Bowman and Stone are not directly comparable, it is apparent that pioneer fringes identified by Bowman in eastern Asia south of parallel 60° north, in central Asia east of the Aral Sea, and along the outer edges of the Amazon Basin have been considerably reduced in extent.

Political frontiers

Political frontiers used to separate neighbouring countries and geographic interest in them is mainly concerned with their physical characteristics, their position, the attitudes and policies of the flanking states, the influence of the frontier on the subsequent development of the cultural landscape, and the way in which boundaries were drawn within the frontier.

Lord Curzon's essay in 1907 contained the seeds of a useful classification which was brought to fruition by East. East distinguished between frontiers of contact and frontiers of separation, and observed, 'states have always sought frontiers which foster separation from, rather than assimilation with, their neighbours' (East 1937). Some frontiers, either by the attraction of their resources or the ease with which they can be crossed, allowed contact between separated political groups. This contact took the form of trade, the payment of tribute, migration, or conflict. On the other hand, the frontier sometimes possessed physical qualities which made it unattractive to exploiters and travellers alike. However, in no case did the geography of the frontier determine the degree of intercourse between states; rather the attitudes and policies of the flanking states were decisive. When Chile achieved its independence, its state limits included the Atacama Desert to the north and the Andes Mountains to the east, both physical barriers which inhibited cultural contacts. Yet during the last century the expansionist policies of the Chilean government carried the country into war with Peru and Bolivia over the Tacna-Arica districts of the Atacama and into a dispute with Argentina concerning the trans-piedmont slopes of the Andes, which in some cases were settled by Chilean emigrants. The successful northward advance against Peru and Bolivia, made for economic and strategic reasons, delivered to Chile the port of Arica and access to the valuable borax, copper, and nitrate deposits (Dennis 1931).

The effects of a policy of isolation are revealed by considering the case of the Kingdom of Benin west of the Niger Delta. Although the forested frontiers of this state were no more difficult to cross than similar frontiers surrounding other indigenous kingdoms, there was practically no contact between European traders and Benin, because traders were not welcome. Eventually it required an expeditionary force to conquer the country and to establish relations between the colonial and indigenous authorities.

International boundaries have replaced international frontiers throughout the world; therefore research into them must have a strong historical and anthropological basis. Political frontiers generally enjoyed less intensive economic development than the territories they separated. This was because the environment was unfavourable, or because the resources of the existing state area were sufficient, or because it was the policy of the state to neglect the frontier thereby enhancing its divisive character. Deserts, mountain ranges, rivers, flood plains, and woodlands have all formed frontiers at some period in history. It follows from this that frontiers were usually less densely populated than the flanking states and that the inhabitants of the frontier if there were any, experienced a lower standard of living than the citizens of neighbouring states. Tacitus the eminent Roman historian of the 1st century described the debased condition of the Slavic Venedi who occupied the wooded and mountainous area between the Peucini and Fenni. A more recent example was cited by Tilho (Ministère des Colonies 1919): the wretched Beddé pagans survived in the swampy areas between the kingdoms of Sokoto and Bornu in the western Sudan only if they could escape, being enslaved by raiding parties from both sides.

Yesner has described the frontier which separated the Aleut and Inuit populations on the Alaskan peninsula:

> The boundary zone between Eskimo and Aleut populations on the Alaska Peninsula was not a resource-rich buffer zone, but a (relatively) resource-poor zone that acted as an isolating barrier between two relatively richer resource zones – one nearly exclusively maritime and one providing a mixed coastal/riverine/tundra suite of resources . . . Once established, the boundary between these two regions remained relatively stable over a long period until the Thule peoples – either because of environmental change, increased technological efficiency, or population pressure – were finally able to penetrate the boundary. (Yesner 1985, p. 84)

Where there was the threat of invasion or trespass, political frontiers were selected for their defensive advantages and this point was thoroughly investigated by Curzon and Holdich. Curzon mentioned that deserts formed the best defensive frontiers, but it seems worth remarking that extensive deserts, such as the Sahara, were often the habitat of mobile, warlike tribes, such as the Tuareg, which plagued the surrounding semi-argicultural tribes. Davies made this precise point when he wrote of the northwest frontier of British India: 'So long as hungry tribesmen inhabit barren and almost waterless hills, which command open and fertile plains, so long will they resort to plundering incursions in order to obtain the necessaries of life' (Davies 1932, p. 179). Linear mountain ranges and rivers had the strategic advantage of allowing the defending forces to focus their strength at passes and bridges. The possession of limiting deserts, mountain ranges, and major rivers is a matter of geographical good fortune, and it seems likely that many of the original frontiers consisted of woodland and marshes. The swamps and forests surrounding Westphalia played a major part in the defeat of some Roman legions. A more recent example is provided by the forested margins of Kikuyuland in East Africa. This forested zone was about two hours' march in width and enabled the Kikuyu to defeat Masai invaders, who seemed invincible on the grassy plains of Masailand (Hohnel 1894, p. 1).

Frontier markers

Many states tried to mark their frontiers in some way. The famous Roman and Chinese walls are the best examples. The Great Wall of China served not only to exclude nomadic barbarians but also to restrict the number of Chinese who adopted a modified agricultural system which made them more difficult to control from the Chinese capital (Lattimore 1940). The modern journey from Peking to the restored portion of the Great Wall passes a number of earlier, local walls which show the same disregard for steep gradients exhibited by their famous national equivalent. The walls of the Roman Empire, unlike the Great Wall of China, did not mark a major environmental divide and seemed to be built for the sole purpose of defending the empire, by permitting some control over, if not total exclusion of, the barbarians. Where clear physical features were not available, the Romans constructed walls such as the well known Hadrian's Wall, linking Solway Firth to the Tyne valley. Two others were built across the re-entrant formed by the upper courses of the Rhine and Danube, and east of the Drava–Danube confluence. The barbarians north of the Roman wall also built earthworks to delimit their territory. It is recorded that the Angrivarii constructed a broad earthwork to mark their boundary with the Cherusci. It might be asked whether these walls were not boundaries rather than frontiers, even though they were generally selected in a unilateral manner. The reply would be that the walls formed the first or last line of a system of defence in depth rather than the limit of national sovereignty (Baradez 1949). The Roman walls were reinforced by establishing farmers on land behind the walls, in a zone called *agri limitanei*. The men of these families were expected to assist in the defence of the wall in time of need, in a way reminiscent of some Israeli Kibbutzim adjacent to Arab territory. However, as an exception to this rule Collingwood (1923) noted that the *vallum* behind Hadrian's Wall marked the limits of Rome's civil government.

The counterparts of the Roman and Chinese walls could be found in Africa until quite recently:

> The kingdom was surrounded, where there were no natural defences by deep and wide ditches defended by tree trunk palisades and crossed at intervals by narrow bridges. The northern frontier was formed by the Gojeb River, called Godafa by the Kafa. Beiber gives the dimensions of the ditches as 6 metres in width and 3 metres in depth; he describes the gates, *kello*, as consisting of circular fenced enclosures entered by drop gates. Customs dues were collected at these gates. Outside the line of fences was a strip of unoccupied land like the *moga* of the Galla states. At points where Galla attacks were expected, the gates were additionally defended by a high rampart and several lines of entrenchment, a form of defence much admired by the neighbours of the Kafa. (Huntingford 1955, p. 116)

The *moga*, or uncultivated strip, of the Galla states of the Horn of Africa was inhabited only by fierce brigands, who were encouraged by the Galla rulers to attack common enemies and recapture escaped slaves.

The reduction of political frontiers

Fischer maintains that a 'rather extensive literature deals with the development of boundary lines out of such [frontier] zones or related features' (Weigert *et al.* 1957, p. 79). However the works which he cites do not treat this aspect of boundaries in detail. The general impression is that as states separated by frontiers extend their territory, the unclaimed land diminishes. Eventually property disputes arise and an attempt is made to resolve these difficulties by delimiting a precise boundary. No doubt this situation has occurred in many cases, but there are some significant variations on this theme, which are examined in the following paragraphs.

We can begin by saying that frontiers normally diminish in width and that frontiers of separation are replaced by frontiers of contact. There have been cases where frontiers have increased in width, although this is not usual. An example of a widening frontier was provided by Hashtadan; it was situated on the borderlands of Afghanistan and Persia south-west of the great northward bend of the Hari Rud at Koshan. This area was investigated by General Maclean in 1888–9 prior to making a boundary award as requested by the Persian and Afghan governments. He found village ruins, faint field patterns, and portions of underground canals which revealed that the area had once been occupied and fairly prosperous.

He established that an epidemic throat disease had decimated the valley's population in 1788 and that subsequent raids by Uzbek Hazarah and Turcoman raiders were responsible for the subsequent devastation of the valley:

> But from the appearance of the ruins and abandoned fields it is quite evident that the valley has been deserted for some generations . . . Upon the whole, looking to the nature of my present information, it seems to me that neither Persians nor Afghans can produce proofs of recent possession in support of their respective claims, neither having felt inclined to stand the brunt of collisions, in such an exposed locality, with the Turkomans. (Prescott 1975, p. 154)

In the more usual process frontiers diminish in width either by incorporation of parts of the frontier by one or both of the flanking states or by the creation of subsidiary political units within the frontier.

Annexation of parts of the frontier might take place because of rising land hunger within the state or through the development of new techniques which enable the frontier resources to be revalued. Fifer (1966) has provided a most interesting account of the extension of political control by Bolivia and Brazil over the 'empty, unknown, formerly negative frontiers of separation' which lay between them. She clearly identifies the desire for new lands on which high-quality rubber could be produced during the second half of the 19th century. This process was accelerated during the 1880s when a severe drought throughout Brazil's eastern province of Ceara encouraged a mass exodus of labourers in search of work: 'not until the rubber boom involved the Amazon headwater region, thus probing beyond Brazil's undisputed territorial claims, did old negative frontiers of separation suddenly assume both economic and hence political significance' (Fifer 1966, p. 361).

If the frontier existed because of the internal weakness of the neighbouring states or their preoccupation with threats from other quarters, the removal of the threat or the resolution of the internal weaknesses might allow the frontier to be appropriated. Alternatively, the frontier may be invaded and annexed for strategic reasons. For example, after the Roman successes in Gaul, the eastern flank of this advance was protected by the annexation of Noricum, Pannonia, Meosia, and Dacia in the Danube Basin. This advance also removed the scene of conflict from the Mediterranean centres of the Empire (East 1962). In some cases annexation for one reason carried additional benefits. The Romans invaded the area between the River Rhine, the River Main, and the Taunus ridge to stop the raids by the Chatti. Later they discovered hot springs and iron and silver deposits in the region.

Political organizations within political frontiers

The subsidiary organizations which can be created within the frontier include marches, buffer states, and spheres of interest or influence. A march is a border territory organized on a semi-permanent military system to defend the frontier. An illustration of the creation of marches was provided by the policies of Charlemagne and Otto:

> From these Marks, intended to safeguard the Frontiers of the Empire from Slavonic or alien contact, and ruled by Markgrafs or Markgraves, sprang nearly all the kingdoms and states which afterwards obtained an independent national existence, until they became either the seats of empires themselves, as in the case of the Mark of Brandenburg, or autonomous members of the German Federation. (Curzon 1907, p. 27)

The German Kingdom was protected from the Slavs and Magyars by a series of marches stretching from the Baltic Sea to the Adriatic Sea. The Markgraves had the responsibility of defending the kingdom and extending the territory subject to it, and the winning of territory led the North Mark and East Mark to become the cores of major states. North Mark, founded in AD 928 by Henry the Fowler, later became known as Altmark, and it acquired Mittelmark, Vormark, Ubermark, and, eventually, Neumark east of the Oder River. This enlarged region was created the Mark of Brandenburg in AD 1157 and subsequently provided the territorial basis for the unification of Germany. East Mark, after being almost submerged beneath Magyar raids, was recreated by Otto the Great in AD 955. The additions of the Marks of Styria and Carinthia by AD 1282 created the core from which the Austro-Hungarian Empire grew (Fig. 2.1).

Buffer states have been constructed in frontiers when two strong neighbours decided to reduce the possibility of conflict between themselves. Britain's prime boundary strategy in the Indian subcontinent involved maintaining a system of small, weak states between British India and the territories of Russia, China, and France. In 1895 Britain and Russia fashioned the Wakhan panhandle which extends Afghan territory to the limits of Chinese territory, in order to avoid direct contact between British India and Russian central Asia. In the Himalayas, Britain encouraged the existence of Nepal, Sikkim, and Bhutan to separate China and India. Britain would have also welcomed an independent Tibet playing the same rôle in Sino-British relations as Mongolia played in Sino-Russian relations. In the East, Britain tried to persuade Thailand or China to accept the equivalent of the Wakhan panhandle, along the Mekong River, in order to separate British Burma and French Indochina:

> We could not have a conterminous frontier with France in Burmah. That would involve vast expenditure on both sides, and lines of armed

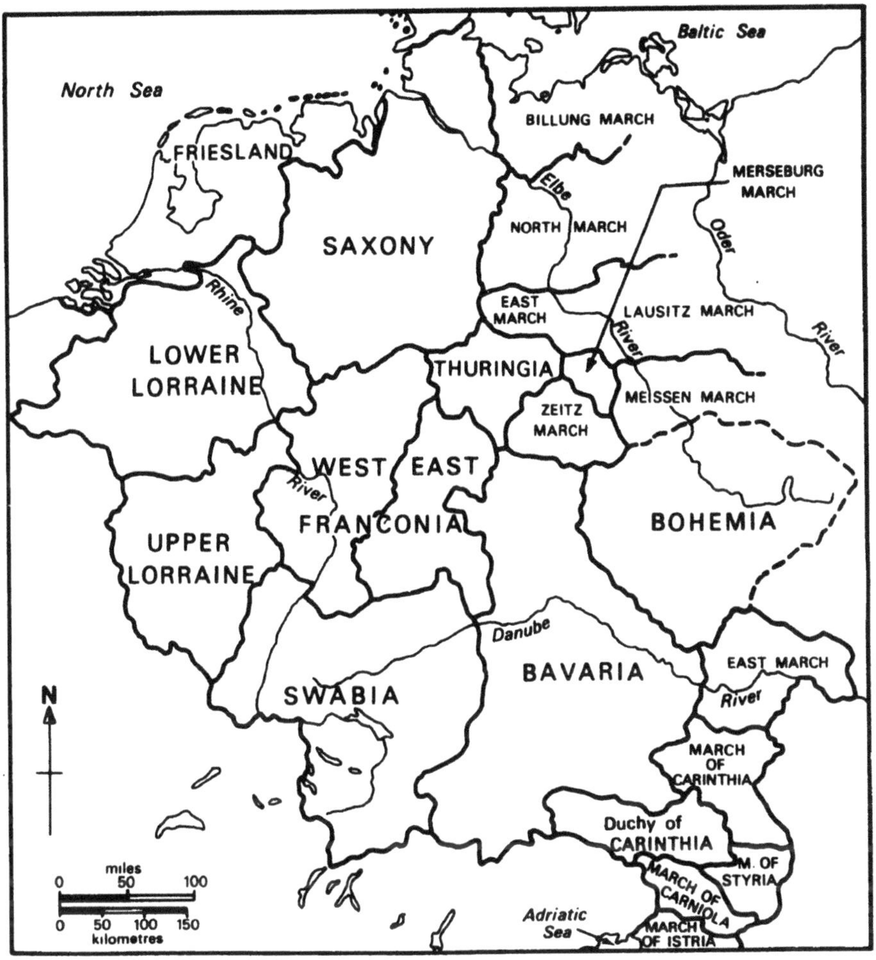

Figure 2.1 European marches.

> posts garrisoned by European troops . . . We had proposed the buffer state in the interests of both countries, for it was evident that if our boundaries were contiguous, any fussy, or ill-conditioned frontier officer, whether English or French, would have it in his power to magnify every petty incident into a grave international question, which would be transferred to Europe, and thus grow into a cause of exacerbation between the two Governments. (*British and Foreign State Papers* 1894–5, pp. 272, 379)

Some colonial powers employed neutral zones to serve the same function as buffer states. Britain and Germany separated their spheres of influence in Togoland and the Gold Coast by a neutral zone in 1887; it lay north of the confluence of the Dakka and Volta rivers. At various times there were also proposals for neutral zones between British and German and between

British and Portuguese territories in southern Africa. In 1965 and 1973 Kuwait and Saudi Arabia eliminated a neutral zone which they had jointly controlled. Neutral zones were often convenient solutions to difficult territorial questions, but they were really only short-term remedies. Usually their continued existence proved inconvenient and a greater source of friction than a detailed debate about their division between the two neighbouring states.

The concepts of spheres of interest and influence developed during the last century, when the major European powers were establishing actual and potential claims to parts of Asia and Africa. At no time have the responsibilities assumed under either concept by the claimant powers been defined. Both concepts are means of reserving a portion of territory from the political interference of another state, and it has been assumed that a sphere of interest is a less significant claim than a sphere of influence. Holdich (1916, p. 96) suggests that a sphere of interest becomes a sphere of influence when there is the threat of competition by another state, but against this it must be said that the formal definitions of spheres of interest and of influence were found usually in bilateral territorial agreements. An example is provided by the second article of the Anglo-French Agreement of 1890, in respect of African territories:

> The Government of Her Britannic Majesty recognises the sphere of influence of France, to the south of her Mediterranean possessions up to a line drawn from Say on the Niger to Barruwa on lake Tchad, drawn in such a way as to comprise in the sphere of action of the Niger Company all that properly belongs to the kingdom of Sokoto. (Hertslet 1909, p. 730)

This boundary stretched for 700 miles and the country through which it passed was largely unknown to the two parties. Indeed Lord Salisbury admitted as much when he commented on the results of the treaty:

> We have been engaged in drawing lines upon maps where no white man's foot has ever trod; we have been giving away mountains and rivers and lakes to each other, only hindered by the small impediment that we never knew exactly where the mountains and rivers and lakes were. (Kennedy 1953, p. 224)

The degree of interference with the indigenous organizations in the sphere of interest or influence varied in almost every case. At one end of the scale, the European power assumed no responsibility but claimed the exclusive right of its nationals to trade in the area; at the other end of the scale, there was a high degree of political control more appropriate to the condition of a protectorate.

Political frontiers and the evolution of landscapes

It was noted earlier that one aspect of geographical research connected with settlement frontiers concerned identifying elements in the landscape derived from frontier origins. Such studies are also a proper facet of political frontier studies, although little has been done in this direction by geographers. Noteworthy studies are by Cornish (1936) and Wilkinson (1955). Cornish traced the evolution of the language borderlands of Europe, such as Flanders, Lorraine, Friuli, Istria, and Macedonia. He found that in each case the language frontier coincided with an earlier political frontier between Christendom and heathen states, which had been static for some time. The growth of polyglot language regions occurred only where the frontier did not coincide with a divisive physical feature. Cornish called such regions 'link-lands', to emphasize their position between larger state areas. The heathen languages were eventually reduced to writing through contact with Christianity, and their traditions were thus preserved. Cornish points out that only during the 19th century, with improved means of mass communication between the larger state areas and the link-lands, did the bonds of language become more important than the regional ties of the link-lands.

Wilkinson shows how the Jugoslav Kosmet, at various times, formed the frontier between the Eastern and Western parts of the Roman Empire, the Bulgar and Byzantine empires, Christianity and Islam, and Yugoslavia and Albania. This situation has resulted in some neglect of the economic resources of the area and hindered its integrated development. Problems have arisen when boundaries have been drawn through upland areas which provide the summer pastures of a transhumance economy.

The following example of the geographical analysis of frontiers in West Africa is based on fieldwork and the review of an extensive literature, of which the most important works were by Barth (1857), Staudinger (1889), and Hogben (1930).

Frontiers in the Niger–Benue area

The largest state in West Africa in the middle of the 19th century was the Sokoto–Gando Empire founded by the Fulani Jihad 50 years earlier. It stretched from Libtako in the west to Adamawa in the east and from Katsina in the north to the latitude of Ilorin in the south. This territory, which was not subject to uniform political authority, was organized into provinces, each having a degree of independence, which varied directly with their distance from Sokoto or Gando. In the provinces of Zaria, Bida, Kontagora, Nassarawa, Kano, and Muri, the Fulani subjugated the indigenous tribes. In other areas such as Bauchi and western Adamawa, enclaves of pagan groups retained their independence on hilltop settlements. Finally, in Libtako and eastern Adamawa, only the main towns on the principal trade routes were subject to Fulani authority. These Fulani towns were exclaves within uncontrolled pagan areas, and might have been described as march towns (Fig. 2.2).

North and west of Sokoto lay the Habe states, organized by Hausa chiefs, who continued the struggle against the Fulani from new capitals. The westernmost Habe state was Kebbi, which had a narrow frontier with Sokoto and Gando, in the neighbourhood of which many raids were carried out and battles fought. The other two Habe states, Gober and Maradi, were separated from Sokoto by a frontier of separation, formed from a devastated zone. The towns of Jankuki, Dankama, and Madawa were destroyed by Fulani attacks. This depopulated zone became more thickly wooded than the rest of the area and served as a refuge for robbers. On the northern fringes of this frontier Maradi established the marches of Gazawa and Tessawa.

North-east and east of Sokoto lay the Bornu Empire and its vassal states, which included Zinder. The reduced power of Bornu after the Fulani conquest and subsequent Bornu revival had increased the degree of autonomy enjoyed by its traditional northern tributary states, including

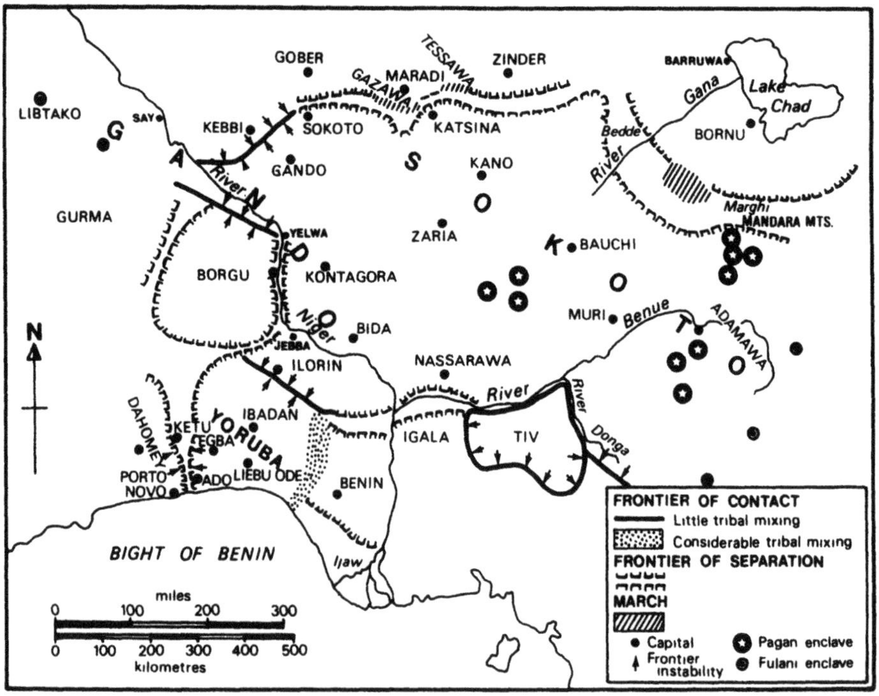

Figure 2.2 Frontiers in Nigeria in the 19th century.

Zinder. Between Kano Province of the Sokoto–Gando Empire and Zinder, there was a deserted frontier of separation resulting from the weakness of both states; there were only occasional raids across this frontier.

The frontier of separation dividing Bornu from Sokoto can be divided into three parts. North of the River Gana lived the Bedde pagans protected by a forested, swampy environment. Armies from both Sokoto and Bornu conducted slave raids against these people. Between the River Gana and the Mandara Mountains the forested frontier of separation was defended on the Bornu side by a series of quasi-independent marches, which had a long history of resistance to the Fulani. The Mandara Mountains themselves formed the third section of the frontier of separation between Bornu and Adamawa. This area was occupied by the Marghi pagans against whom the Fulani exerted intermittent pressure. The continued independence of the Marghi was advantageous to Bornu, since it prevented possible collision with the Adamawa Fulani and discouraged slaves from escaping southwards.

The southern frontier of the Sokoto–Gando Empire marked the broad division between the states of the Sudan and those of the forested zone. In the west Gando had a common frontier with the kingdom of Borgu. Westwards from Yelwa, on the south side of the River Niger, there was a narrow frontier of contact against which the Fulani exerted continuous pressure unsuccessfully. Between Yelwa and Jebba the River Niger flows through a series of deep gorges which effectively separated the two states. This frontier was continued westwards from Jebba into a hilly, forested zone.

There was an unstable frontier of contact between Ilorin province of the Gando Empire and the Yoruba states of the south. Both states maintained permanent armies against each other, and the position of the frontier depended upon their relative strength at any time. Eastwards this frontier broadened into one of separation between the Fulani of Kabba and the Benin Kingdom, resulting partly from weakness of the Fulani and partly from the isolation policy of Benin. At intervals both states raided the frontier for slaves, further fragmenting the small Yoruba groups living there. This frontier of separation was continued east of the River Niger between Nassarawa Province of the Gando Empire and the Igala tribes. The frontier zone lay generally south of the Benue and was flooded with refugees from the north bank, which was effectively conquered by the Fulani. The tribes of the south were protected by the river, except at periods of low flow when the Fulani raiders could easily cross. The Benue formed the frontier of separation between the Tiv and Fulani states, expect for a small holding which the Tiv maintained on the north bank of the river. The stability of this frontier resulted partly from the sturdy independence of the Tiv and partly from their traditional friendship with the Fulani. The other frontiers of the Tiv group were remarkably unstable frontiers of contact, resulting from the outward migration of the Tiv, which involved the absorption of the farmlands of the Igala and Ogoja tribes. This continued migration caused trouble for the colonial authorities. In 1912, after a Tiv raid had dispossessed the Gabu of their land, the authorities built an earth wall, 5 feet high and 34 miles long, to restrict the Tiv. The plan was not successful and today there are many Tiv south of the wall, which has now fallen into serious disrepair.

It now remains to describe the common frontier of the four recognizable forest-states. The weakness of Borgu, together with the conflicts of the Yoruba Confederation with the Ilorin Fulani and Dahomey, resulted in Borgu being limited to the south by a wide forested frontier of separation, which was unpopulated expect for some brigands. Between Egba, the westernmost Yoruba state, and Dahomey a frontier of separation narrowed towards the coast, which was the target for both states seeking to dominate trade between the Europeans and the interior. Both armies made frequent raids into the frontier during the dry season when rivers posed no obstacles. The Ewe-speaking refugees from the west and the Egbado refugees from the east formed a complex ethnic mixture in the frontier. The distinction between Yoruba and Benin territory was not a sharp one. The peaceful frontier contained a complex intermixture of both groups, gradually shading to Yoruba dominance in the west and Benin dominance in the east:

> It is impossible, at the present time, to determine the extent of the Benin Empire at any particular period of the past. The frontiers were continually expanding and contracting as new conquests were made and as vassals in the border rebelled and were reconquered. (Bradbury 1957, p. 21)

To the south of Benin the delta tribes, such as the Ijaw, preserved their independence largely as a result of the defensive character of the swamps and creeks. By the middle of the last century the policy of isolation had caused Benin to withdraw its authority from the western bank of the Niger. East of the Niger the political organization of the Ibo did not rise above the level of the clan or family. Although some of these groups must have been surrounded by areas of unclaimed forest, their distribution cannot be reconstructed on the present scale of inquiry.

Conclusions

There seem to be three main results of the former location of past frontiers in the present landscape. First, the colonial boundaries which were superimposed on the indigenous political fabric did coincide to some extent with the indigenous frontiers. The Anglo-French boundary between Dahomey and Lagos was drawn within the frontier between Dahomey and Egba. The Anglo-French boundary between the Niger and Northern Nigeria showed some correspondence with the devastated sections of the frontier between Sokoto and the northern Habe states of Maradi and Zinder. The former federal boundary between Western and Northern Nigeria is clearly related to the northern boundary of the former Yoruba and Benin kingdoms. Secondly, pressure from both flanks of some frontiers of separation has created ethnic shatter zones. The ethnic complexities of the areas between the former Dahomey and Egba kingdoms, between the Kabba and Benin kingdoms, and between Nassarawa and the Igala tribes are revealed in the striking variations over short distances in house types and agricultural methods. Thirdly, these shatter zones, marginal to the cores of the original states and their colonial successors and lacking unified political control have remained underdeveloped and have not shared in the extension of services which have characterized other areas.

References

Adams, J. T. 1943. *Atlas of American history*. New York: Scribner.

Anderson, R. L. 1981. A government-directed frontier in the humid tropics: Para, Brazil, 1870–1920. *Agricultural History* **55**, 392–406.

Baradez, J. 1949. *Fossatum Africae*. Paris: Arts et Métiers Graphiques.

Barth, H. 1857. *Travels and discoveries in North and Central Africa*. New York: Appleton.

Becker, H. 1980. Siedlungsgrundungen des ausgehenden 19. Jahrhunderts an die kanadischen Bergbaufrontier im Klondike-Goldfeld (Settlement foundations of the Canadian mining frontier in the Klondike gold field at the end of the 19th century). *Die Erde* **111**, 329–52.

Billington, R. A. 1960. *Westward expansion*, 2nd edn. New York: Macmillan.

Bowman, I. 1931. *The pioneer fringe*. American Geographical Society Special Publication No. 13, American Geographical Society, New York.

Bradbury, R. E. 1957. *The Benin Kingdom*. London: International African Institute.

British and Foreign State Papers 1894–5, Vol. 87. London: HMSO.

Burt, A. L. 1957. If Turner had looked at Canada, New Zealand and Australia when he wrote about the West. In *The frontier in perspective*, W. D. Wyman & C. B. Kroeber (eds), 59–77. Madison: University of Wisconsin Press.

Clark, T. D. 1959. *Frontier America*. New York: Scribner.

Christopher, A. J. 1982. Towards a definition of the 19th century South African frontier. *South African Geographical Journal* **64**, 97–113.

Collingwood, R. G. 1923. *Roman Britain*. London: Oxford University Press.

Cornish, V. 1936. *Borderlands of language in Europe and their relation to the historic frontier of Christendom*. London: Sifton Praed.

Curzon of Kedleston, Lord 1907. *Frontiers*. The Romanes Lecture. Oxford: Oxford University Press.

Davies, C. C. 1932. *The problem of the Northwest Frontier 1890–1908*. Cambridge: Cambridge University Press.
Dennis, W. J. 1931. *Tacna and Areca*. New Haven: Yale University Press.

East, W. G. 1937. The nature of political geography. *Politica* **2**, 259–86.
East, W. G. 1962. *An historical geography of Europe*, 4th edn. London: Methuen.

Fifer, J. V. 1966. Bolivia's boundary with Brazil: a century of evolution. *Geographical Journal* **132**, 360–72.

Harrison, J. A. 1953. *Japan's northern frontier*. Gainesville, Fla.: University of Florida Press.
Hertslet, Sir E. 1909. *Map of Africa by treaty*, Vol. 2, London: HMSO.
Hogben, S. J. 1930. *The Muhammedan Emirates of Nigeria*. London: Oxford University Press.
Hohnel, L. von 1894. *The discovery of Lakes Rudolph and Stefanie*. London: Longman.
Holdich, Sir T. H. 1916. *Political frontiers and boundary making*. London: Macmillan.
Huntingford, G. W. B. 1955. *The Galla of Ethiopia*. London: International African Institute.

Kennedy, A. L. 1953. *Salisbury 1830–1903: portrait of a statesman*. London: Murray.
Kerr, D. G. G. 1961. *An historical atlas of Canada*. Toronto: T. Nelson.
Konrad, V. A. 1981. An Iroquois frontier: the north shore of Lake Ontario during the late seventeenth century. *Journal of Historical Geography* **7**, 129–44.
Kristoff, L. A. D. 1959. The nature of frontiers and boundaries. *Annals, Association of American Geographers* **49**, 269–82.

Lattimore, O. 1940. *Inner Asian Frontiers of China*. New York: American Geographical Society.
Lawless, R. I. 1974. Uranium mining at Arlit in the Republic of Niger. *Geography* **59**, 45–8.

Ministère des Colonies 1910. *Documents scientifiques de la mission Tilho*. Paris: Government Printer.
Mitchell, R. D. 1972. The Shenandoah Valley frontier. *Annals, Association of American Geographers* **62**, 461–86.

Paullin, C. O. 1932. *Atlas of the historical geography of the United States*. Baltimore: Carnegie Institution of Washington.
Porter III, F. W. 1979. The Maryland frontier 1722–1732: prelude to settlement in western Maryland. *University of Maryland, Occasional Papers in Geography* **4**, 90–107.
Prescott, J. R. V. 1975. *Map of mainland Asia by treaty*. Melbourne: Melbourne University Press.

Staudinger, P. 1889. *Im Herzen der Haussaland* (In the heart of Hausaland). Berlin: A. Landsberger.
Stone, K. H. 1979. World frontiers of settlement. *GeoJournal* **3**, 539–54.
Straight, S. 1978. The American frontier did not stop at the Canadian border. *Bulletin of the Illinois Geographical Society* **20**, 47–53.

Turner, F. J. 1953. *The frontier in American history*. New York: Holt.

Weigert, H. W., H. Brodie, E. W. Doherty, J. R. Fernstrom, E. Fischer & D. Kirk 1957. *Principles of political geography*. New York: Appleton-Century-Crofts.

Whittlesey, D. 1956. The United States: expansion and consolidation. In *The changing world*, W. G. East & E. A. Moodie (eds), 261–84. London: Harrap.

Wilkinson, H. R. 1955. Jugoslav Kosmet. *Transactions and Papers*. Institute of British Geographers. 171–93.

Yesner, D. R. 1985. Cultural boundaries and ecological frontiers in coastal regions: an example from the Alaska Peninsula. In *The archeology of frontiers and boundaries*, S. W. Green & S. M. Perlman (eds), 51–91. Orlando, Fla.: Academic Press.

3 *The evolution of boundaries*

> The missionary, the conqueror, the farmer and, of late the engineer, have followed so closely in the traveller's footsteps that the world, in its remoter borders, has hardly been revealed before one must record its virtually complete political appropriation. (Mackinder 1904, p. 421)

Mackinder was speaking at the close of the most intensive period of boundary construction in the world's history, a period which had created a closed political system enclosing even barren tropical deserts and unexplored equatorial forests. International boundaries have now replaced frontiers in all the continents, including Antarctica. The previous chapter showed some of the ways by which frontiers were reduced in width; this chapter examines the ways in which boundaries are created by states.

The onset and progress of boundary negotiations

Boundary negotiations between states usually originated once a conflict of interest developed or seemed imminent, and they were usually designed to promote peace and better administration. Vattel regarded boundary delimitation and demarcation as a useful cure for international disputes: 'To remove every subject of discord, every occasion for quarrel, one should mark with clarity and precision the limits of territories' (Vattel 1758, p. 137).

The use of boundary treaties to promote peace is sometimes recorded in the treaty's text. In 1864 Russia and China signed the Protocol of Chuguchak regarding the boundary south-west of modern Mongolia, which recorded that it was for 'the promotion of the good understanding between the two Empires' (Inspector-General of Customs 1917, Vol. 1, p. 144).

The conflict of interest sometimes followed direct contact between forces or citizens of the two countries, and in such cases the boundary was determined subsequent to established patterns of occupance, which at the least would involve identification and use of principal routes and the construction of camps or forts. In 1891 a collision occurred between Portuguese forces marching inland from the coast of Mozambique and police of the British South African Company engaged in settling Rhodesia. There was some confusion over which nationals had valid prospecting treaties with the local rulers of Manica, and on 11 May 1891 the battle of Macequece was fought between 48 police and volunteers on the British side and 359 Portuguese troops, of whom 300 were Africans from Angola. There were no casualties after two hours and the Portuguese withdrew during the night. The British booty consisted of 'nine machine guns [which] were in position on field carriages, thousands of rounds of ammunition of all descriptions, officers and men's baggage, stores of all kinds, and one pair of ladies drawers' (Rhodesian Archives, CT 1/12/1–7, Report by Lt Col. E. G. Pennefather, 13 May 1891). Exactly one month after this clash Britain and Portugal signed a treaty which defined their respective spheres of influence in the hinterland of the Mozambique coast.

When there was direct contact between the citizens of the competing states the boundary often had to be determined within confined limits, because territorial questions were inextricably linked with issues of national pride. The direct conflict of interests often involved the use of land for grazing and cultivation, mining claims, the use of rivers and passes for travel and trade, and exclusive commercial treaties signed with indigenous chiefs. Governments recognized that, apart from the serious risk of military conflict between zealous officers, so long as no precise boundary was defined there was the critical problem of encouraging commercial activities in an uncertain political climate.

Conflicts of interests which did not involve direct contact were usually of a strategic nature. For example, in 1727 China agreed to further negotiations with Russia about the westward extension of their boundary through the Amur valley because China wanted to avoid the possibility of any Russian interference amongst tribes in the Chinese borderland. In 1899 Britain proposed various boundaries to China in the area of Hunza and the Aksai Chin, in efforts to secure a good defensive line and to thwart any chance of a Russian advance through the Pamirs. The Chinese neither accepted nor rejected these proposals. Fifteen years later, by some devious diplomacy, Britain attempted to secure a sound defensive boundary in northern Assam through the negotiations which produced the McMahon Line.

Many of the boundary negotiations in Africa were commenced long before the colonizing movements of Britain and France or Germany and Portugal came into contact. They were initiated in an effort to realize some grand design, such as France's determination to link her Mediterranean and West African possessions and Portugal's hope to unite Angola and Mozambique. Although there was greater scope for selecting boundaries through areas which had not been explored or exploited, problems often arose because countries were anxious not to forgo any of the valuable resources which might exist in the borderland. In the 1880s Britain and Germany were arguing about the boundary proceeding inland from the coast between Nigeria and Kamerun; a British official recorded that neither side was disposed to make concessions because the disputed area 'might prove to be an Eldorado or a worthless swamp'. Rudin (1938) noted that negotiations over this area were finally encouraged because neither administration could persuade firms to develop areas in the disputed zone or near it. Commercial firms regarded such areas as poor investment risks because political expediency might transfer their area of operation from one country to another.

The presence of no-man's-land between states facilitated escape by individuals from financial and juridical responsibilities, and sometimes such areas became refuges for brigands. Countries often acted quickly to eliminate such inconvenient areas. In 1899, Britain and Germany agreed to divide the neutral zone, which had separated their colonies of the Gold Coast and Togo, by a river which could be clearly identified by the two administrations.

There were three basic situations in terms of the relative power of states under which negotiations might be launched. First, states of comparable strength may decide to enter negotiations in order to solve administrative problems and avoid risks of a serious clash. Secondly, a stronger state may propose negotiations to a weaker state in order to gain sovereignty over areas not previously held. In 1858 and 1860 Russia was much stronger than China, which was racked by internal rebellion and external threat; by boundary negotiations during that period Russia forced China to cede the trans-Amur and trans-Ussuri territories. In 1893 France exerted considerable pressure against Thailand and whittled away large areas which today form part of Cambodia. The British Ambassador in Paris described the French tactics in the following terms: 'The Siamese Government were now in possession of an ultimatum, a pen-ultimatum and an ante-penultimatum. In fact the word "ultimatum" had completely lost its meaning, for each new one seemed to procreate a successor' (Prescott 1975, p. 432). The third situation arises when a weaker state proposes negotiations in order to try to protect the sovereignty which it possesses. Mexico in 1848 welcomed boundary negotiations as a device to stop further erosion of its territory by the United States. A quarter of a century later Afghanistan was encouraged by Britain to try to settle a boundary with Russia in a bid to halt that country's rapid advance across central Asia towards India.

Once two states commence boundary negotiations they may follow a number of different courses. States will often engage in preliminary, sometimes informal, contacts to make an estimate of the difficulties concerned. Then, before formal talks are held, public servants from the relevant departments gather information which can be used to devise a number of options for the state's negotiators. If the negotiations are successful, a boundary will be selected and will probably eventually be demarcated. If that outcome is impossible to achieve, the parties have three other choices, unless they decide to leave the matter unresolved (Fig. 3.1).

First, a provisional line can be drawn to reduce the level of

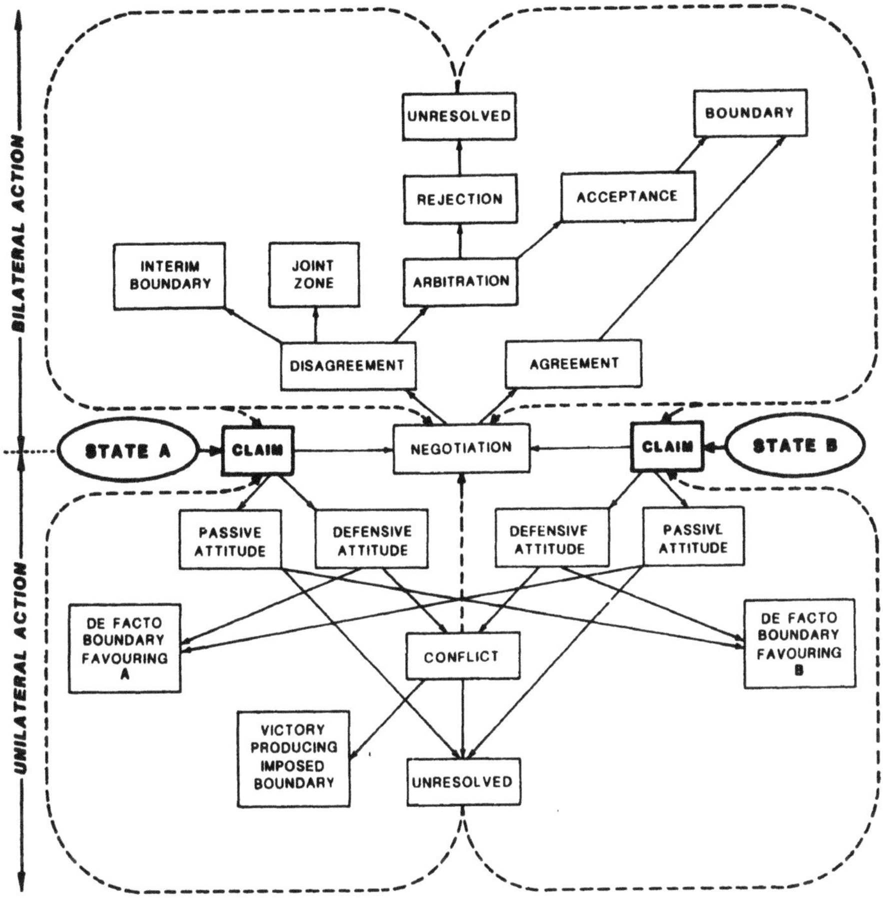

Figure 3.1 The procedures of boundary negotiations.

administrative inconvenience while the search for a final boundary continues. Provisional lines are fairly rare, except at the end of a war, when they are normally called cease-fire lines. After the partition of British India in 1947 fighting occurred in Kashmir between forces representing India and Pakistan. By the middle of 1948 a static front had developed leaving Pakistan in control of Gilgit, Baltistan, and a narrow strip of the western part of the Vale of Kashmir, Punch, and Jammu. India occupied Ladakh and the remainder of the Vale of Kashmir, Punch, and Jammu. On 27 July 1949 a treaty was agreed fixing the cease-fire line, and its military origin was evident in its description. Renewed fighting in 1965 resulted in the 1949 line being confirmed as a cease-fire line. The third contest in this area began in December 1971, and both sides made gains across the 1949 line. On 3 July 1972 a new line of control was agreed and still marks the boundary between the two countries.

Secondly, the two states may decide to make the disputed area into a neutral zone in which they have equal rights. After World War I, Kuwait and Saudi Arabia were unable to agree on the division of an area of about 2500 square miles near the coast of the Persian Gulf, and in their treaty of 2 December 1922 it was agreed that this region would be left as a neutral zone. Neither side showed any interest in dividing this neutral zone until oil was discovered in the Al Burqan fields north of the zone in 1938. Negotiations were delayed until the early 1960s and they were completed by a treaty signed on 7 July 1965 and ratified on 18 December 1969. The neutral zone was divided into two parts, almost equal in extent, by a single straight line measuring 34 miles in length.

Recourse to arbitration is the third solution states may adopt if they cannot agree about a boundary. On 11 June 1891 Britain and Portugal agreed on a boundary to separate their spheres of influence west of the Zambezi River above the Victoria Falls. By this agreement the Barotse Kingdom was reserved for Britain. Since the two countries could not agree on the extent of this kingdom, it was decided to establish a provisional boundary during negotiations in May and June 1893. Further negotiations proved fruitless, and it was decided by an agreement dated 12 August 1903 to invite King Victor Emmanuel III of Italy to determine the extent of the Barotse Kingdom on 11 June 1891. The King announced his award on 30 June 1905 and both sides accepted it. Unfortunately it was discovered later that this arbitral award contained an ambiguous expression and this new difficulty was not resolved until 21 October 1964. If one or both sides reject the award of the arbiter, then the matter remains unresolved and a new round of negotiations may begin, as indicated in Figure 3.1.

If the countries decide to pursue this conflict of interest through unilateral action, there are a number of possible outcomes. Each country can choose from two basic options. First, it can adopt a passive attitude to the dispute and refuse to protect its citizens if they seek to operate in the disputed zone or to interfere with the activities of aliens in the same area. Secondly, a country can adopt a defensive attitude towards its claim; this involves defending the rights of its citizens in the disputed zone and seeking to deny any rights to aliens in that region. If both countries adopt passive stances, the problem will remain unresolved and any administrative inconvenience will remain and perhaps become worse. If one adopts a passive attitude while the other takes a defensive posture, a *de facto* boundary will develop favouring the defensive state.

If both countries adopt defensive attitudes, conflict may result. Such conflict may end inconclusively, leaving the matter unresolved, or one side might be victorious and have the power to impose a boundary on the defeated state. The history of Europe provides many examples of powerful states imposing boundaries on weaker neighbours; the northern boundary of China with the Soviet Union was effectively determined by a powerful Russia in 1858 and 1860 when China was beset by internal problems and unable to resist.

This analysis of how territorial conflicts may be resolved applies equally well to disputes over maritime zones, with one important exception: there is not yet any significant history of maritime boundaries being imposed by military strength to divide a disputed area near the coast. However, there have been cases where states have defended claims into areas which other countries consider to be the high seas; for example, Iceland's expanded claims to fishing zones in 1952, 1958, and 1973 provoked the so-called Cod Wars with Britain.

Aspects of boundary evolution

Three aspects of boundary evolution are appropriate to geographical analysis – evolution in definition, evolution in position, and evolution in the state functions applied at the boundary. Examination of these three aspects will illuminate the two main lines of geographical research into boundaries – the influence of geographical factors on the location of the boundary, and the reciprocal influence of the boundary, once established, on the development of the landscape through which it is drawn.

Geographical knowledge is one of the fundamental factors influencing boundary location and an indication of that knowledge is often contained in boundary definitions. Geometric boundaries in Africa usually meant that little reliable information was available about topography and drainage. Successive boundary definitions often record advances in exploration and cartography as geometric boundaries are exchanged for lines coincident with rivers and relief features or lines of cultural differentiation.

The way in which a boundary influences the development of a border landscape and the lives of its inhabitants is likely to be a function of the accuracy with which that boundary is defined and can be located and the number and quality of state functions applied at the boundary. Often the most striking influences upon the border landscape and its inhabitants will result from changes in boundary position which transfer areas from one state to another. One example will serve at this point. After World War II 2.8 million Germans moved out of former German territory east of the Oder–Neisse Line when it was transferred to Poland. The agricultural activities of the Polish immigrants in the transferred area, on both peasant and communal farms, have produced significant landscape changes (Wiskemann 1956). In order to measure the geographical significance of the boundary, it is necessary to know the relationship which the original boundary bore to the landscape at the time when the boundary was drawn. This clearly involves the application of the methods of historical geography in order to discover the original correspondence between the boundary and the cultural patterns, such as the distribution of population groups, the location of economic activities, and the direction and volume of trade.

Research on the evolution of boundaries

It is now necessary to consider how these avenues of research can be followed, apart from studying secondary sources produced by other scholars. Information about boundaries may be gathered by studying relevant documents and published works, by analysing maps of the area traversed by the line, and by undertaking fieldwork in the borderland.

The documentary material may be classified into three sections. First, there are the copies of correspondence conducted between the negotiating powers and between representatives of the same government. Such records are only rarely published and it is generally necessary to consult them in archives, which will normally restrict access to those files which are more than 30 years old. These sources are invaluable and alone will provide the detailed considerations which led to the selection of a particular boundary. In the letters, minutes, and reports can be discovered the geographical, political, economic, ethnic and legal factors that played an important part in producing agreement on the general location and specific site of the boundary. An analysis of the Anglo-French correspondence dealing with the settlement of the inter-Cameroons boundary in 1920 revealed the following points which could not have been derived in any other way. First, the location of the boundary in Bornu resulted from the incorrect decoding of a British telegram in Lagos in 1916. France, which benefited from this error, refused to allow the line to be corrected, but this obstinacy permitted Britain to press successfully for the reunification in the British Cameroons of the Holma, Zummu, and Higi pagans, who had been divided by the original Anglo-German boundary. Secondly, France was anxious to secure Garua, a port on the Benue River, and the land route from Douala on the coast to Garua. Meanwhile, Britain was anxious to reunite the former Emirate of Yola which had been split between Britain and Germany in the 1880s. Thirdly, the use of inaccurate maps in delimiting the boundary resulted in two disputes. The first, at the southern end of the boundary, involved rich plantations, whereas the other, in the north, related to swamplands which were used for cotton production and winter grazing. Fourthly, the negotiations were conducted with regard to boundary arrangements being made in respect of other German colonies (Prescott 1971, pp. 45–62).

The second class of documents and published material consists of the boundary treaties which are eventually agreed between states. Most countries will publish their own treaty series in parliamentary records and many treaties are recorded in the *United Nations Treaty Series*. Sir E. Hertslet has published collections of treaties dealing with Africa, Europe, and China, and Martens edited a very useful general treaty series. Indeed, the bibliography of publications containing the texts of treaties would be a very long one, and scholars are recommended to consult university and national libraries. The treaties are important because their letter is the final basis on which the boundary is demarcated. If the treaty is not carefully drafted, ambiguous phrases may lead to future boundary disputes. Boundary treaties will often include information about the conduct of affairs in the borderland as well as the definition of the line. For example, the fifth clause of the agreement relating to the boundary between Afghanistan and British India dated 26 February 1895 contains the following provisions:

> We have also jointly agreed on the following matters relating to the portion of the boundary line defined in this clause: Firstly, that the rights attaching to the Psein land which is within Afghanistan and close to and to the west of boundary pillars Nos. XVI, XVII, and XVIII of water from the Kakars, who own the right to the water of the Loe Wuchobai nullah above that, will remain as hitherto.
>
> Secondly, that the Kakar Tribe should continue to enjoy the rights of grazing, as hitherto, throughout the country lying between the Kand river, and Loe Wuchobai nullah, and Babakr Chahan and Sam Narai.
>
> Thirdly, that the Pseins should continue to enjoy the right of grazing, as hitherto, in the tract of land commonly known as Psein Dagh, which is situated on the south of the Psein Lora.
>
> Fourthly, that the water of the Psein Lora and Kadanai river belongs jointly to the people residing on both banks of the river.
>
> If any of the subjects of the British or Afghan Government wish to construct a new water channel leading from the Psein Lora or Kadanai river, they must first obtain the permission of the district officers concerned of both Governments. (Prescott 1975, p. 194)

According to the fifth and sixth Articles of the Anglo-Tibetan Convention of 1904, Tibet undertook to keep roads between the boundary and Gyantse and Gartok free from obstructions and in a state of repair suited to the needs of trade and to raze all fortifications along these routes which might impede communications (Inspector-General of Customs 1917, Vol. 2, pp. 656–7). Lamb (1964) gives a lead to scholars, who may be looking critically at modern treaties, in his analysis of the Sino-Pakistan boundary treaty of 1963.

One of the most fruitful sources of information related to treaties concerns the decisions of arbitrations. There have been many cases, in the last century and in recent times, when judgments have been made either by a single arbitrator acceptable to both sides or by a judicial tribunal. For example, in 1875 France ruled in Portugal's favour when Britain disputed Portugal's ownership of Delagoa Bay, immediately south of Maputo. In 1897 Italy adjudicated a compromise boundary between the conflicting claims of Britain and Portugal on Rhodesia's eastern border; eight years later it performed the same service for the same two countries along the border between Angola and Zambia. In 1911 the disputed boundaries of Walvis Bay were settled by Spain, which dismissed the German arguments and found in favour of Britain. In 1962 the International Court of Justice ruled on the alignment of the boundary between Thailand and Cambodia near the temple of Preah Vihear, and in 1968 a judicial tribunal resolved the dispute between India and Pakistan about the boundary through the Great Rann of Kutch. Whenever arbitration or judicial processes are involved the countries submit as much evidence as they possibly can, and the records of these activities contain great stores of useful information on the evolution of the earlier boundaries. In the Rann of Kutch hearings the two sides presented evidence which occupied 10 000 pages of typescript and illustrated their arguments with 350 maps.

The third class of documents and published works comprise the reports and personal accounts of the individuals who demarcated the boundary or were involved in its creation. Probably because so much of the demarcation work involved travel in remote areas and in some cases primary exploration, many boundary commissioners published accounts in the journals of learned societies such as the Royal Geographical Society. These accounts often include detailed descriptions of the borderland's physical geography and precise accounts of the nature of indigenous societies near the boundary and of the attitude of those populations to the presence of the new limit. For example, the papers by Nugent (1914) and Detzner (1913), who were joint leaders of the Anglo-German demarcation team betwen Nigeria and Kamerun, give a clearer impression of the problems faced and the dislocation caused by the boundary to the economic and political life of local tribes than any other source. The memoirs of men such as Holdich (1916) and Ryder (1925) who were practically involved in identifying and marking boundaries in the landscape contain much more detail about the boundaries than the memoirs of statesmen such as Sir Mortimer Durand (Sykes 1926), who, perhaps understandably, were more concerned with the broad questions of international diplomacy rather than the precise alignment of boundaries. Sir Mortimer Durand negotiated the boundary between India and Afghanistan in 1893, which was later known as the Durand Line.

The study of maps by scholars interested in the evolution of boundaries is an essential undertaking for a number of reasons. First, it will in many cases never be possible for the scholar to undertake fieldwork in the borderland because countries such as China and the Soviet Union or Iran and Pakistan do not allow such activities. Secondly, it is important to study the maps available to the negotiators because these maps will give a good indication of the perception of the area held by them. Many of the decisions of negotiators, which subsequently created problems, are inexplicable if analysed on modern maps but thoroughly comprehensible when related to the maps available during the negotiations. For example, the Anglo-Russian Convention of 1825 stated that the boundary between Alaska and British Canada should follow 'the summits of the mountains situated parallel to the coast', providing that line shall be no more than 30 miles from the coast (Davidson 1903, p. 81). Commentators have shown that such a boundary cannot be located (Davidson 1903, Hinks 1921). This boundary definition was accepted by the negotiators because their work was related to maps based on Vancouver's explorations of 1792–4, which were published in an atlas in 1798. Vancouver represented a range of mountains along the whole length of the Pacific coast of North America, located 10 to 24 miles from the coast. Even in 1867, when Russia transferred Alaska to the United States, the official American charts were still based on Vancouver's maps.

Maps of the period when the boundary was drawn will also be very helpful in identifying places which have changed their names in the subsequent period. Problems connected with place names can be particularly difficult to solve. I recall being uncertain for a number of weeks about the exact location of the Shabina Pass which marked the western terminus of the 1727 Sino-Russian boundary in central Asia; the answer was found after consulting dozens of maps of various ages.

Finally, maps will sometimes provide useful information about the cultural landscape at the time the boundary was established and will therefore provide a benchmark from which ensuing change in the borderland can be measured. Such maps will often record the patterns of settlement, communication, and land ownership, which may not be described in any other form. Obviously, maps on the largest possible scale should be used.

If it is possible, fieldwork is a most valuable source of information. Simply by travelling through the borderland it is possible to gain a familiarity with the topography and drainage pattern which will illuminate analysis of the boundary negotiations several decades ago. The techniques of observation and local inquiry do not differ from fieldwork in other branches of geography. It is unlikely that a scholar undertaking such fieldwork would restrict the inquiry only to those aspects useful in understanding the boundary's evolution; when in the region it would be sensible to accumulate information about the current importance of the boundary as a factor influencing the cultural landscape and the lives of people living close to it. For example, it would be useful to acquire information on the state of the boundary's demarcation and whether or not it agreed with the published description. It would also be helpful to identify any significant changes in the cultural landscape near the boundary and the impact of the boundary on the lives of local residents. The major crossing points should be listed and, if possible, the volume of traffic at each over a period of time should be measured.

During fieldwork on part of the Nigeria–Benin boundary only 3 of the 20 monuments marking the section could be found. All three were found near villages; none was complete as each one had been used to sharpen cutlasses or axes. There were no significant changes in the cultural landscape within 20 miles of the boundary; the lives of the people seemed unaffected by the boundary, though some Nigerian farms were lying partly in Benin, in accordance with the provision of the Anglo-French Agreement of 1906:

> The villages situated in proximity to the boundary shall retain the right to arable and pasture lands, springs and watering places, which they have heretofore used, even in cases where such arable and pasture lands, springs and watering places are situated within the territory of one Power, and the village within the territory of the other. (Hertslet 1909, Vol. 3, p. 861)

The two crossing points at Ijoun and Idiroko were 32 miles apart, and most of the border inhabitants crossed the boundary by uncontrolled paths. At times of tax collection there was some movement across the boundary in order to escape responsibilities. During an interview with the Aleketu, who is a Yoruba chief in Benin separated from the majority of his tribe in Nigeria, he said, 'We regard the boundary as separating the French and the English, not the Yoruba.' All this information was valuable in understanding the present condition and significance of the boundary and could be obtained only by fieldwork.

Evolution in definition

Geographers have propounded several systems of boundary evolution. Brigham (1917) employed a threefold division – tribal, transitional, and ideal. The tribal boundaries were primitive and were not defined in any document. Such divisions should be described as frontiers, since they had a zonal quality however clearly the last lines of defence were marked in the landscape. Brigham envisaged the transitional stage as being one when the boundary was likely to change its position, carrying the implication that the boundaries were finding a position where the forces from either side were neutralized. Finally, in the ideal stage the boundary became permanently fixed, and a gradual diminution of functions applied at the boundary reduced its significance as a landscape element. This altruistic concept of boundary evolution probably owed much to the world situation when it was published, and the ideas have not been further developed by other workers.

Lapradelle (1928) distinguished three stages of boundary evolution – preparation, decision, and execution. He emphasized the tentative nature of the first stage compared with those that follow by using *le trace*, which means 'outline' or 'sketch', instead of *la limite* meaning 'boundary'. Jones (1945) follows Lapradelle closely in suggesting four stages of boundary evolution – allocation, delimitation, demarcation, and administration. Allocation refers to the political decision on the distribution of territory; delimitation involves the selection of a specific boundary site; demarcation concerns the marking of the boundary on the ground; and administration relates to the provisions for supervising the maintenance of the boundary. Nicholson (1954, p. 116) tried to marry the schemes of Jones and Brigham by carrying the process through from the tribal stage to the demarcated boundary. However, he admits that the only correlations between the first frontiers and the final boundaries in Canada were fortuitous and that there was no continuous development. There seems to be no reason why his ideas should not apply where there is a continuous history of indigenous development. In cases of widespread colonization, as in America, Africa, and Australia, the extent to which colonial boundaries are drawn within indigenous frontiers will depend on the extent to which the colonizing state considered existing political structures.

It must not be presumed that all boundaries have passed through the stages of allocation, delimitation, and demarcation in an orderly sequence. In some cases the original allocating boundary has been demarcated with no intervening delimitation. In other cases there has been more than one delimitation before demarcation occurred. Finally, there are many boundaries in the world which are still undemarcated. In the following discussion of boundary evolution the stages suggested by Lapradelle and Jones are used.

Allocation

When a boundary was created in a frontier where the geographical facts were well known and where the population density was moderate to heavy it was sometimes possible to select a boundary site, and in such cases the stage of allocation was omitted. In areas which were less well known, often supporting low population densities, the stage of allocation provided the first formal political division. The boundaries which resulted were often arbitrary and consisted of two main kinds. The first type was made up of straight lines connecting prescribed co-ordinates or points in the landscape which had been identified, such as a waterfall or a village.

The Portuguese–German Declaration of 1886 described such a line allocating territory to Angola and South West Africa:

> The boundary follows the course of the river Kenene from its mouth to the waterfalls which are formed south of the Hunbe by the Kenene breaking through the Serra Canna. From this point the boundary runs along the parallel of latitude to the river Kulingo [Okavango], then along the course of that river to the village of Andura which is to remain within the German sphere of influence, and from thence in a straight line eastwards to the rapids of Catima on the Zambezi. (Hertslet 1909, Vol. 3, p. 703)

The quotation shows how the series of straight lines or stream courses linked up the waterfall, village, and rapids which had been identified and approximately located on maps. This boundary has been preserved intact to the present time because the two sides were unable to agree on any alterations. From time to time there were problems about identifying the exact point where the boundary leaves the Cunene, the location of Andara village, and the exact point where the boundary intersects the Katima Rapids. The problem over the village concerned the fact that the village was moved 3 miles downstream at the turn of the century. In some cases meridians and parallels were selected as boundaries, but, of the two lines, parallels were more reliable until accurate radio time signals could be sent and received. An official of the Royal Niger Company made this point forcefully in September 1893. 'Meridians moved around Africa like mountains. An error of a degree or even half a degree might cost England Kukawa, and therefore all Bornu' (Prescott 1971, p. 34). Kukawa, situated west of Lake Chad, was the capital of Bornu which Britain claimed by virtue of a treaty. The company was afraid that if some meridian was accepted which was thought to lie east of Kukawa it might turn out eventually to be west of that settlement. Accordingly, the British government persuaded Germany to define the boundary's terminus as the meridian which intersects the southern shore of Lake Chad 35′ east of Kukawa. It was considered by both sides that such a point coincided with the meridian 14° east because the value ascribed to the meridian through Kukawa, by Vogel, in 1853, was 13°24′ east. This had been altered to meridian 13°25′ east on Kiepert's map in Deutscher Kolonial Atlas. In fact the demarcation commission of 1903 found that the meridian passing through Kukawa was 13°33′ east, and so the terminus was set on the shore of Lake Chad at 14°8′ east.

A straight line connecting known points was often hard for administrative officers to determine when they were some distance from either point. This would be particularly true when the terrain through which the line passed was forested or hilly. It was probably even harder for the local population, in colonial situations, to know exactly where the unmarked straight line traversed their region. Such uncertainties led to serious administrative problems in parts of east Africa, including Rhodesia and Mozambique, where the 1891 Anglo-Portuguese Convention defined meridian 33° east as the boundary between the Mazoe River and the parallel 18°30′ south. In the three years following that Convention there were several exchanges of letters regarding allegations, by officers of both sides, of trespass and the illegal collection of hut taxes on the wrong side of the boundary.

The second type of boundary defined in principle the division of territory. It has been usual in earlier studies, including the important contribution by Jones (1945), to refer to definition in principle as one of the methods by which boundaries are delimited, but it seems more appropriate to include such definitions in the stage of allocation. Definition in principle means that there is a statement about the basis of territorial division and the result desired. For example, the boundary drawn by Russia and China in 1689 between the Little Gorbitsa River and the coast was defined in the following terms:

> The territories of the two Empires should be divided in such a manner that all land and rivers both great and small which flow from the south side of this range into the River Saghalien Vla [Amur] will come under the sway of the Chinese Emperor, while all lands and rivers which lead in a northerly direction on the other side of the watershed will remain under the control of the Russian Empire. (Prescott 1975, p. 13)

In short, the boundary had to follow the northern watershed of the Amur River Basin. The two countries had a vague idea of the location of that watershed but in some areas they would not have known its exact location within 20 or 30 miles. Where there is such uncertainty about the location of the boundary it seems proper to treat it as allocating territory, because delimitation refers to the selection of a specific site. It could be argued that the selection of a watershed or a river implies the selection of specific line, but only in exceptional situations would there be no disagreement possible about the alignment of a watershed or the headwater tributary which represents the source of a particular river.

It is beyond question that definition in principle related to physical features involved considerable uncertainty about the course the boundary would follow. For example, in 1885 Britain and Germany defined their common boundary between Nigeria and Kamerun as following the right bank of the Rio del Rey from its mouth to its source. According to Sir Claude Macdonald it was considered that the river was a long one, as shown on the Admiralty charts of the day (Nugent 1914, p. 647). Unfortunately the chart was wrong as the German Consul Baron von Soden pointed out to his British counterpart in 1888: 'You know as well as I do there is no Rio del Rey, at least no source of such a river. I do not know if the rapids of the Cross river can be more easily found, or whether they have the same mythical existence' (Prescott 1971, pp. 13–14). The problem was that the Rio del Rey was only 4 miles long and it was fed by two main tributaries, the Akpayafe and Ndian rivers which diverged sharply. Britain regarded the Ndian as the proper continuation of the Rio del Rey and, predictably, Germany was equally certain that the Akpayafe River was the correct boundary. At least the Cross River rapids did exist, but instead of being located at 9°8′ east and 5°40′ north, as shown on the chart, they occurred at 8°50′ east and 6°10′ north.

When the British and Belgian governments concluded an agreement defining their spheres of influence in east and central Africa in 1894 the boundary between the Sudan and Belgian Congo was stated to follow the watershed between the Nile and Congo rivers. This watershed is about 400 miles in length and the country is not particularly hilly; it has never been demarcated and it is predictable that any attempt to demarcate the line will provoke debate between the Sudan and Zaire over its precise alignment.

Definition in principle can also refer to cultural features, which generally are even more likely to cause disagreement than physical features. In September 1885 Britain and Russia defined Afghanistan's northern boundary, and east of Hauzi Khan the line had to be 'fixed in a manner which leaves to Russia the land cultivated by the Saryks and also their pastures' (Prescott 1975, p. 124). The demarcation commission which started work within two months of the agreement was unable to agree on the line limiting the Saryk areas and a further set of protocols had to be negotiated at St Petersburg in 1887.

In 1878 the Treaty of Berlin allocated territory between Montenegro and Albania by a line which was defined in principle according to the location of specified tribes:

> It then coincides with the existing boundaries between the tribes of the Kuci-Drekalovici on one side, and the Kucka-Krajna, as well as the tribes of the Klementi and Grudi on the other, to the plain of Podgorica, from whence it proceeds towards Plavnica, leaving the Klementi, Grudi and Hoti tribes to Albania. (Hertslet 1891, Vol. 4, p. 2782)

The last boundary treaty between Italy and Ethiopia, in May 1908, which is the most recent negotiated agreement regarding this bitterly disputed boundary between Ethiopia and Somalia, defined the boundary as the line which separated the Rahanwein tribes in the Italian area from all other tribes to the north which remained in Ethiopia. Definition in principle is a clear indication that the exact distribution or location of the physical features of cultural attributes is unknown. Such imprecise lines concern the first stage of boundary definition and the second stage of delimitation is designed to eliminate this inconvenient uncertainty.

Delimitation

The allocation of territory by arbitrary straight lines or by lines related to the uncertain distribution of physical and cultural features generally solved immediate territorial conflicts of interest and allowed governments to plan the development of territory with a sense of security, which encouraged some commercial firms. The delimitation of the boundary requires the selection of a specific boundary site and was usually only undertaken when the borderland possessed some intrinsic economic value, or if the interests or antagonisms of the two states required the rigid application of state functions at a specific line.

The retention of the arbitrary straight lines occurred when one or more of the following conditions applied. First, straight lines were preserved if the borders lacked any economic or strategic value and if the surveying of the boundary would have been an unnecessary and unjustifiable expense. An examination of the political map of the world reveals that many geometric boundaries are located in tropical deserts and Antarctica. Secondly, straight lines persisted when the two countries concerned were unable to agree on any alteration. This condition explains the continued use of straight sections in the boundary between Angola and Namibia. The inability of the Italians and Ethiopians to reach any modifications after 1908 underlies the persistent trouble along the Ethiopian–Somali boundary. Thirdly, it was usual for straight lines to be maintained when the same colonial power came into possession of the separated territories. This situation applied between Egypt and the Sudan during the period of British paramountcy, and between Kenya and Tanganyika and Botswana and South West Africa when German authority was eliminated after 1918.

Two types of boundary definitions can also be distinguished during the process of delimitation. The first involves complete definition, which requires the surveyors to proceed into the field and trace the line which had been so closely described. The second type of definition gives the surveyors power to vary the line, usually by a set distance on either side, in order to make the identification and administration of the boundary easier.

There are three common ways of delimiting boundaries. The first records the alignment of the boundary in the same fashion as the track of a ship by means of bearings and distances. It is important to specify whether the bearings are measured from true or magnetic north and to note precise distances. Further, it is useful to identify by other means some of the most important turning points on the boundary, otherwise any errors made during the demarcation of the line will accumulate. MacLean used this method in delimiting the boundary through Hashtadan when he had to arbitrate between Persia and Afghanistan in 1891. He recorded the forward and back bearings between each pair of consecutive pillars, and as an additional check he also gave bearings from each of the 39 pillars to other prominent features of the landscape, including villages, mills, and hills. He was able to specify the distances for 24 of the pillars; the remainder would have to be calculated by resection (Prescott 1975, pp. 159–70).

The second method simply records the turning points of the boundary and requires that they be connected by straight lines. The turning points may be physical features, such as the confluence of rivers, a peak, or the head of a ravine, or cultural features, such as crossroads, houses, or bridges. There have even been cases where letters of place names shown on a nominated map have been used as turning points. Problems arise when the selected turning point refers to an area rather than a single point. Physical features, such as hilltops, often possess an area within which the exact turning point must be located, and there is scope for disagreement about the exact confluence of many rivers and their tributaries.

The third method makes the boundary coincide with linear features in the landscape such as roads, rivers, or crests. Once again many of these linear features, especially rivers and crests, possess areas within which a line has to be selected.

It is during the delimitation stage that many of the seeds of future boundary disputes are sown. During the stage of allocation it is recognized that the boundary is arbitrary and likely to be modified to produce a more practical line, but the delimitation of the boundary is expected to produce a definitive line. Disputes of this nature are considered in the next chapter, but it will be useful to indicate here the traps which exist for the unwary boundary architect. Most subsequent difficulties arise because the individuals who define the line have not visited the borderland; instead they rely on descriptions from travellers and especially on maps. It is very easy to draw boundaries on maps because the scale representation of the landscape can be easily comprehended. In addition, because the cartographer has to be selective in the features he records they are given a prominence which may not be so obvious in the field. Rivers and crests stand out on many maps, especially some of the beautifully hand-coloured maps of the second half of the last century which can be found in archives illustrating reports of explorers and administrative officers. It may seem sensible when looking at the representation of a river on a map to nominate a bank as the boundary, but different surveyors and geomorphologists would interpret the bank of a river in different ways.

Other problems arise because the treaty draftsmen try to make assurance doubly sure by defining a single point in two ways. This would be helpful if the two ways were always in total agreement but unfortunately they sometimes contradict each other. For example, the first boundary between Peru and Bolivia included as a turning point the confluence of the Lanza and Tambopata rivers, which was clear enough, apart from the need to select a single point at the confluence. But in an effort to remove any possible doubt the treaty specified that the confluence lay north of the parallel 14° south. In fact the confluence lay south of that parallel and therefore the two administrations argued whether it was the confluence which was important or a location on the river north of parallel 14° south.

It must be noted that some international boundaries are not created by negotiations which produce an allocation of territory, followed by the delimitation of a line fitted to the landscape. Some international boundaries are created by the promotion of internal boundaries to international status. The dissolution of the British and French empires caused internal boundaries to achieve international rank in French West Africa and French Indochina and in British India. Usually internal boundaries are not described with the rigour demanded for international boundaries, and therefore it was not surprising that positional boundary disputes arose between countries such as Benin and Niger, Upper Volta and Mali, Cambodia and Vietnam, India and Pakistan, and India and Bangladesh.

Demarcation

Demarcation involves the identification of a delimited line in the field, the construction of monuments or other visible features to mark the line, and the maintenance of the markings. Generally the instrument of delimitation will define the composition of the demarcation commission and the distance by which the final line may vary from the delimited line in the interests of clear marking and good administration. Often demarcation does not follow promptly after delimitation; in fact there are many boundaries which have never been demarcated. Sometimes new boundary agreements render demarcation of the earlier line unnecessary, or matters of greater priority may make it impossible to spare survey teams for the work. The commencement of war in 1914 cut short a number of demarcation agreements, which were not renewed after the war. Laws (1932) and Peake (1934) have described how the boundaries separating the former Belgian Congo from Northern Rhodesia and Tankanyika respectively remained undemarcated until copper and tin mining made demarcation essential if major disputes were to be avoided and large companies encouraged to invest further capital.

When many of the world's international boundaries were demarcated the commissioners faced serious problems from hostile tribes, dangerous fevers, slow travel, and adverse climates. Those problems have now almost been eliminated by improved means of travel and communication and the use of aerial photographs and new survey methods. However, modern demarcation commissioners still face the other set of problems which faced their predecessors. These problems concerned the ambiguities and inconsistencies contained in the text delimiting the boundary.

Sometimes the delimitation document was so unsatisfactory that the commissioners were unable to reach agreement and had to pass the problem back to the diplomats. Such an event has already been mentioned in connection with the Russo-Afghan boundary in 1887; five years later the British and Portuguese commissioners could not agree on a single interpretation of the text governing the boundary between Rhodesia and Mozambique and the matter was passed to Italian arbitration, which was handed down in 1897.

The classical case of a contradictory definition concerned the boundary between Argentina and Chile promulgated in 1881. A section of the boundary was defined as 'the most elevated crests of said Cordillera that may divide the waters' (Varela 1899, Hinks 1921). Unfortunately, the process of headward erosion by the rivers flowing eastwards into Argentina had pushed the watershed west of the crest of the Andes, so Chile pressed for the crest and Argentina argued for the water divide.

The boundary may be marked in a variety of ways. In semi-arid areas pillars of stone are often constructed, but in sandy deserts materials have to be imported to construct monuments. At least in deserts there is usually the possibility of long lines of sight between pillars, whereas in forested or scrub areas the clearing of vistas is often a long and difficult undertaking. In addition to erecting pillars along the Sino-Burmese boundary, which was settled in 1960, the demarcation commission also planted flowering trees near the boundary to make its location more obvious.

It is regrettable that many of the boundaries which were demarcated at the beginning of this century in Africa and Asia were not maintained by regular inspections. Various natural processes tend to obliterate boundaries. Vegetation grows up in the cleared lines, plants break down pillars and cover them, large animals may knock them over, and the elements of weather can cause deterioration. There is the added problem that the local population will sometimes move the boundary markers because they do not agree with them, or steal the materials of which the markers are constructed for their own purposes. Clifford (1936) and Ryder (1925), working in Somalia and Turkey respectively, described how nomads destroyed boundary pillars within 24 hours of erection in the belief that sovereignty was vested in the people rather than in governments. In 1907 a demarcation commission placed 226 beacons along the 270 miles of boundary between Rhodesia and Botswana. Each beacon consisted of an iron pole sunk 3 feet into the ground and surrounded by a pile of rocks. In 1959 a second demarcation commission worked on this line and found only 105 of the original beacons, and some of them were only found after a long search. Lambert (1965) has written an interesting account of the maintenance of the boundary between Canada and the United States by the International Boundary Commission appointed in 1925. This commission supervises the inspection of the entire boundary once every decade and arranges for damaged boundary monuments to be repaired and overgrown vistas to be cleared.

Evolution in position

Analysis of the evolution of the boundary's position is important for three reasons. First, the operation of state functions at the boundary and of state policies in the borderland can influence the development of landscapes along the boundary and the life of communities in the borderland. This process may produce recognizable landscape differences along the boundary and those differences will be a function of the time the process has operated. It is therefore important to know how long the boundary has occupied particular sites. Secondly, as the boundary changes its position areas will be transferred from one authority to another. Such changes in sovereignty may set in train changes in settlement patterns, the volume of migration, and the orientation of regional economies, and these are all important subjects for consideration. Thirdly, the transfer of territory from one country to another may provide the cause for subsequent irredentist movements within the detached areas and may underlie territorial claims at a later date by the country which lost the territory.

Before looking at these points in more detail it is useful to relate the scale of change in boundary position to the stages of evolution in definition. The areas transferred by changes in position usually decrease as the definition proceeds from allocation to demarcation. This point is illustrated by the history of the Anglo-French boundary between the River Niger and Lake Chad. When the second allocating boundary, drawn in 1898, is compared with the first, drawn eight years earlier, it is evident that the maximum movement of the boundary was 90 miles and that Britain gained 14 800 square miles and lost 4550 square miles. This situation was reversed by the delimitation of the boundary in 1904 when the maximum movement of the boundary was 70 miles and France gained 19 960 square miles. When this delimited boundary was demarcated in 1907 the commission made only nine small changes, the largest of which involved 17 square miles.

The geographical effects of changes in a boundary's position

It is proposed to leave consideration of the rôle of boundaries in landscape construction to Chapter 6, but brief reference is made here to the possible effects which the transfer of territory from one country to another may produce.

Several studies of this subject have been made. Pallis (1925) studied national migrations in the Balkans during the period 1912–1924 when the boundaries within the area significantly altered. He concluded that the movements were the largest since the breakup of the Roman Empire. The movement of people from territory ceded to neighbouring states formed a significant part of the migrations. For example, in 1913 the total Greek population, numbering 5000, left the *qazas* of Jam'a-i-Bala, Melnik, Nevrokop, and Stromitsa when they were ceded to Bulgaria by the Treaty of Bucharest. In 1914, approximately 100 000 Moslems left the portions of central and eastern Macedonia, which had been ceded to the Balkan states by the peace treaty with Turkey, and settled in eastern Thrace and Anatolia. This scale of movement was exceeded after World War II when the Polish boundary moved westwards to the Oder–Neisse Line at the expense of Germany. Wiskemann (1956, p. 118) estimates that, in 1946, 1 460 621 Germans left Polish-occupied territory and settled in British-occupied Germany and that a further 600 000 moved into the Soviet sector. In the next three years a further 800 000 moved into the Soviet zone. By 1954 the number of Poles living in the Polish-occupied territory had risen from the prewar figure of 1 million to 7 million (Wiskemann 1956, p. 213). The changes in population structure in the former German areas were accompanied by alterations in the pattern of agriculture. All holdings over 100 hectares were confiscated and much of this land was redistributed to Polish peasants as farms having an average size of 12 hectares. Altogether, 3.6 million hectares were distributed to 605 000 families. In addition, some collective farms were organized which bore a closer resemblance to some of the former German estimates.

Economic changes resulting from boundary changes have also been studied by geographers, including Schlier (1959) and Weigend (1950). Schlier contrasts the spheres of influence of Berlin before and after World War II in respect of administration, services, food supplies, and employment. His maps show how Berlin's areas of influence in all respects have been truncated by the movement of international boundaries and how links with the Federal Republic are restricted to a few well defined roads, railways, and air-corridors. Weigend (1950) examined the changes which occurred in the area of the South Tyrol which was transferred from Austria to Italy in 1919. He points to the striking proportional increase of Italians in the population and makes some interesting comments on the economic changes. The fruit and wine producers of the transferred area continued to export their products to their traditional markets, which now lay across the boundary. Because the Italian producers were now competing on equal tariff terms, it was necessary for the producers in South Tyrol to improve the quality of their products. Other farmers in the transferred area adjusted production to the requirements of the population of the Po plain, which had become their obvious market. The constant demand for seed potatoes and Swiss Brown cattle led to their import into, and production in, the transferred areas. Although the tourist trade suffered because of the transfer, this disadvantage was partially offset by the establishment of some industries including hydroelectricity generation and aluminium refining at Bolzano.

A review of the available studies of the effects of boundary changes suggests the general conclusion that the effects will be less severe when one or more of the following situations exists:

(1) The altered boundary has existed for only a short time.
(2) Few state functions have been applied at the boundary.
(3) The groups formerly separated by the boundary have a cultural similarity.
(4) The economy of the transferred area was formerly oriented across the boundary.
(5) The economy of the transferred area is of a self-contained subsistence nature.

The transfer of the Juba strip to Italisn Somaliland from the Protectorate of Kenya after World War I met the second, third, and fifth conditions outlined above. Neither the Italian nor British governments had rigorously applied state functions at the boundary, and the Somali groups from either side were free to cross the boundary to find pastures during their subsistence stock movements. For all these reasons the transfer took place smoothly without any dislocation to the lives of the borderland inhabitants.

The converse of this argument is that boundary changes are likely to be most severe in their effects upon the population and the landscape when the boundary has existed for a long time, when the population of the transferred area is ethnically dissimilar from the state in which it is incorporated, when the states have applied many fiscal and security functions at the boundary, and, finally, when the economy of the transferred area has been closely integrated with the core region of the state from which it is removed.

An example of a country reclaiming territory which it lost through the relocation of a boundary is provided by Portugal. In 1886 Germany claimed the Kionga Triangle south of the Rovuma River and this area was detached from Mozambique and added to Tanganyika. In 1918, after Germany's defeat, Portugal successfully reclaimed the area at the Versailles Peace Conference. A more recent example is provided by the claim of Burkina Faso to the narrow strip of territory known as Oudalan, which is held by Mali. This region, which has a plentiful and permanent supply of water and rich pastures, was part of Upper Volta from 1919 to 1932. In 1932 France reorganized the administration of French West Africa and Upper Volta was divided amongst the neighbouring administrative regions of Mali, Niger, and Ivory Coast. The territory was reconstituted in 1947, but the Oudalan was left in Mali on the grounds that it had been administered from Tombouctou in 1911.

Evolution in state functions

The only function of a boundary is to mark the limits of sovereignty. The nature of the boundary's definition and the condition of the demarcated boundary will determine the effectiveness with which the boundary serves this function. In certain areas where international boundaries are located in deserts or tropical forests, it may often be difficult for the traveller to know when he passes from one state to another. This was one of the reasons why the Sino-Italian border situation became difficult in 1961. In most parts of the world, however, it is now difficult for travellers to stray inadvertently across international boundaries and the local residents usually know exactly where the territorial limit lies.

Usually a state will find it convenient to carry out some of its functions at the boundary. At points of entry passports are inspected and customs and health regulations are enforced. These points of entry are likely to be at the boundary on land routes and at airports and seaports for other travellers. In the last two cases the checking is done at the first convenient point after the traveller enters the state's territory, which extends to the outer edge of the territorial waters and to an unspecified height in the atmosphere. Boggs (1940) has listed state functions applied at the boundary, although he ascribes them to the boundary itself. There does not appear to be any study of the order in which state functions are applied at the boundary and it is unlikely that one will ever be made. Such a study would have been interesting in the evolution of colonial boundaries in Africa and Asia as administrations played an increasingly important rôle in the conduct of commercial and strategic relations. Restrictions on the import of weapons and the export of ivory and slaves were often some of the earliest acts of colonial governors. Gradually, other regulations were added to control the circulation of people, goods, and ideas until, by the end of the colonial period, some of these limits were difficult to cross unless an extensive array of forms had been correctly filled in and certified. Once independence was achieved the new governments acted promptly to ensure they they exercised all necessary functions at the boundary. The supervision of movement across some of these international boundaries was very strictly controlled at times when fighting in neighbouring states produced streams of refugees, as in Zaire in 1960, and when any threat to the established government was detected, as in Benin in 1975. In 1983 Nigeria created 12 new border control posts in an effort to reduce the influx of illegal immigrants.

It must also be noted that sometimes countries have closed boundaries in order to register displeasure with neighbouring states. For example, in 1964, 12 boundaries were closed by African states during quarrels with neighbours, and more recently, in 1983, Somalia closed its boundary with Djibouti.

Of course there are also cases where there is a reduction in the number of state functions applied at the boundary. The best example in recent years is provided by the greater ease in circulation within the European Economic Community. It seems likely that the most fruitful avenue for research into the evolution of functions will concern the way in which the application of state functions at the boundary influences the development of the border landscape.

To illustrate some of the points made in this chapter it is proposed to consider the evolution of the boundary between Mexico and the United States of America.

Evolution of the boundary between Mexico and the United States of America since 1847

The long previous history of the region through which this boundary was drawn has been described and interpreted by Bancroft (1884), Bannon (1970), Bustamante (1979), and Spicer (1962). House (1982) provided a useful summary of the earlier history of this region when it was either a political or a settlement frontier.

A state of war legally came into existence between Mexico and the United States on 13 May 1846. In less than a year the American forces had made considerable advances and secured the Mexican provinces of New Mexico, Upper and Lower California, Coahuila, Tamaulipas, Nuevo Leon, and Chihuahua. Accordingly the American government decided to appoint a commissioner who would remain with the army and be ready to accept any opportunity for negotiating a satisfactory peace (Miller 1937, p. 261). The conditions which the United States government would find satisfactory were carefully laid down in a draft agreement given to the commissioner; we are concerned here only with the territorial provisions.

At that time the *de jure* boundary between the two states was that promulgated in 1819 and coincident with the Sabine, Red, and Arkansas rivers and latitude 42° north. Under Article IV of the draft treaty, the United States sought a southward extension of the boundary to include all Texas, which had joined the Union in 1845, New Mexico, and Upper and Lower California (Fig. 3.2):

> The boundary line between the two Republics shall commence in the Gulf of Mexico three leagues from land opposite the mouth of the Rio Grande, from thence up to the middle of that river to the point where it

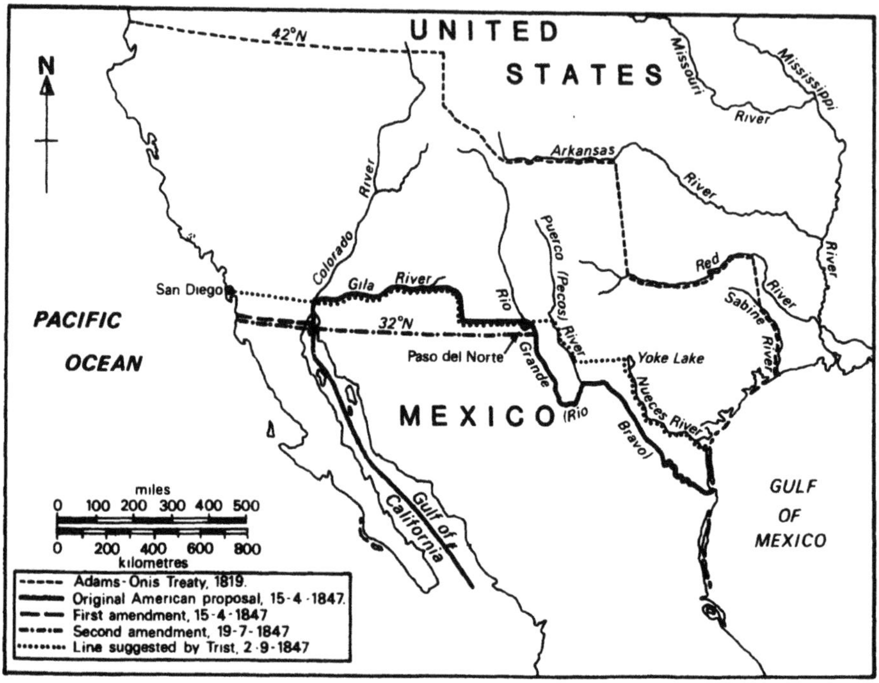

Figure 3.2 American boundary proposals.

> strikes the Southern line of New Mexico, thence westwardly along the Southern boundary of New Mexico, to the South Western corner of the same, thence Northward along the Western line of New Mexico until it intersects the first Branch of the river Gila, or if it should not intersect any branch of that river, then to the point on the line nearest to such branch and thence in a direct line to the same and down the middle of said branch of the said river until it empties into the Rio Colorado, thence down the middle of the Colorado and the middle of the Gulf of California to the Pacific Ocean. (Miller 1937, p. 265)

In addition to this territorial gain, the United States sought to secure transit rights for American citizens and goods across the Istmo de Tehuantepec. The draft agreement represents the maximum concessions America hoped to gain; however, the government indicated that it would be satisfied with less and outlined a series of payments which could be authorized to Mexico, depending upon the territory and rights secured:

(1) Up to $30 million would be paid for Upper and Lower California and New Mexico, together with transit rights over the Tehuantepec peninsula.
(2) Up to $25 million would be paid either for the three Mexican

provinces alone or Upper California and New Mexico, together with transit rights.

(3) Up to $20 million would be paid for Upper California and New Mexico.

If it proved impossible to secure Lower California, the conclusion of the boundary description would be altered to read as follows: 'to a point directly opposite the division line between Upper and Lower California; thence, due West, along the said line which runs north of the parallel of 32° and South of San Miguel to the Pacific Ocean' (Miller 1937, p. 263).

The intention of the boundary definition was clear, but it contained the seeds of disputes. Boggs (1940) has shown the difficulty of identifying 'the middle' of any river. Further, the description assumes that there was no uncertainty about the position of the southern boundary of New Mexico. Lastly, considerable difficulty may have been attached to locating the point on the western boundary of New Mexico nearest to any tributory of the River Gila, and the actual situation might have involved a considerable northward extension of the boundary.

Second thoughts by the United States government resulted in further choices being suggested to the commissioner. In order to gain the Paso del Norte and the whole of the Gila valley, which had been identified as a favourable route to the Pacific Ocean, it was suggested that the boundary should follow the Rio Grande to the parallel 32° north and along that latitude to the middle of the Gulf of California. This line could be extended across the Californian peninsula if Lower California could not be obtained, but it was essential that the Americans should have uninterrupted access through the Gulf of California and that San Diego be secured. This course was recommended since it would prevent any dispute about the southern boundary of New Mexico which, so far as America knew, had never been 'authoritatively and specifically determined' (Miller 1937, p. 770).

At the first meetings between American and Mexican commissioners the latter revealed their government's proposals. No doubt, like the American draft, there were several possibilities. The most important point which emerged from the first exchange was that the Mexican government laid down two conditions as *sine qua non* which prevented even the minimum American demands from providing a basis for discussion. First, the Mexicans required a neutral strip of territory adjacent to the north bank of the Rio Grande in order to afford military protection against the United States and restrict the incidence of smuggling which would reduce Mexico's revenue and injure their manufacturing industries. Secondly, Mexico required a land connection between Lower California and Sonora around the head of the Gulf of California (Fig. 3.3).

Instead of breaking off the negotiations, the American commissioner exceeded his instructions and submitted to his government for consideration

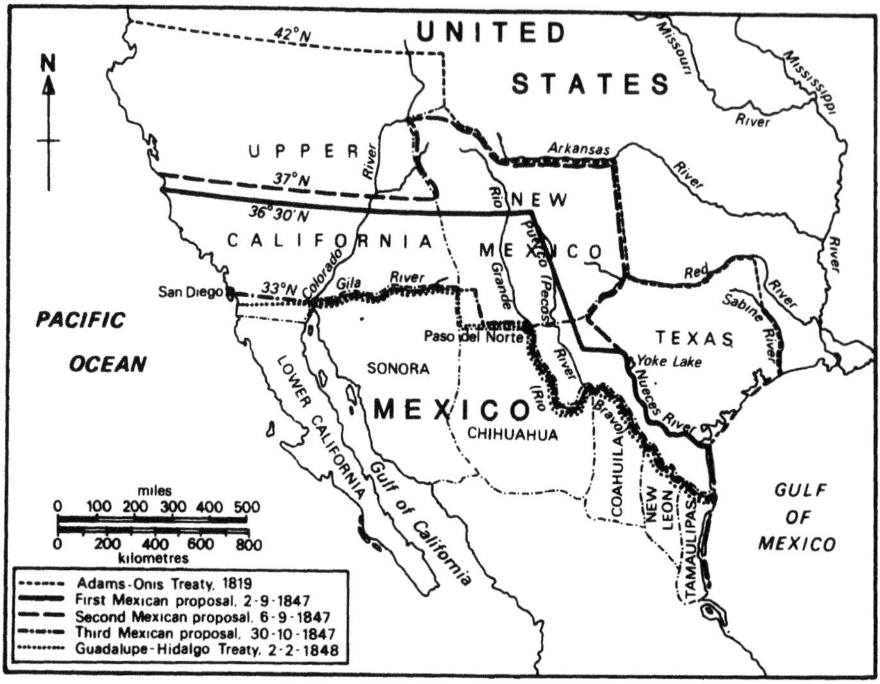

Figure 3.3 Mexican boundary proposals.

a line which met the Mexican conditions. Historians have undoubtedly judged the commissioner: geographers can be concerned only with the results of this action. The recommended boundary was defined as follows:

> The boundary line between the two Republics shall commence at a point in the Gulf of Mexico, three leagues from Land, opposite to the middle of the Southernmost inlet into Corpus Christi Bay; thence, through the middle of said inlet, and through the middle of said bay, to the middle of the mouth of the Rio Nueces; thence up the middle of said river to the Southernmost extremity of Yoke Lake, or Oagunda de las Yuntas, where the said river leaves the said Lake, after running through the same; thence by a line due west to the middle of the Rio Puerco, and thence up the middle of said river to the parallel of latitude six geographical miles north of the Fort at the Paso del Norte on the Rio Bravo; thence due west, along the said parallel to the point where it intersects the western boundary of New Mexico; thence northwardly along the said boundary, until it first intersects a branch of the River Gila; (or if it should not intersect any branch of that river, then to the point on said boundary nearest to the first branch thereof, and from that point in a direct line to such a branch) thence down the middle of said branch and of the said Riva Gila, until it empties into the Rio Colorado,

> and down or up the middle of the Colorado, as the case may require, to the thirty-third parallel of latitude; and thence due west along the said parallel, into the Pacific Ocean. And it is hereby agreed and stipulated, that the territory comprehended between the Rio Bravo and the above defined Boundary, from its commencement in the Gulf of Mexico up to the point where it crosses the said Rio Bravo, shall for ever remain a neutral ground between the two Republics, and shall not be settled upon by the citizens of either; no person shall be allowed hereafter to settle or establish himself within the said territory for any purpose or under any pretext whatever; and all contraventions of this prohibition may be treated by the Government of either Republic in the way prescribed by its laws respecting persons establishing themselves in defiance of its authority, within its own proper and exclusive territory. (Miller 1937, p. 288).

The form of this description implies that the neutral strip lay within Mexico, although the sense of the description is that it would be the responsibility of both governments to restrict settlement there. It is not clear why, in order to give a land connection between Lower California and Sonora, the boundary had to be drawn along the parallel 33° north, which would deny the United States access to San Diego and San Miguel. The 'parallel six geographical miles north of the fort at Paso del Norte on the Rio Bravo' was coincident with the southern boundary of New Mexico shown on Disturnell's map, and this avoided any dispute about the position of that provincial boundary.

There was no chance of America accepting this boundary since it would compromise Texan sovereignty and exclude America from the two main ports of Upper California and the Gulf of California. The American commissioner was recalled before the resumption of hostilities, which were to force Mexico to sue for peace on American terms (Miller 1937, pp. 289–93). The American commissioner continued to make history by ignoring his recall and remaining in Mexico to negotiate a treaty, although by then he lacked authority.

Before examining further negotiations it may be recalled that up to this stage the process of boundary evolution had been normal. Both states had proposed lines which would allocate territory between the states to their greatest advantage. The descriptions revealed the generalized nature of geographical knowledge of the area and were drawn in response to broad strategical motives. Mexico's proposal for a neutral zone is the transparent device of a weak state trying to limit its territorial concessions to a stronger neighbour.

The final round of negotiations began in December 1847. The Mexicans gave up the idea of a broad neutral zone and instead sought to draw the boundary parallel to, and one league north of, the river. Further, they introduced a claim calling for part of the boundary to coincide with the summits of the Sierra de los Mimbres which would have preserved the south-west quadrant of New Mexico. The Mexican government did not give up its claims for a land connection between Lower California and Sonora which would include San Diego.

It was only this last point which prevented rapid agreement, for it will be recalled that the American Commissioner had been instructed in the first draft to secure a boundary which was to be defined in the west as follows:

> . . . down the middle of the Colorado river and the Gulf of Mexico to a point opposite the division line between Upper and Lower California; then due west along said line which runs north of the parallel of 32° and south of San Miguel to the Pacific Ocean. (Miller 1937, p. 263)

The American Commissioner found himself in some difficulty for three reasons. First, some cartographic authorities showed San Miguel to be south of latitude 32° north. Secondly, the Mexican government, and other authorities, represented the political division between Upper and Lower California as being north of San Diego. Thirdly, it was suspected, correctly, that the mouth of the Colorado River lay south of parallel 32° north.

Eventually, after several proposals and counterproposals, the commissioners agreed to a boundary which coincided with the original American draft, except in the extreme west where the boundary followed a direct line from the confluence of the Gila and Colorado rivers to a point on the Pacific coast named Punto de Arena, which was south of San Diego (Miller 1937, p. 325).

This suggestion was transmitted to both governments and, not surprisingly, the American government decided to accept it, although their agent lacked authority in the final negotiations. The treaty was endorsed by the American Senate with certain amendments which did not relate to the territorial provisions. The fifth Article defined the boundary in the following way:

> The Boundary line between the two Republics shall commence in the Gulf of Mexico, three leagues from land, opposite the mouth of its deepest branch, if it should have more than one branch emptying directly into the sea; from thence, up the middle of that river, following the deepest channel, where it has more than one, to the point where it strikes the south boundary of New Mexico; thence, westwardly, along the whole southern boundary of New Mexico (which runs north of the town called Paso) to its western termination; thence, northward, along the western line of New Mexico, until it intersects the first branch of the river Gila; (or if it should not intersect any branch of that river, then, to

> the point on said line nearest to such branch, and thence in a direct line to the same); thence down the middle of the said branch of the said river, until it empties into the Rio Colorado; thence, across the Rio Colorado, following the division line between Upper and Lower California, to the Pacific Ocean.
>
> The southern and western limits of New Mexico, mentioned in this Article, are those laid down in the Map, entitled 'Map of the United Mexican States, as organized and defined by various acts of the Congress of said Republic, and constructed according to the best Authorities. Revised Edition. Published at New York in 1847 by J. Disturnell:' of which Map a Copy is added to this treaty, bearing the signatures and seals of the Undersigned Plenipotentiaries. And, in order to preclude all difficulty in tracing upon the ground the limit shall consist of a straight line, drawn from the middle of the Rio Gila, where it unites with the Colorado, to a point on the coast of the Pacific Ocean, distant one marine league due south of the southernmost point of the Port of San Diego, according to the plan of said port, made in the year 1782 by Don Juan Pantoja, second sailing master of the Spanish fleet, and published at Madrid in the year 1802, in the Atlas of the voyage of the schooners *Sutil* and *Mexicana*. (Miller 1937, pp. 213–15)

There are two points to notice. First, the description was similar to that originally proposed by America and it continued to reflect the generalized topographical knowledge available about the area in question. Secondly, the definition hoped to avoid the two main points of controversy by specifying the maps which were authorities for fixing the southern and western boundaries of New Mexico and the terminal point on the Pacific Ocean. Subsequent events showed that controversy was not avoided.

In 1849 a commission tried to determine the Pacific coast terminus of the boundary, which was defined as follows: 'one marine league due south of the southernmost point of the port of San Diego, according to the plan of the said port, made in the year 1782 by Don Juan Pantoja' (Miller 1937, p. 214). The commission found little correspondence between the map and the actual coastline. They did find one point near the port which appeared to coincide with the present coast. Accordingly, they measured the distance on the map between this point and the southernmost point of the port. This distance was then laid off on the ground and the marine league measured from there.

In 1852 the United States government made financial provision for the commission appointed to demarcate the southern and western boundary of New Mexico. The availability of the money depended upon the following condition being met:

> That no part of this appropriation shall be used or expended until it shall be made satisfactorily to appear . . . that the southern boundary of New

> Mexico is not established . . . farther north of the town called 'Paso' than the same is laid down in Disturnell's map, which is added to the treaty. (Miller 1937, p. 369)

Now, according to the treaty map the latitude of El Paso was 32°15′ north and the boundary intersected the Rio Grande eight miles north of the town and extended 3½° of longitude west of that river, which was shown to be in longitude 27°40′ west of Washington. The surveyors quickly found that the correct latitude of El Paso was 31°45′ north and the boundary intersected the Rio Grande at longitude 29°40′ west of Washington. The Mexican members of the commission pressed for a boundary starting on the Rio Grande at 32°22′ north and proceeding westwards for 1½° of longitude. This means that they wished to accept the latitude of the boundary shown on the map but to correct the longitudinal error in the river position. For them the position of El Paso was unimportant, provided it remained Mexican. However, the American government insisted on following the map and using the position of El Paso as the datum from which measurements in the field would be made. The area between the two interpretations of the boundary was about 11 000 square miles (Fig. 3.4).

Neither side was prepared to concede the area nor to compromise, and therefore new boundary negotiations were started in 1853. After extravagant proposals by both sides, a new boundary definition crystallized out of the discussions. The new boundary followed the middle of the Rio Grande to latitude 31°47′ north, which it followed westwards for 100 miles before turning due south to the parallel 31°20′ north. This parallel was followed westwards as far as longitude 111° west of Greenwich. The boundary then followed a straight line to the Colorado River 20 English miles below its confluence with the River Gila and then upstream along the middle of the Colorado to the line agreed in 1848 (Malloy 1910, p. 1122). This line ceded about 24 000 square miles to the United States and secured for that state the entire catchment of the River Gila (Fig. 3.4).

No problems were experienced in demarcating this second allocating boundary between the Rio Grande and the Colorado River. The only problems associated with the boundary since 1853 have all been concerned with the Rio Grande, which proved to be most unstable in position in the section which marked the boundary. Changes in the river's course occurred gradually by accretion and suddenly by avulsion. The most enduring problem concerned El Chamizal, a tract of 630 acres opposite El Paso on the north bank of the river which forms the boundary. Mexico claimed that this area was south of the boundary in 1853, when the agreement came into force, but was transferred to the north bank of the river when the course was suddenly changed in the floods of 1864. The two states could not solve the dispute and arbitration by Canada in 1910

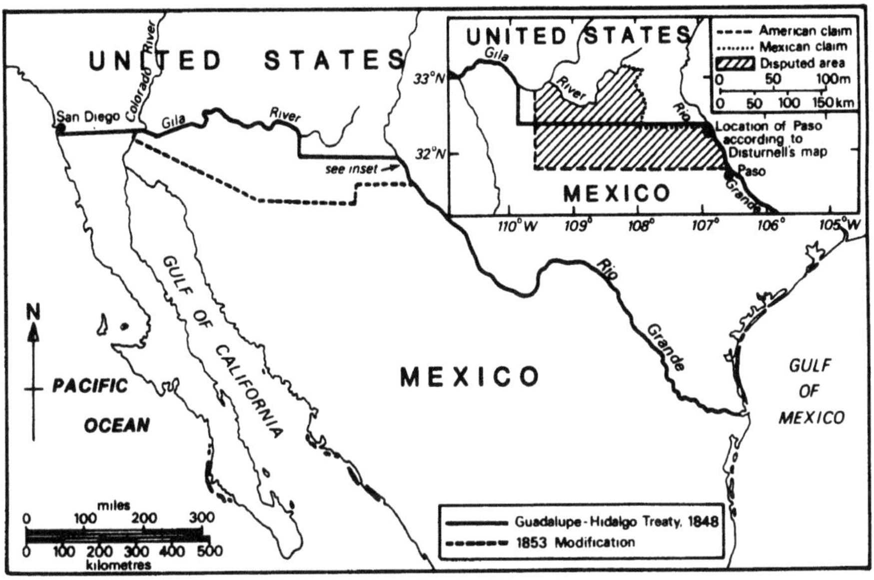

Figure 3.4 The disputed zone and agreed boundary.

failed to produce an answer. The Canadian decision was that the boundary should follow the course of the river as it existed before the floods of 1864. It did not prove possible to establish this line through this territory to the satisfaction of both sides until 1964 when it was agreed to divide the disputed area. Hill (1965) has written an excellent account of this settlement.

In 1884 the two states agreed that in future the boundary would coincide with the centre of the normal channel of the river and continue to follow changes in the river's course resulting from accretion. The boundary would follow the abandoned river course when changes resulted from avulsion. This means that the area transferred from one bank to the other, when a meander neck was severed, locally known as *bancos*, would remain under the sovereignty of the original state (Malloy 1910, p. 1159). In 1905 the governments agreed to minimize the difficulties resulting from avulsive river changes by exchanging all *bancos* other than those having an area of more than 617 acres or a population of 200. A permanent commission was responsible for this mutual exchange, which simplified boundary administration.

Further complications resulted from the fact that the Rio Grande is a valuable resource used by nationals of both states. To regularize the use of the water, Mexico and America signed a Convention in 1906 to ensure equitable division. Under the terms of the Convention the United States government contracted to deliver 60 000 acre feet to the head of a Mexican canal 1 mile below the point where the river became the internal boundary. The United States would also be responsible for the distribution of water as far as Fort Quitman in Texas. In order to prevent damaging floods, a dam was built in 1916 at Elephant Butte in Texas. This effectively stopped the flood waters derived from above the dam, but the reduction in the flow rate of the river resulted in the deposition of alluvium, which had previously been scoured by floods. At El Paso the river bed was 12 feet higher in 1933 than in 1907. To remedy the situation the government agreed to construct a rectified canal from El Paso-Juarez to the mouth of the Box Canyon below Fort Quitman. The canal of 88 miles shortened the river course by 67 miles and the increased gradient prevented alluvial accumulation. The canal was 590 feet wide and was aligned along the boundary axis, requiring both sides to exchange 3500 acres. In addition to making the adjoining farmlands more secure from flood, the new canal increased the area of land available for cultivation and simplified boundary maintenance and control.

This account of the evolution of the boundary between the United States and Mexico reveals four points. First, the American motives in concluding the initial boundary agreement were to secure at least the former Mexican territory of New Mexico and Upper California and to end the war without the need to occupy all Mexico and maintain a military administration. Mexico agreed to negotiate in order to maintain sovereignty over the area south of the ceded portion. Secondly, at no time was the boundary delimited. The Guadalupe–Hidalgo Treaty laid down a boundary which allocated territory between the two states, and this line was modified by the Gadsen Treaty when the first boundary definition proved impossible to apply on the ground. In both cases the boundary was defined by imprecise physical features and straight lines linking known points or coinciding with parallels of latitude or meridians. The descriptions reflected the generalized nature of existing knowledge about the area. Thirdly, the demarcation of these allocating boundaries was hindered by the lack of correspondence between the maps named in the treaty and the actual landscape and by the nature of the Rio Grande. Fourthly, state controls over immigration and trade were applied as soon as the boundary could be identified.

References

Bancroft, H. H. 1884. *History of the northern Mexican States.* San Fransisco: A. L. Bancroft.

Bannon, J. F. 1970. *The Spanish Borderlands frontier 1513–1821.* Albuquerque: University of New Mexico Press.

Boggs, S. W. 1940. *International boundaries: a study of boundary functions and problems*. New York: Columbia University Press.

Brigham, A. P. 1917. Principles in the delimitation of boundaries. *Geographical Review* **7**, 201–19.

Bustamante, J. A. 1979. El estudio de la zona fronteriza Mexico–Estados Unidos (The study of the Mexican–American frontier zone). *Foro Internacional* **19**, 471–516.

Clifford, E. H. M. 1936. The British Somaliland–Ethiopia boundary. *Geographical Journal* **87**, 289–337.

Davidson, G. 1903. *The Alaska boundary*. San Francisco: Alaska Packers Association.

Detzner, H. von 1913. Die Nigerische Grenze von Kamerun zwischen Yola und dem Cross-Fluss (The Nigerian boundary of Kamerun between Yola and the Cross River). *Mitteilungen aus den deutschen Schutzgebieten* **26**, 317–38.

Hertslet, Sir E. 1875–91. *The map of Europe by treaty* (4 volumes). London: HMSO.

Hertslet, Sir E. 1909. *Map of Africa by treaty* (3 volumes plus atlas). London: HMSO.

Hill, J. 1965. El Chamizal: a century old boundary dispute. *Geographical Review* **55**, 510–22.

Hinks, A. R. 1921. Notes on the techniques of boundary delimitation. *Geographical Journal* **58**, 417–43.

Holdich, Colonel Sir T. H. 1916. *Political frontiers and boundary making*. London: Macmillan.

House, J. W. 1982. *Frontier on the Rio Grande: a political geography of development and social deprivation*. Oxford: Clarendon Press.

Inspector-General of Customs 1917. *Treaties between China and Foreign States* (2 volumes). Shanghai: Statistical Department of the Inspectorate of Customs.

Jones, S. B. 1945. *Boundary-making, a handbook for statesmen, treaty editors and boundary commissioners*. Washington, DC: Carnegie Endowment for International Peace.

Lamb, A. 1964. The Sino-Pakistan border agreement of 2 March 1963. *Australian Outlook* **18**, 299–312.

Lambert, A. F. 1965. Maintaining the Canada – United States boundary. *Cartographer* **2**, 67–71.

Lapradelle, P. de 1928. *La frontière: étude de droit international* (The boundary: a study of international law). Paris: Les Éditions Internationales.

Laws, J. B. 1932. A minor adjustment in the boundary between Tanganyika Territory and Ruanda. *Geographical Journal* **80**, 244–7.

Mackinder, H. J. 1904. The geographical pivot of history. *Geographical Journal* **23**, 421–44.

Mallow. W. M. 1910. *Treaties, conventions, international acts, protocols and agreements between the United States of America and other powers 1776–1909*, Vol. 1. Washington, DC: Government Printing Office.

Miller, H. 1937. *Treaties and other acts of the United States of America*, Vol. 5. Washington, DC: Government Printing Office.

Nicholson, N. L. 1954. *The boundaries of Canada, its Provinces and Territories.* Memoir 2. Department of Mines and Technical Surveys, Geographical Branch, Ottawa.

Nugent, W. V. 1914. The geographical results of the Nigeria–Kamerun boundary demarcation commission. *Geographical Journal* **43**, 630–51.

Pallis, A. A. 1925. Racial migrations in the Balkans during the years 1912–24. *Geographical Journal* **66**, 315–31.

Peake, E. R. L. 1934. Northern Rhodesia – Belgian Congo boundary. *Geographical Journal*, **83**, 263–80.

Prescott, J. R. V. 1971. *The evolution of Nigeria's international and regional boundaries 1861–1971.* Vancouver: Aldine.

Prescott, J. R. V. 1975. *Map of mainland Asia by treaty.* Melbourne: Melbourne University Press.

Rudin, H. R. 1938. *Germans in the Cameroons.* London: Cape.

Ryder: C. H. D. 1925. The demarcation of the Turco-Persian boundary in 1913–14. *Geographical Journal* **66**, 227–42.

Schlier, O. 1959. Berlins Verflechtungen mit der Umwelt fruher und Heute (Berlin's relations with the surrounding region yesterday and today). *Geographische Rundschau* **11**, 125–51.

Spicer, E. H. 1962. *Cycles of conquest: the impact of Spain, Mexico and the United States on the Indians of the Southwest 1533–1960.* Tuscon: University of Arizona Press.

Sykes, Sir P. M. 1926. *The Right Honorable Sir Mortimer Durand.* London: Cassell.

Varela, L. V. 1899. *La République Argentina et le Chilli: histoire de la démarcation de leurs frontières depuis 1843 jusqu'a 1899* (Argentina and Chile: the history of the demarcation of their boundaries 1843–1899) (2 volumes). Buenos Aires: M. Biedma.

Vattell, E. de 1758. *The law of nations or the principles of natural law*, Vol. 3. Washington, DC: published by the Carnegie Institution of Washington in 1916 after translation by G. G. Fenwick.

Weigend, G. G. 1950. Effects of boundary changes in the south Tyrol. *Geographical Review* **40**, 364–75.

Wiskemann, E. 1956. *Germany's eastern neighbours.* London: Oxford University Press.

4 *International boundary disputes*

> The relations between modern states reach their most critical stage in the form of problems relating to territory. Boundary disputes, conflicting claims to newly discovered lands, and invasions by expanding nations into the territory of their weaker neighbours have been conspicuous among the causes of war. (Hill 1976, p. 3)

Boundary disputes have long been a popular subject for research among political geographers, lawyers, political scientists, and historians. Their research has a refreshing topicality and often results in governments making available information which would otherwise have remained buried in correspondence files and secret reports.

This chapter begins by describing the aspects of boundary disputes on which geographers can profitably concentrate and provides an example of the logical arrangement of those aspects in the analysis of boundary disputes. The next section describes four types of boundary disputes and each type is then examined separately. The rôle of maps in boundary disputes is then considered and, finally, the main points of this chapter are summarized in the conclusion.

Aspects of boundary disputes

Geographers are not alone in studying boundary disputes; they have also been a profitable field of research for other scholars. For example, Hsu (1965) published a detailed historical account of the boundary dispute between China and Russia in the Ili Valley between 1871 and 1881; Johnson (1966) prepared an interesting account of the legal aspects of the dispute over the Columbia River between the United States and Canada; and Touval (1966) presented a useful study of the boundary dispute between Morocco and Algeria from the viewpoint of a political scientist. However, the facility of geographers with maps and their understanding of regional characteristics have given them an advantage in such studies. There are clearly some aspects of boundary disputes which a geographer is not competent to consider, such as the involved decisions about the legality of treaties and the rôle of individual persons in successfully pressing arguments in favour of one or another case. There still remains a great deal the geographer can study in making a distinct contribution to an understanding of the situation.

The analysis of any boundary dispute should provide information on the following aspects. First, it is necessary to uncover the *cause* of the dispute. In most cases the cause will be found in the boundary's history, which will reveal that the evolution of the boundary is incomplete. For example, the quarrel between China and the Soviet Union over the island at the confluence of the Amur and Ussuri rivers rests ultimately on the ambiguous boundary definition of the 1861 treaty, which failed to specify the course of the boundary in the vicinity of the island.

Secondly, it is important to identify the *trigger action* which created a situation where one side judged it necessary to argue in favour of rectification of the boundary. Clearly, many of these disputes, which are rooted in history, have periods of quiescence and periods of sometimes intense activity. Thus the dispute over the Falkland Islands, which has a history over nearly two centuries, flared into a fierce war when the Argentinian authorities believed that the time was ripe for a diversion from domestic problems and that the risk from distant Britain was acceptable. This example leads into the third important aspect which concerns the *aims* of governments initiating boundary disputes. In many cases the government will be seeking additional territory or relief from some unacceptable administrative irritation connected with the boundary. However, in some cases disputes over boundaries may be launched for other reasons. For example, it is generally believed that the Philippines' claim to northern Sabah in the early 1960s was designed to delay or prevent the incorporation of Sabah and Sarawak into Malaysia.

The *argument* used by governments to justify the position adopted in any dispute is the next important aspect to consider. Arguments based in history and geography are common in the scholarly and official publications dealing with boundary disputes, but there is also recourse to issues of strategy, economics, and anthropology in some cases. Regardless of whether a dispute is settled quickly or allowed to drag on for many years, there will usually be some consequences. The *results* may be confined to the borderland if the dispute is settled quickly and the changes are minor. For example, the realignment of a boundary to take account of changes in the course of a river may mean that some farmers now own land on both sides of the boundary and need travel documents to conduct their operations. The movement of the boundary between Iran and Iraq from the east bank of the Shatt el Arab to the median lines in the river appeared to be a reasonable compromise in 1975 and was expected to result mainly in improved access for Iranian vessels to the ports of Khorramshahr and Abadan. Five years later, at a time of political turmoil in Iran, the government of Iraq launched an attack to fix the boundary on the east bank of the river again. This example illustrates the need to consider the effect of settlements and continued disputes on the international relations of countries directly and indirectly concerned.

Aspects of the boundary dispute between Afghanistan and Pakistan

Afghanistan claims that the Pathan tribesmen in Pakistan should be allowed to form a state with their fellow tribesmen in Afghanistan; Pakistan in reply denies that the Pathans desire the establishment of Pushtunistan, as the proposed state is known. The basic *cause* of the situation can be found in the Anglo-Afghan Agreement of 1893, which delimited, by means of a map, the boundary between the spheres of influence of Britain and Afghanistan (Sykes 1940, Vol. 2, p. 353). This agreement was confirmed by further treaties in 1905, 1919, 1921, and 1930 (Qureshi 1966). This boundary divided the territory occupied by the Pathans in such a way that 2.4 million remained within British territory. The boundary was the result of a British dilemma. Britain wanted friendly relations with Afghanistan and a stable boundary behind which British subjects would be safe. There was no obvious boundary and when Britain took action against aggressive hill tribes relations with Afghanistan worsened; if this was avoided by a British retreat, sooner or later the hill tribes followed and started to raid British areas.

The *trigger action* which encouraged Afghanistan to make this claim was the partition of India in 1947 when Pakistan, facing internal difficulties and external pressure from India, replaced Britain as the sovereign neighbour of Afghanistan. After a period when the dispute was dormant, it was revived when the Pakistan authorities were beset with problems of internal revolt in Baluchistan after the secession of Bangladesh and after the revolution, which established a republican government in Afghanistan in July 1973.

The *arguments* advanced by Afghanistan fall into three categories. First, it is argued that the 1893 treaty was not legally binding since Afghanistan signed it under duress; that in any case the tribal territories between Afghanistan and the administered territories of the British sphere formed independent territory; and that Pakistan cannot inherit the rights of an 'extinguished person', namely the British in India (Fraser-Tytler 1953, p. 309). These are legal and moral arguments which the geographer can note but not evaluate. Secondly, it is claimed that historically Afghanistan controlled much of India, and certainly the present area of western Pakistan (Taussig 1961). Recourse to historical political geography shows that the State of Afghanistan was formed in 1747. At the maximum extent of the Durrani Empire in 1797, the area controlled reached eastwards almost to Delhi and Lahore. Lahore was ceded to the Mogul Empire in 1798 and Peshawar was lost in 1823. If territorial control for 76 years 160 years ago was accepted as a strong argument in favour of territorial reversion, the world map would be liable to dramatic change! Thirdly, it is claimed that the Pathans in Afghanistan and Pakistan form a single ethnic unit which should be united in one state. This argument is undoubtedly stronger, although Caroe (1961) has shown that the eastern Pathans have enjoyed close economic and political ties with the major states of the Indus valley and have developed linguistic differences with the western Pathans. Further, the area claimed for Pushtunistan stretches from the Pamirs to the Arabian Sea and is bounded on the west by Afghanistan and Iran and on the east by the Indus River. This includes large areas where there are few Pathans, such as Chitral, Gilgit, Baltistan, and Baluchistan. When the strongest argument is exaggerated so remarkably as to weaken its force one suspects the altruism of Afghanistan and seeks the real aim of this dispute. The claim to Baluchistan suggests that Afghanistan is hoping to control the proposed state and use it for an outlet to the Arabian Sea. Hasan (1962) has also suggested that the ruling Pathan dynasty in Afghanistan is seeking to bolster its position with regard to the Persian and Turki-speaking Afghans who form two-thirds of the state's population.

The *results* of the dispute have been significant on the regional scale, and in international relations. Diplomatic relations were broken off between the two countries in September 1961 and were not resumed until May 1963. During this period the boundary was closed, except for a short period in January 1962 when American foreign aid supplies were allowed through from Pakistan. Since Afghanistan's major markets were in India and most of its supplies came from Japan, America, and India there was a reorientation of Afghanistan's trade. The Soviet Union assumed a dominant commercial rôle in Afghanistan's international commerce nearly a decade before it started to assert its political dominance. Even before Soviet intervention in Afghanistan in December 1979 there had been friction along the Afghan–Pakistan border. This continuing problem encouraged the authorities in Pakistan to improve roads and airfields so that their forces could operate more efficiently near the boundary.

The closing of the boundary also affected the traditional transhumance movements of the Powindas of Afghanistan. Normally, about 100 000 Powindas, with their large herds, migrated to Pakistan's lowlands as winter settled over the Afghan uplands. These people supplemented their incomes in Pakistan by manual labour, especially in harvesting sugar cane, and money lending. After wintering in the lowlands they returned to the new summer pastures in Afghanistan. The Pakistan authorities began to insist on travel documents for the people and certificates of health for the stock on the grounds that grazing was scarce in Pakistan and the Powindas' herds were suspected of carrying disease. The transhumance movements have not been resumed on the former scale, and the Powindas have been resettled in parts of Afghanistan, including Sistan.

Since the entrance of Soviet forces into Afghanistan the number of border incidents has increased. This is mainly due to the large flow of refugees from Afghanistan into Pakistan. By the end of 1983 it is estimated that there were 3 million refugees on the Pakistan side of the boundary. Some of the refugee camps have provided recruits for the forces opposing the Soviet army in Afghanistan, and these forces crossed into Pakistan's territory when they needed sanctuary. Not surprisingly, the Soviet and Afghan forces have struck at camps across the boundary, which they believe are occupied by the rebels. Pakistan recorded 400 violations of its territory in the period from December 1979 to April 1982. Afghanistan has also complained about incursions from Pakistan, such as the one on 17 December 1983 which resulted in the destruction of the customs post at Torkham on the boundary between Paktia in Afghanistan and Waziristan in Pakistan.

One important development has occurred in the Wakhan valley. This narrow region was ceded to Afghanistan by Britain in 1893 because Britain did not want to share a common boundary with Russia. In June and July 1981 Soviet forces occupied this valley and drove out the rebels who had occupied it. The boundary with Pakistan was closed and Soviet officers were installed in the various border posts. On 18 November 1981 the United States' delegate at the United Nations, Mrs Jeane Kirkpatrick, alleged that the Soviet Union was guilty of a *de facto* annexation of the Wakhan valley (*Keesing's Contemporary Archives* 1982, 28, p. 31544). Although Afghanistan and the Soviet Union signed a border agreement after these events, its terms have never been published. If there is any truth in the allegation of annexation, then Pakistan has acquired a new neighbour and the connection between China and Afghanistan has been broken.

On 31 August 1976 Afghanistan celebrated Pakhtoonistan Day and 18 months earlier President Daud had outlined his country's position to the Secretary-General of the United Nations:

> Afghanistan has, since the time the British divided our land by force of arms and annexed part of our territory to their empire, supported the lawful right of these people [Pathans and Baluchis] living on the borders of Afghanistan and Pakistan, and will continue to do so until they are restored. (*Keesing's Contemporary Archives* 1976, 22, p. 27851)

It seems likely that this is a dispute which Afghanistan will revive from time to time as convenient circumstances arise. An irridentist movement on behalf of Pathans in Pakistan might be one policy which would promote the unity of Afghanistan after the civil war is ended.

This brief account indicates the aspects of boundary disputes on which political geographers can most profitably concentrate and the remainder of the chapter examines the four types of boundary disputes in detail.

Types of boundary disputes

The general term 'boundary dispute' includes four quite different types of disagreements between countries.

The first type of dispute may be described as a *territorial* boundary dispute and this results from some quality of the neighbouring borderland which makes it attractive to the country initiating the dispute. The second type of boundary dispute concerns the actual location of the boundary and usually involves a controversy over the interpretation of terms used in defining the boundary at the stage of allocation, delimitation, or demarcation. This type may be called a *positional* boundary dispute. Both territorial and positional disputes can only be solved in favour of the claimant state by altering the position of the boundary; the two remaining types can be solved without altering the boundary's location. The third type arises over state functions applied at the boundary, and they may be described as *functional* boundary disputes. Such disputes might arise because states are unreasonably diligent in applying regulations or because they are negligent in enforcing rules. The last type of dispute concerns the use of some transboundary resources such as a river or a coalfield. Disputes of this kind usually have as their aim the creation of some organization which will govern use of the particular resource and they may be called disputes over *resource development.*

Territorial boundary disputes

Although it must be accepted that if a government feels sufficiently strong it might press territorial claims which have no obvious basis, nevertheless in most cases some arguments, no matter how weak, are raised.

In most borderlands between countries it would have been possible to draw a variety of boundaries based on different features, even if the work was done by an impartial commission. For example, the boundary drawn to separate linguistic groups probably did not coincide with the line separating religious factions; neither of these limits might have coincided with boundaries which distinguished areas with particular economic interests; the boundaries of former political units might have provided another set of unique lines, and unfortunately these cultural divides could have followed a different alignment to boundaries which separated the physical regions of the borderland based on morphology, drainage, and vegetation. The boundary which was finally selected after a process of political negotiations might have corresponded with none of the lines mentioned or with different sections of the various lines. In the latter case the boundary would lack any uniform basis. It will therefore always be possible for a government to make a territorial claim by emphasizing some pattern in the borderland which was discounted during the boundary's construction.

It can therefore be expected that most boundaries were drawn as a compromise between the strategic, economic, and ethnic requirements of the two states and will therefore have some degree of unconformity with features in the borderland. In most cases these discrepancies will not be serious enough to provoke a territorial dispute. In those cases where territorial boundary disputes develop, it is possible to identify three processes by which the boundary's unconformity might have arisen.

The origin of territorial disputes

First, the boundary might have been drawn without full knowledge of the distribution of people or topographical features. Many of the boundaries of Africa were drawn through areas for which no precise information was available; this was particularly the case during the stage of allocation, as the last chapter shows. Unfortunately, it is also true, in Africa, Asia, and South America, that some boundary decisions were based on inaccurate information. Tribal chiefs would frequently exaggerate the extent of their territory in the hope that the colonial powers would confirm their authority over such extended kingdoms. For example, on 4 April 1903 the District Commissioner in Barotseland was instructed to obtain answers in very great detail to 16 questions regarding the extent of the Barotse Kingdom in 1891. This information was necessary because Britain had a disagreement with Portugal about the interpretation of their 1891 treaty regarding the boundary between Angola and Northern Rhodesia. The answers were required within one week! The commissioner duly replied and reported on the problem and technique for solving it:

> It is extremely difficult to collect reliable evidence and to obtain proof of statements as to the distance Lewanika's influence [Lewanika was Paramount Chief of the Barotse] extended in 1891, since practically the only white men who had lived in Lealui at that time were the Revd Mr Arnott and the Revd Mr Coillard, who were both missionaries and who therefore had no interest in determining the boundaries of the Barotse influence . . . it is almost necessary in order to form a correct idea of the position in 1891, to find out the actual truth at some later date, which would permit of no doubt, and then to work back on the most probable assumptions to what the position would have been at the earlier period; here we know to a large extent the position in 1897, and six years earlier in such large areas as these, the position could not evidently have been very different. (File on Barotse Boundary in the Archives of Rhodesia)

It was precisely because the colonial authorities were aware of the problems created by boundaries which divided tribes that they made considerable efforts to obtain reliable information. Unfortunately, the task was sometimes beyond them and in other cases overriding strategic interests caused them to disregard reliable information about tribal distributions. In other cases it proved impossible to reconcile the data regarding tribal distributions which were acquired by both sides.

Secondly, at the conclusion of a war, new boundaries sometimes were forced on the defeated country which did not correspond with established patterns. For example, when the Hapsburg Empire was dismembered in 1919, Italy was awarded the area of Tirol lying south of the Brenner Pass. Nearly 70 per cent of the population in the transferred area spoke German and had strong cultural affinities with the Austrians on the other side of the boundary. Following the surrender of Italy in 1943 and the defeat of Germany in 1945, Austria attempted to reclaim the southern Tirol but was unsuccessful.

Thirdly, it is possible for new distributions of population to develop after the boundary is drawn and to give rise to territorial claims. This is especially possible following the establishment of an antecedent boundary. After the states of Peru, Chile, and Bolivia had been established, valuable guano and nitrate deposits were discovered in the coastal areas of all three states, with the richest deposits being located in Bolivia. In Peru the deposits were developed under a government monopoly, but in northern Chile and southern Bolivia the smaller deposits of guano were developed by private contractors, from whom the government derived duty when the material was exported. The manual work was done by Chilean peasants in the entire borderland and their presence in Bolivia, close to Chile, encouraged that country to make a successful claim to the Bolivian littoral, which helped to transform that country to a landlocked state (Dennis 1931, pp. 37, 73).

The timing of territorial disputes

Even if a boundary separates a state from a neighbouring area which has certain attractive qualities, there is no certainty that a dispute will develop. Clearly, a definite act is required by the claimant state to initiate the dispute. This action will be taken in the most favourable circumstances and therefore it will often be noticed that a boundary which has created no problems for a very long time will suddenly become the subject of dispute. Generally the trigger action which creates a favourable situation is related to some change in the government or the relative strengths of the states concerned. It was noted earlier that the Pushtunistan dispute was revived in July 1973, when a republican government replaced the traditional monarchy. The two most common circumstances which have promoted the occurrence of boundary disputes are the conclusion of wars and decolonization.

At the conclusion of a war it often happens that there is a change in the relative strengths of states which allows territorial claims to be advanced by the country which has been strengthened. In 1919 and 1945 European boundaries were altered in favour of states such as Poland, Czechoslovakia, and Yugoslavia at the expense of the defeated countries.

Civil wars may promote territorial disputes in two ways. First, when a country is involved in a civil war it is unlikely to engage in territorial claims which might invite attack. However, once that civil war is concluded and the country's strength renewed it may begin actively to prosecute claims to external territory. After 1949 China began to assert itself in the borderlands and demand the renegotiation of treaties which were alleged to have been forced on China during a period of weakness. This development has caused serious disputes with the Soviet Union and India. Once the Cambodian revolution was completed in 1975 it began to press claims against Thailand, and in the same fashion the emergence of a communist government in Laos in May 1975, which marked the end of internal strife, was quickly followed by militant policies towards the boundary with Thailand along the Mekong River. These involved artillery duels and the sinking of a Thai patrol vessel. Secondly, when a country is engaged in a civil war its capacity to resist external aggression is reduced and territorial claims may be pressed against it. The Somali government obviously decided that the serious fighting in Eritrea Province of Ethiopia in June 1975 provided an excellent opportunity to try to settle the Somali claim to the Haud and Ogaden by force. Libya annexed the northern areas of Chad in 1975 when that government was beset by internal revolt.

The transfer of power to indigenous governments during the process of decolonization has caused a number of territorial disputes to flare. The new governments have often undertaken a much more diligent surveillance of their boundaries than the colonial administrations. In some cases this has led to a recognition that the boundary does not coincide with tribal distributions or prominent physical features, and boundary issues which were debated and settled by the colonial powers have been revived by the new authorities. Lesotho's claim to much of the Transvaal and the Orange Free State, Uganda's claim to western Kenya, and Tanzania's claim to part of Lake Tanganyika are typical of such territorial disputes. Decolonization is sometimes the signal for neighbouring states to take advantage of the withdrawal of colonial armies. In November 1971, when Britain withdrew from the Persian Gulf, Iran seized three islands in the Strait of Hormuz which had been controlled by Britain on behalf of the Trucial States. It was noted earlier that Afghanistan's claims to Pushtunistan coincided with Britain's withdrawal from the Indian subcontinent and with fighting between religious groups in Pakistan. In February 1976, when Spain withdrew its forces from Spanish Sahara, the territory was annexed by Morocco and Mauritania.

In some cases it is the action of one state that induces another to make a territorial claim. In 1915 and again in 1927 Guatemala made grants of land to the American Fruit Company in the area between the Matagua River and the Meredon Mountains. This immediately prompted Honduras to raise claims which had been dormant for a long time. In a similar fashion, the granting to the United States of a 99-year lease on the Great and Little Corn Islands by Nicaragua encouraged Colombia to launch a claim that the islands were formerly part of the Province of Veragua and therefore properly part of Colombia under the principle of *uti possidetis*.

States' aims in territorial disputes

When the aims of the state initiating the dispute are considered they can be divided into two classes. First there are those claims when the state genuinely wants the territory claimed and believes that it has some chance of obtaining it. In such cases the aim involves strengthening the state by the accretion of territory. The increased strength may come from resources found in the area, from the population which lives there, from the improved access the claimed territory gives to the sea or to major lines of communication, or because a better strategic situation is created. For example, the South-West African People's Organisation, which is the main nationalist movement in Namibia, has demanded that the territory of Walvis Bay be transferred from South Africa to the control of authorities in Windhoek. Walvis Bay is the only port on the coastline of South-West Africa capable of accommodating large vessels, and it is sought to improve the country's transport facilities. Somalia's latent claim against Djibouti would strengthen Somalia by giving it an economic lever which could be used against Ethiopia, which conducts much of its trade through that Red Sea port. On the other side of the continent it seems certain that Morocco's annexation of the northern part of Spanish Sahara was directly connected with the rich phosphate deposits of Bou Craa. Iraq's claim to Kuwait, six days after that country became independent, was connected with the desire to acquire an area with considerable potential for producing petroleum and at the same time to extend Iraq's coastline on the Persian Gulf.

The second class of aims applies to those claims which are apparently made without any hope of successful outcome. In such cases it appears that the dispute is initiated to serve some domestic or international policy. For example, it is generally considered that the Philippines claimed northern Sabah at the time the Federation of Malaysia was being formed in an effort to postpone or prevent the emergence of that political association. The claim was based on the most flimsy ground and was apparently abandoned in June 1977. Some countries have brought forward territorial claims to distract attention from internal difficulties, and it seems likely that the Chinese claims against Russian territory are made less with the hope of regaining territory and more with the intention of scoring points in the ideological debate, by portraying Russian leaders as latter-day czars, who have precisely the same foreign policy aims and territorial ambitions as their royal predecessors.

It is worth noting that sometimes countries will encourage secessionist movements in neighbouring countries rather than claim territory for themselves. Of course if the secession is successful the new state affords fresh opportunities for diplomatic and commercial influence, and the state from which the territory is detached has been weakened. India's encouragement of secession in Pakistan and Zaire's encouragement of secessionist forces in Cabinda were not altruistic. India was convinced that Bangladesh would be a better neighbour that East Pakistan and that Pakistan would be made weaker by the loss of its eastern territory. Zaire concluded that its interests would be best served if Cabinda could be detached from Angola because the small oil-rich state would give opportunities for trade and political influence and because Angola, from which an invasion of Zaire's Shamba Province was launched in 1977, would be weakened by the loss of a large slice of its export revenue.

It is rarely possible to gain definite confirmation of this second category of aims, except by research in archives long after the event. However, that should not dissuade scholars from making guesses about policy aims in contemporary situations.

Arguments supporting territorial claims

When the arguments in favour of any territorial claim are considered it is useful to follow the division suggested by Hill (1976, p. 26). He distinguishes legal arguments relating to a statement that the territory should belong to the claimant state from all other arguments designed to show that it would be more appropriate if the territory was ceded to the claimant state but where there is no claim that the territory is illegally held.

Although geographers cannot make a major contribution to the analysis of legal arguments in territorial disputes, which remain the proper preserve of the lawyer, it is useful to examine the legal basis of claims to territory; the following discussion of this aspect follows the excellent exposition by Hill (1976, Ch. 10). Occupation of territory is one of the soundest bases on which to mount legal claims. Such claims were especially characteristic of the 19th century, when colonial powers were disputing territory in Africa, Asia, and North America. It was generally considered that a statement of intention to occupy territory must be supported by a physical presence in the area. That was certainly the position adopted by the United States in disagreements with Russia over territorial questions in north-west America. According to the General Act of the Congress of Berlin, legal rights to colonies in Africa could be secured only through effective occupation. Article 35 stated that the claimant state had to demonstrate satisfactorily to other states that the claim should be respected because a sufficient degree of authority was established throughout the area. This Article was generally obeyed on the coast, but in the interior it was honoured more often in the breach than in the observance. In 1933 Denmark was able to validate its claim to Greenland against Norway because of its occupation of part of the east coast and the enactment of regulations covering the whole area. Triggs (1983, pp. 19–89) has provided a detailed account of the concept of effective occupation in a study dealing with Australian sovereignty in Antarctica.

Claims were also made to territory during the last century on the grounds of contiguity and territorial propinquity. Such arguments maintain that the closest authority should exercise sovereignty over adjacent unclaimed land. For example, in 1834 Peru claimed the island of Lobos because although it was 35 miles from the shore it was more distant from other countries. The hinterland doctrine advanced by many colonial powers in Africa was related to these concepts. It was alleged that any country that had occupied a section of coastline was entitled to the hinterland of that shore. It was recourse to this argument that persuaded the British government not to restrict Germany's claim in South-West Africa to a strip 20 miles wide along the coast when it was urged to do so by the government of Cape Province in 1884. In a sense, various sector claims to Antarctica are also based on concepts of propinquity. Hill (1976, p. 153) asserts that such claims have no strength in international law, and in modern times, when the whole land surface of the earth has been politically appropriated, it is difficult to see such arrangements being entertained. However, it is interesting that the concept of propinquity underlies the use of lines of equidistance in settling disputes over claims to areas of the sea and continental shelf. This point is considered in detail in the next chapter.

According to some authorities on international law, it is recognized that legal claims to fragments of territory could be based on symbolic acts of possession. Such a concept is particularly important in supporting claims to islands in the Pacific Ocean and the South China Sea. For example, there is a current dispute between China and Vietnam over ownership of the Paracel Islands, which China occupied with a military force in January 1974. According to Vietnamese sources, the Emperor Gia Long had made a symbolic act of acquisition in 1816: 'In 1816, he [Emperor Gia Long] went with solemnity to plant his flag and take formal possession of these rocks, which it is not likely any body will dispute with him' (Taberd 1837, p. 738). Then, in 1833, the Emperor Minh Mang ordered trees to be planted on the islands for the benefit of navigators. In 1836 markers were placed on the islands bearing the following inscription:

> In the year Binh Than, 17th Year of the reign of Minh Mang, Navy Commander Pham Huu Nhat, commissioned by the Emperor to Hoang So islands [Paracel islands] to conduct map surveyings, landed at this place and planted this marker so to perpetuate the memory of the event. (Vietnam Foreign Ministry 1975, p. 31)

In 1930 the crew of the French vessel *La Malicieuse* erected 'sovereignty columns' on some of the Paracel islands. Eight years later further monuments were erected recording French authority and the date 1816, which was the first recorded symbolic act of possession by Vietnam. It is of course entirely possible that the Chinese authorities can produce earlier evidence to support its claim to these islands.

Vanuatu, which became independent in July 1980, claims Matthew and Hunter Islands which are considered by France to form part of New Caledonia. Although these tiny islands have no intrinsic value, they do permit claims to be made to 59 400 square nautical miles of sea (Office of National Assessments 1984, p. 18).

Disputes over islands were common during the search for guano during the last century, and an American statute enacted in 1856 determined that any citizen who discovered a guano island and took possession of it in the name of his government provided grounds for it to be claimed by the President as an American possession. Such claims created some problems when the United States negotiated maritime boundaries in the south-west Pacific Ocean. Its agreement with the Cook Islands in June 1980 included a provision that the United States abandon claims to Pukapuka, Manihiki, Penrhyn, and Rakahanga.

The legal basis of claims in South America is usually the principle of *uti possidetis* (Ireland 1938). This principle means that the new states in the post-colonial period accepted the same boundaries as the colonial territories they replaced. This was designed to ensure that the European colonial powers, such as Britain, France, and Germany, were prevented from making claims to uncontrolled areas on the borders of the new states. The principle is apparently derived from the same term in Roman law, which applied to an edict that preserved the existing state of possession of an immovable object, such as a house or a vineyard, pending litigation. The principle established in 1810 in fact caused some conflict because of confusion over its interpretation. Some states regarded the rule as applying to the limits legally in force when the act of decolonization occurred. Others believed that the rule was concerned with the boundaries observed for practical administration by the colonial authorities. These two lines did not always coincide and competing states would urge the interpretation which suited themselves best. For example, the Venezuelan

Constitution in 1830 proclaimed the state as being coincident with the area previously known as the Captaincy-General of Venezuela. It was then discovered that the Spanish administrators had governed beyond this legal limit in good faith. Venezuela then espoused the second interpretation. In 1964 the Organization of African Unity passed a resolution by which all members solemnly declared that they would respect the borders existing 'on the achievement of national independence'. This resolution was not designed to forestall any acts by the colonial powers; it was aimed at preventing any territorial disputes between African countries. Unfortunately it has not succeeded in that intention.

Conquest is another means by which legal title to territory is acquired. As Hill (1976, p. 161) notes, this involves actual possession based on force, an announcement of the intention to hold the territory, and an ability to make good that declaration. This is usually a unilateral action in which no treaty is involved. For example, on 17 December 1961 Indian troops invaded the Portuguese colonies of Goa, Damao, and Diu, and these territories were formally incorporated into the Indian state on 19 December 1962. It also seems likely that some territory captured in the Golan Heights and on the West Bank of the Jordan by Israel in 1967 will be retained by that country.

Territory is often claimed on the grounds that it was ceded by treaty. Sometimes the cession may be voluntary, as when Britain ceded the Wakhan Strip to Afghanistan in 1893; at other times it may be involuntary. Involuntary or forced cessions usually occur at the end of a war or after the threat of force; they are typified by the cession of the south part of Sakhalin to Japan by Russia in 1905 and the cession of the trans-Ussuri area to Russia by China in 1860.

The last legal claim mentioned by Hill is based on prescription, which is defined in the following terms:

> . . . the acquisition of sovereignty over territory through continuous and undisturbed exercise of sovereignty over it during such a period as is necessary to create under the influence of historical development the general conviction that the present condition of things is in conformity with international order. (Hill 1976, p. 156)

India uses arguments based on prescription as part of its case against China in the disputed areas of the Himalayas. India maintains that British authority was exercised for a long, continuous period without Chinese objection.

A recent example of a territorial dispute argued on purely legal grounds involved the Philippines and Malaysia. The Philippines in 1963 began to assert a claim to parts of Sabah on the island of Borneo. In 1877–8 a British syndicate secured the transfer to themselves of the rights of the Sultan of Sulu over the territories and adjacent islands in return for the payment of a pension. According to the English translation, the treaty stated that the land was ceded and granted forever and in perpetuity. In 1881 the syndicate was taken over by the British North Borneo Company which received a royal charter in the same year. In 1883 the area was made a British protectorate, and in 1903 a confirmatory deed was signed by the Sultan specifying the islands not individually named in the original treaty. In 1946 the area became a British colony. The Philippine claim rests firstly on the ground that the Sultan was not empowered to sign the treaty since Spain was the sovereign power. That argument did not impress Malaysia because Britain did not recognize the Spanish treaties with the Sultan in 1836, 1851, and 1864 since Spain could not control him. Furthermore, in 1885 Spain renounced its rights in favour of Britain in return for recognition in the Sulu archipelago. America replaced Spain as the dominant power in the area in 1898, and Britain secured specific American recognition of the British claims in 1930. The second argument used by the Philippines to support its claim questioned the correct interpretation of the Malay word *padak*, which is used in the treaty. Britain had interpreted the word to mean 'granted and ceded', but the Philippines insisted that it was more accurately interpreted as 'leased'. In 1977, to demonstrate its goodwill towards Malaysia the Philippines abandoned its claim at a meeting of the Association of Southeast Asian Nations. However, the Philippines appeared to be in no hurry to confirm this new attitude by passing the necessary legislation.

Territorial disputes based solely on legal arguments that the territory ought to belong to the claimant state are comparatively rare. It is usual for the legal arguments to be underpinned by other assertions founded in history, geography, strategy, and economics. Indeed, the largest number of territorial disputes lack any significant legal component and instead are based on the view that for a variety of reasons the claimed territory should belong to the claimant state.

In most cases the historical arguments refer to periods that are not well defined and before the period when legal titles may have been gained. In 1919 France claimed the Saar on historical grounds, but its case was weak. Saarlouis alone had been founded by the French, under Louis XIV in 1680, but even this area had not been under French control for more than 23 years (Temperley 1920, Vol. 2, p. 177). Italy's claim to part of the Dalmatian coast at the same conference was based on strategic and historical arguments. The Italian Premier expressed the historical claim in the following sentences:

> And can one describe as excessive the Italian aspiration for the Dalmatian coast, this boulevard of Italy throughout the centuries, which Roman genius and Venetian activity have made noble and great, and whose Italianity, defying all manner of implacable persecution

> throughout an entire century, today shares with the Italian nation the same feelings of patriotism? (Temperley 1921, Vol. 5, p. 404)

This statement is typical of many historical claims which often appear as padding for the more pertinent arguments. However, a more specific historical argument is provided by assertions that the original colonists of an area have a prior right, which takes precedence over claims by subsequent migrants. This was one of the arguments used by the Serbs to Banat and by Italy to the Julian Venetia.

Geographical arguments are normally designed either to show the desirability of extending a state's territory to make the boundary coincide with some physical feature or to demonstrate the basic unity of an area which is divided or threatened with division. The Banat was one such area at the end of World War I. The territory was contested by Rumania and Yugoslavia and is bounded by the Danube, Tisza, and Murec rivers. As in the Drava valley to the west, the population is of mixed Magyar, Serbo-Croat, and Rumanian origin. The Yugoslavs claimed the lower central and western parts and Rumania claimed the entire area as a geographical unit. This view was based on the 'natural frontiers' which the rivers provide, the complementary nature of the products of the plains and the hills to the east, and the opportunity the plains afforded to the hill-dwellers for employment. The published reports of the Sino-Indian boundary talks make it clear that India rests much of its case on the 'natural boundary' provided by the main watershed of the Himalayan system:

> In the discussions on the location and natural features of alignment, the Indian side demonstrated that the boundary shown by India was the natural dividing line between the two countries. This was not a theoretical deduction based on the rights and wrongs of abstract principles. The fact that this line had received the sanction of tradition and custom was no matter of accident or surprise because it conformed to the general development of human geography and illustrated that social and political institutions are circumscribed by physical environment. It was natural that peoples tended to settle up to and on the sides of mountain ranges; and the limits of societies – and nations – were formed by mountain barriers. The Chinese side recognised this fact that high and insurmountable mountain barriers provided natural obstacles and suggested that it was appropriate that the boundary should run along such ranges. But if mountains form natural barriers, it was even more logical that the dividing line should be identified with the crest of that range which forms the watershed in that area. Normally where mountains exist, the highest range is also the watershed; but in the few cases where they diverge, the boundary tends to be the watershed range . . . it is now a well-recognised principle of customary international law that when two countries are separated by a mountain range and there

> are no boundary treaties or specific agreements, the traditional boundary tends to take shape along the crest which divides the major volume of the waters flowing into the two countries. The innate logic of this principle is self-evident. The inhabitants of the two areas not only tend to settle up to the intervening barrier but wish and seek to retain control of the drainage basins. (India, Ministry of External Affairs 1961, pp. 235–6)

> Similarly it is manifest that there are passes all along the high mountains and that there are always contacts across the ranges. But this does not invalidate the general conclusion that the watershed range tends to determine the limits of the settlements of the inhabitants on either side and to form the boundary between the two peoples. Neither the flow of rivers through the ranges nor the contacts of peoples across them can undermine the basic fact that a high watershed range tends to develop into the natural, economic, and political limits of the areas on the two sides. (India, Ministry of External Affairs 1961, p. 237)

It is difficult to know where to start challenging this statement since it contains so many concepts with which a political geographer must disagree. The concept of 'natural boundaries' which is explicit in this statement was once popular in political geography. Simply, it was assumed that the main, linear, physical obstacles in the world set the limits within which nations fashioned their political life. It was the political aspect of determinism which conferred on the environment and the physical landscape the dominant rôle in shaping man's economic and social characteristics and activities. Pounds (1951, 1954) has published excellent accounts of the origin of the idea of the natural frontiers of France, which were considered to be the sea, the Alpine watershed, the Pyrenees, and the Rhine. But the idea of 'natural boundaries' has been discredited for decades. Writers such as Hartshorne (1936), East (1937), Boggs (1940), and Jones (1945) have clearly shown that all political boundaries are artificial because they require the selection of a specific line within a zone where change in the physical characteristics of the landscape may be more or less rapid. Thus in a mountain range there is not one line along which all physical characteristics change sharply. First, all physical changes occur over a zone which may vary considerably in width, from the very narrow arête which marks a watershed to the zonal change from one dominant type of vegetation to another. Secondly, even if the changes in vegetation, climate, drainage, elevation, structure, morphology, and geology could each be reduced to a single line, these lines would not coincide with each other. It must also be added that the response of different human communities to the same environment varies. It is unwise for lowlanders to assume that high mountains mark absolute barriers to highland communities. Kirk (1962) has made the point that there are distinct communities in the Himalayas following a complex transhumance economy, which carries them over crests and watersheds as they use pastures and camping grouds which to lowlanders appear to be uniformly barren. Kingdom Ward described a similar situation earlier:

> But obviously a pass of 15,000 feet is nothing to a Tibetan who habitually lives at 10,000 or 12,000 feet altitude. The Tibetan is not stopped by physical but climate barriers, and no boundary pillars are needed to make him respect these. His frontier is the verge of the grassland, the fringe of the pine forest, the 50-inch rainfall contour beyond which no salt is (until indeed you come to the sea) or the 75 per cent saturated atmosphere. The barrier may be invisible; but it is a more formidable one to a Tibetan than the Great Himalayan ranges. If he crosses it he must revolutionize his mode of life. (Ward 1932, p. 469)

Ryder (1926) has described the problem of marking a boundary between Turkey and Iran through mountain communities, who removed the pillars as soon as they had been erected because they intersected routes to traditional pastures. In the reports of the men who marked the Durand Line, there are dozens of cases where tribal limits did not coincide with obvious watersheds. It is also clear that communities will occupy both banks of major rivers and will exist in desert environments, both of which are physical features which the proponents of natural boundaries believed marked the limits of nation building. Finally, it must be noted that 'natural boundaries' are always the limits to which a state wishes to expand. There is no recorded case of a state wishing to withdraw to 'natural boundaries'.

Turning from a general criticism of 'natural boundaries', it is now necessary to focus on the specific physical limit which provides the basis of the Indian claims. A watershed is the line that separates areas in which the flow of water, after precipitation or the melting of snow and ice, is in different directions. In this situation only surface drainage is considered; the complication of underground drainage, especially in limestone areas, need not be considered here. Now, just as there are first-, second-, and third-order rivers there must be first-, second-, and third-order watersheds. For example, in this region, the primary watershed would be between the rivers flowing into the Bay of Bengal and those that flow intermittently into interior drainage basins of Tibet. The secondary watersheds would separate any two river basins which drain to the Bay of Bengal; thus there would be a secondary watershed between the Tsangpo and the Subansiri rivers. The tertiary watersheds separate the adjacent tributaries of any river; such a watershed would separate the Ange and Tangon tributaries of the Dibang River. This identification of a hierarchy of watersheds can continue until the smallest rivulets are reached. The problem for the Indian government is to justify the selection of one watershed rather than another. Where the watershed and the crest coincide there is no difficulty, but in many parts of the Himalayas the crest and the watershed do not coincide. Rivers, through the process of headward erosion, or because they are antecedent, have cut through the crest, displacing the watershed. Unfortunately for the Indian argument the McMahon Line does not consistently follow primary, secondary, or tertiary watersheds, or the crests where they form watersheds. It is thus difficult to avoid the conclusion that the alignment of the McMahon Line was the result of a series of *ad hoc* decisions, which the Indian government has tried to mask by the uniform gloss of the watershed principle. The Indian statement also identifies the watershed by 'the major volume of the waters flowing into the two countries', but it is hard to understand how this can be calculated when a single river basin, such as the Subansiri or Luhit, is divided between the two countries.

Strategic arguments in favour of the transfer of territory usually have one of two aims. In some cases they are designed to deprive a country with a history of aggressive policies of territory from which attacks can be easily launched. In other instances the arguments support territory being given to a country which has a history of being attacked so that its security is increased by the territorial buffer. The Greek and Yugoslav claims against Bulgaria at the Versailles Peace Conference provided examples of the first situation. In each case the territory claimed had been used by German and allied forces to launch rapid and successful attacks. France's claim to the Saar Basin in 1919 was based partly on the desire to move the German boundary farther away from the iron-ore field of Briey and Thionville. These deposits had been quickly overrun in 1914, and their loss had severely handicapped the French production of armaments.

The economic arguments in favour of territorial claims are usually designed to show the economic integration of the claimed area with a zone already held, the need for the area as a routeway for supply of raw materials, or the value of the region as reparations for damage suffered during a war. Claims by Czechoslovakia to the Teschen district of Silesia provide examples of the first two types of economic argument. First, the Freistadt area was regarded as being inextricably linked with the industrial complex of Ostrava, where metal foundries depended upon the Karvina coking coal. The coal was also needed to a lesser extent in Bohemia and Moravia. Secondly, the Czechoslovakian government claimed that the Olderberg–Jablunka–Sillein railway was of vital importance since it formed the arterial line connecting Slovakia with Bohemia-Moravia. The railway through the Vlara Pass, which Poland claimed could be further developed, was not considered suitable by the Czechs because of the steep gradients and sharp curves. Further, the only other line from Breclava to Bratislava was too far to the south. Poland reversed this economic argument to the east in Zips and Orava when certain highland areas, occupied by people speaking a dialect transitional between Czechoslovakian and Polish, were claimed on the ground that because of easy communication they were more closely attached with Cracow than with Kralovany, the Czech county town.

Economic claims to territory often focus on access to the sea. Bolivia claimed the Gran Chaco from Paraguay because it wanted access to the sea by way of the Paraguay River, and there was fighting over this region in the period 1932–5. The issue was eventually resolved through arbitration which awarded part of the region to Bolivia.

Compensatory claims for property and population losses during the war were made against German colonies. Referring to Belgium's claim to Runda-Urundi, Temperley (1920, Vol. 2, p. 243) notes that 'no one wanted to refuse the insistent claim of a state which had suffered so seriously from Germany's aggression in Europe'. The extension of Poland's territory to the Oder–Neisse Line was also seen as compensation for losses to Russia and as the need to secure a better strategic boundary. Allied to such arguments are the cases where states secure territorial promises for co-operation with another state. A good example of this is provided by the 1915 Treaty of London between Italy and the Allied Powers under which Italy agreed to merge her forces in the general war effort. Under this treaty certain territorial promises were made:

> In the event of the total or partial partition of Turkey Italy was to obtain a just share of the Mediterranean region adjacent to the province of Adalia.
>
> In the event of France and Great Britain increasing their colonial territories in Africa at the expense of Germany those Powers agree in principle that Italy may claim some equitable compensation, particularly as regards the settlement in her favour of the questions relative to the frontiers of the Italian colonies of Eritrea, Somaliland and Libya. (Temperley 1921, Vol. 4, p. 290)

A settlement under the terms of this treaty transferred 36 000 square miles of northern Kenya to Italian Somaliland on 15 July 1924. The boundary was shifted west of the Juba River, which had previously formed the boundary, to the present line, which was demarcated by a commission in 1930.

Territorial claims based on ethnic arguments refer to the human qualities of nationality, race, language, culture, history, and religion. From time to time countries have focused on one of these characteristics as being the definitive guide to the political predilection of a region's population. During the Versailles Peace Conference some delegations laid great emphasis on language as a reliable guide to national aspirations of people and therefore to the proper distribution of territory. Ancel (1936,

p. 76) noted that if this argument was followed to its logical conclusion linguistic divides would become political boundaries and language would become 'the symbol of the nation'. At that conference efforts were made to draw boundaries which reduced the number of minorities to a minimum. Fisher (1940, Ch. 1) praised the settlements because they left only 6 per cent of the population of Europe under alien rule. But the records of the conference show that many of the decisions were inconsistent and that tests of nationality were not uniformly applied. It is difficult to see how any such principles could have been uniformly applied, given the complexity of population distributions in Europe. The intermingling of population in the European borderlands often made it impossible to draw boundaries which precisely separated different ethnic groups. In some regions, such as the Western Banat, there was an intricate mixture of Yugoslavs, Rumanians, Magyars, and Germans (Bowman 1923, p. 272). In another situation Greece claimed the entire Argyro-Castro area because of the larger Greek rural population, which surrounded the towns occupied by Yugoslavs. Yugoslavia used this argument to its own advantage in the Klagenfurt Basin, where the rural population was Yugoslav and surrounded towns occupied by German-speaking people, who formed the basis of Austria's claim to the region.

In Africa territorial claims based on the division of tribes by boundaries drawn during the colonial period are much easier to define because tribal mixing has been on a comparatively small scale. Claims by Somalia to areas of Ethiopia, Kenya, and Djibouti and by Togo to areas of Ghana have a high degree of precision. It would be much easier over wide areas of Africa to make boundary adjustments without creating new minority problems than it would in Europe.

However, it seems certain that no single characteristic or even group of characteristics can be used with confidence to allocate people to different states. The important criterion must be the wish of the people, and the concept of self-determination, which seemed such an attractive solution to European problems in the 1920s, continues to be resuscitated again and again in modern times. But self-determination is a two-edged sword and the concept has never been applied uniformly; it is invariably subject to decisions of political expediency. For example, the Organization of African Unity called for the self-determination of Rhodesian citizens, but it condoned the partition of Spanish Sahara between Morocco and Mauritania without any consultation with the local population. The people of Eritrea were not consulted before they were handed over to Ethiopia in 1952, and therefore it is not surprising that the secessionist movement started ten years later when the Ethiopian government unilaterally abolished Eritrea's federal status.

In an effort to determine the wishes of a group of people regarding alternative nationalities, plebiscites have sometimes been held, especially where a solution was being imposed on the area as in the Klagenfurt Basin. In many cases, however, a plebiscite has not been found satisfactory because the state controlling the area enjoys an important advantage in securing a favourable result. In 1883 the treaty of Ancon, which terminated the war between Chile and Peru, stipulated that the provinces of Tacna and Arica be held by Chile for a period of ten years, after which a plebiscite would decide by popular vote whether the area should remain as part of Chile or part of Peru (Dennis 1931, p. 297). No attempt was made to hold the plebiscite until 1925, by which time Chile had sufficiently nationalized the area to make a decision in favour of Chile certain. But, even so, the plebiscite was never held and the commission appointed to conduct it noted that Chile's failure to provide free voting conditions was the major obstacle. It is interesting that at the Versailles Peace Conference ethnic claims were sometimes made, not because a particular cultural group formed the majority but because the actions of a government had prevented them from forming a majority. For example, Italy insisted that it was the policy of the Austrian government which was responsible for the Slav majority on the Dalmatian coast:

> This is the result of the most outrageous violence that the political history of Europe records during the last century. Austria did not recoil before any form of artifice or violence in Dalmatia in order to repress Italian feelings, after 1866 in order to check any movement toward annexation to Italy, and after 1878 and 1882 in order to carry out her Balkan schemes. (Hill 1976, p. 126)

Unfortunately for Italy the American delegation reversed this argument in respect of Fiume and asserted that it was the policies of the Austro-Hungarian authorities which had allowed an Italian majority rather than a Slav majority to develop in the town.

The results of territorial disputes

When the results of territorial boundary disputes are considered it is useful to classify them into two groups. First, there will be the consequences which flow from the dispute being initiated. Secondly, there will be results which follow any transfer of territory. Claims to the territory of a neighbour will normally cause a deterioration in relations between the two countries concerned, and this may be reflected in a reduced level of commercial contact across the disputed boundary. Within the claimed area the authorities in control will probably seek ways to increase their defences. Such measures might include the removal of people who might sympathize with the proposed transfer; the construction of new strategic roads and airfields; the improvement in the services provided in the region, to win support from the local people. Both countries are also likely to publish their versions of the compelling reasons why the issue should be decided in their favour. This is a bonus to the interested scholar.

If the matter is settled by the transfer of territory, a wide range of consequences are possible. Some people in the transferred region who opposed the change may cross the new boundary to remain citizens of their original country; new patterns of administrative organization might be established; the orientation of the region's economy may be altered towards the country which has acquired the land; different forms of production might be encouraged by the new regulations which apply to the area and by access to a different market. It would require careful fieldwork in each case to establish the range of changes which occurred following the alteration of the boundary. Sanguin (1983) described the problems in that part of Italy which was transferred to France in 1947 and showed how some of the present problems can be traced to that change. In some cases the dispute will enter a dormant phase without any transfer of territory, and this may be encouraged by an agreement between the two countries regarding their border relations. For example, in 1967 Somalia and Ethiopia reached an understanding, which, though not ending the dispute, did encourage better relations between the two countries, promote easier travel across the boundary, and guarantee some safeguards for the civil rights of Somalis living in the Ethiopian borderlands.

Positional boundary disputes

Although the basic cause of territorial boundary disputes is superimposition of the boundary on the cultural or physical landscape, positional boundary disputes arise because of incomplete boundary evolution. It is not the quality of the borderland but rather the defect of the boundary which is crucial. Positional disputes will usually arise at one of two stages. Most of them will arise during the demarcation of the boundary because the commission will be faced with the problem of matching the boundary definition to the landscape. However, it is also possible that positional disputes will arise at a much later date if the demarcation commission makes an error. There is probably an important legal distinction between these two situations because the existence of a demarcated boundary for a considerable period may be used as justification by the satisfied party for maintenance of that location, but essentially they arise from the same causes. These causes are found in the boundary definition and can be grouped into two categories. First, some of the terms will be imprecise because they will allow more than one interpretation; secondly, some of the turning points along the boundary will be defined in two ways which are contradictory.

An examination of imprecise terms leading to positional disputes reveals a basic dichotomy between the legal and geographical definitions of phrases in the text. Generally, the problems of geographical interpretation are more common. An example of problems over a legal definition is provided by the Anglo-German agreement of 1886 regarding the boundary between Nigeria and Kamerun between the coast and the River Benue. The northern terminus was defined as follows: 'a point on the right bank of the River Benue to the east of and as close as possible to Yola as may be found on examination to be practically suited for the demarcation of the boundary' (Hertslet 1909, Vol. 3, pp. 880–1). Leaving aside the error in designating the south bank as the right bank, the real dispute centred on the term 'practically'. When the time came to select the point Britain argued that the term had both economic and political connotations. Politically Britain would find it inexpedient to draw a boundary within sight of the walls of Yola since the Emir of that important town was losing a considerable slice of his territory to Germany. Economically, it was demanded that the boundary should be drawn to allow the free circulation of the people of Yola, which was later clarified to mean that sufficient area would be left to the east to provide satisfactory supplies of firewood and enough pasture. For Germany of course the term 'practically' had only a technical meaning in respect of boundary demarcation.

The literature is replete with examples of problems of geographical definitions and they can generally be divided into three groups. The first involves the identification of a geographical feature such as the bank of a river, the edge of a plateau, or the crest of a range. The second concerns disagreements about place names, and the third results from debate over the location of some provincial or tribal boundary which is being elevated to international status. In several cases the problem is complicated because the demarcation is not undertaken until long after the boundary's delimitation, which may make it hard to reconstruct the exact intentions of the people who drafted the boundary's description.

The Anglo-German treaty of 1890, dealing with various African territories, provides one good example of an indefinite geographical term which still has not been satisfactorily interpreted. It defined the southern limit of German South-West Africa as 'the north bank' of the Orange River. In some sections of the course of the Orange River it was difficult to be certain where the bank was located, especially in view of the fact that the flow exhibited wide seasonal fluctuations. Those variations have now been reduced but not eliminated by the construction of dams in the Orange Free State. The Germans took the view that the boundary coincided with the river's waterline on the north shore, and as the river level fell or rose so they increased and reduced their area of control. Britain regarded such a concept as intolerable and wanted a fixed line. Debate between the two sides continued throughout the early years of this century and by 1909 the Cape Colony government was prepared to offer Germany the thalweg of the river provided Britain continued to own all the islands, British subjects had the exclusive authority to operate ferries, and there were safeguards regarding the construction of weirs from the German bank. The thalweg of a river is the deepest continuous channel. War intervened before the matter was settled, and the question has not been resolved. Hinks (1921) records that a positional dispute arose on the boundary between Peru and Bolivia north of Lake Titicaca because it proved impossible, before the advent of aerial photography, to identify 'the western source of the River Heath', which was one of the turning points of the boundary.

An interesting positional dispute has arisen between China and the Soviet Union over the location of the boundary at the junction of the Amur and Ussuri rivers (Prescott *et al.* 1977, pp. 12–13). The treaty in 1860 defined the boundary as following the Amur River to its junction with the River Ussuri and then turning south along the Ussuri River. Unfortunately, at the junction of the two rivers there is a triangular island with an area of 128 square miles. The triangular island is bounded by three waterways (Fig. 4.1). To the north lies the Amur River, or the Hei-lung Chiang according to China; it is a large river with an average width of 1.4 miles. The south-west coast of the island is washed by the Protoka Kazakevicheva (K'o-tsa-k'ai-wei-ch'ai-wo Schui-tao in Chinese), a narrow watercourse which is 18 miles long and 1000 yards wide. The south-east margin of the island is bounded by a channel which the Chinese regard as the Wu-su-li Chiang (Ussuri in Russian) and which the Soviets regard, together with the Protoka Kazakevicheva, as the southern arm of the Amur River. This section of river is 22 miles long with an average width of 1300 yards. The crux of the problem is to decide the confluence of the Amur and Ussuri rivers. The Soviet authorities place it at Kazakevichevo, but their Chinese counterparts are sure it should be at Khabarovsk. In short, the Soviets believe the island is located between two arms of the Amur River, but the Chinese are equally certain that it is between two arms of the Ussuri River. Even if the Soviet Union accepted that the island was in the Ussuri River, the Chinese would still have the difficulty of persuading the Soviets that the confluence intended was the major one at Khabarovsk and not the minor confluence south of Verkhne-Spasskoye.

Imprecise place names have often been used in treaties. Hinks (1921) notes that the interpretetion of the term 'Barraca of Illampu' created problems for the demarcation commission on the boundary between Peru and Bolivia. The term can either mean the whole estate or the main house on the estate. In 1872 the Indian authorities, when assisting the British Government to negotiate a boundary between Russia and Afghanistan, noted that Afghanistan's claims did not extend north of 'the ford of Kwaja Salar' on the Amu Darya. In the agreement with Russia which was signed

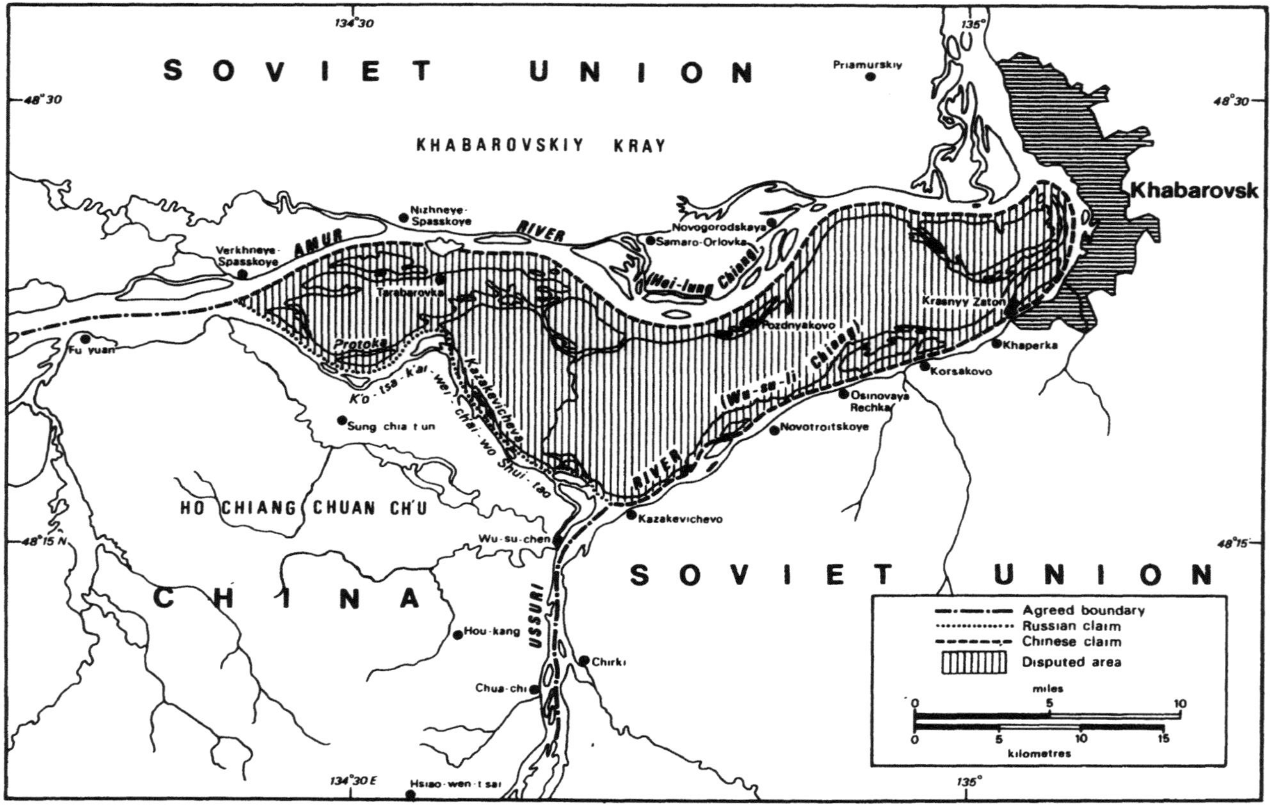

Figure 4.1 The Amur–Ussuri junction.

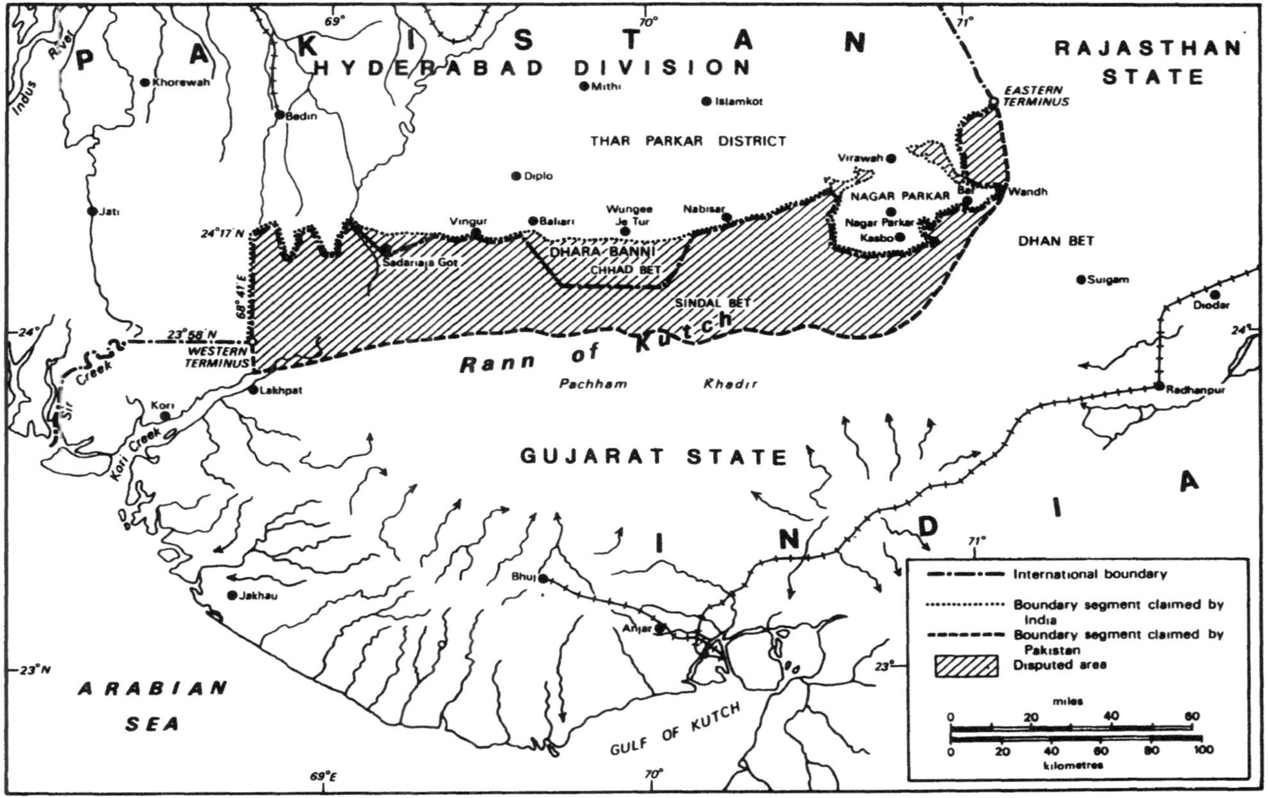

Figure 4.2 The Great Rann of Kutch.

later in 1872 one of the turning points was 'the port of Kwaja Salar' or, as some authorities record it, 'the post of Kwaja Salar'. When the boundary was being demarcated 16 years later it was discovered that Kwaja Salar was a name which applied to a district, a fort, a house, a ferry, and a tomb. Britain identified Islim as the proper terminus intended in the treaty; Russia selected instead the tomb which was 9 miles upstream from Islim. Because the sector of the boundary was defined between two points, of which Kwaja Salar was one, the different termini created a triangular area of dispute.

The area west of Kwaja Salar also affords an example of the boundaries of districts being used to identify the international boundary over which a dispute has been created because of the uncertainty attached to these district boundaries. The boundary was drawn along the northern edges of the Afghan districts of Aksha, Seripool, Mainmenat, Shibberjau, and Andkoi. Needless to say, when the time came to settle the boundary there was long debate over the actual location of these limits (Prescott 1975, p. 108). However, it was not only in historic periods that such confusion existed.

According to the Indian Independence Act of 18 July 1947, the province of Sind was awarded to Pakistan. South of Sind there were a number of suzerainties, including Kutch, Sulgam, Tharad, Wav, and Santalpur, which subsequently acceded to India. This meant that the boundary between Sind and Kutch became the international boundary between India and Pakistan. It soon became evident that India and Pakistan had differing views about the location of that boundary through the Great Rann of Kutch. It was agreed by both sides that the boundary extended from the mouth of Sir Creek in the west to the eastern terminus at the tri-junction of Gujarat, Rajasthan, and Hyderabad. Further they agreed that the western sector had been defined along the Sir Creek to latitude 23°58′ north and then eastwards along that parallel for 22 miles to its intersection with meridian 68°41′ east (Fig. 4.2). The land section of this boundary segment was demarcated by sandstone pillars in 1923 and a further 66 pillars were then placed along the meridian of 68°41′, apparently carrying the boundary northwards to parallel 24°17′ north. This segment of the boundary was disputed between India and Pakistan. India maintained that it was a proper boundary and that it was only necessary to draw the boundary between the northern terminus of this extension and the eastern tri-junction. The Indian authorities pointed out that the northern edge of the Rann, which is a salt-impregnated alluvial tract, would be a convenient and direct boundary. The Pakistan government argued that the boundary drawn along meridian 68°41′ east had no validity and that the boundary should connect the point where that meridian intersected the boundary drawn along the parallel 23°58′ north. Pakistan recommended a boundary which connected this point with the eastern tri-junction via the middle of the Rann. The two countries were unable to resolve the dispute and it was settled by arbitration in 1968 which defined a compromise line between the two extreme claims.

Hinks (1921) described a problem over the location of a point on the boundary between Bolivia and Peru which was defined in two contradictory ways. The delimitation described a turning point at the confluence of the rivers Lanza and Tambopata as being north of parallel 14° south. When this confluence was identified and its co-ordinates fixed it was found to lie south of parallel 14° south. This and other problems eventually were solved by arbitration by Argentina.

Finally, mention must be made of a particular kind of positional boundary dispute which arises because the feature with which the boundary was made coincident changes position. The most common case concerns rivers. Rivers shift their course within their flood plain, sometimes imperceptibly as the meanders widen and proceed downstream, and sometimes dramatically when the neck of a wide meander is cut through. It is this second situation that provides difficulties if the boundary agreement contains no provision for such a situation. This problem underlay the Chamizal dispute between the United States and Mexico which was discussed earlier. The French and Thai authorities tried to avoid similar problems along the Mekong River between Laos and Thailand by a treaty in 1926. The treaty specified that where there was only a single channel the boundary would follow the thalweg and where there was more than one channel, due to the presence of islands, the boundary would follow the channel closest to the Thai bank. It was also stipulated that if the channel nearest the Thai bank dried up the boundary would continue to follow it unless the Joint Permanent High Commission for the Mekong River decided to move the boundary to the nearest water channel.

There was an interesting case of a telegraph line which was supposed to have changed its position on the border between Cambodia and Vietnam. In 1873 the southern section of this boundary was agreed between the two authorities and part of the boundary was fixed along the telegraph line between Giang Thanh and Ha Tien. This boundary was demarcated in the period 1873–6. In 1891 the governor of Cochin China shifted the boundary away from the telegraph line which lay close to the Giang Thanh River. He made the boundary coincide with the Mandarin's Way, which transferred 8 square miles of marshy land occupied by 50 people to Cochin China (Prescott 1975, pp. 471–2). He justified this move on the ground that in 1873 the telegraph line lay along the Mandarin's Way and not along the river road. After complaints from the Cambodian authorities a report was prepared on the matter by the French administrators in the cantons called Ha Thanh and Thanh Gi which occupied the borderland of Cochin China. This report justified the governor's decision and provided the information that the telegraph line had been moved from the Mandarin's Way to the river road between the definition of the boundary in 1873 and the demarcation of the boundary two years later (Fig. 4.3). This alteration was made because the activities of Cambodian rebels had made the Mandarin's Way insecure. The Cambodian complaints continued and a further commission was appointed by the governor-general of Indochina in 1896. This body included surveyors and administrators from both sides of the boundary. It concluded that the telegraph had been established along the river road in 1870 and had not been moved subsequently. That settled the matter, but it is interesting to record that the authorities in Cochin China used this loss of territory, which had never belonged properly to them, to justify the alteration of the boundary around Ha Tien in 1914 when 3 square miles were transferred from Cambodia to Cochin China.

Positional boundary disputes will usually involve smaller areas than those involved in territorial disputes, although positional disputes in two situations may involve considerable areas. The first will occur when a boundary is only defined by a few points and the location of one of those points is disputed. The example of the Russo-Afghan dispute over the proper location of Kwaja Salar was noted earlier. It follows that that relocation of a turning point will affect the alignment of the boundary sectors on both sides; if those sectors are long, large areas may be involved. The second situation occurs when boundaries are defined as following previously existing boundaries such as the southern boundary of Kutch. In such cases there is sometimes scope for disagreement over major areas. Generally, positional disputes create fewer problems for border residents and internatic ıal relations than territorial boundary disputes.

Functional boundary disputes

A functional boundary dispute arises when one government believes that it has been adversely and unfairly affected by the functions of a neighbouring government along the boundary. Unlike territorial and positional disputes, it is possible to solve functional disputes to the satisfaction of both sides without any alteration in the boundary's position. House (1959) has described an interesting functional dispute following the adjustment of the boundary between Italy and France in the Alps Maritimes in 1947. As a result of the boundary change and the state functions applied by both governments, problems had arisen for the inhabitants of two communes. The people in a French commune lacked sufficient summer pasture for their stock and the people in an adjoining Italian settlement lacked sufficient spring and autumn pastures.

Since positional changes in the boundary were not possible, the Swiss

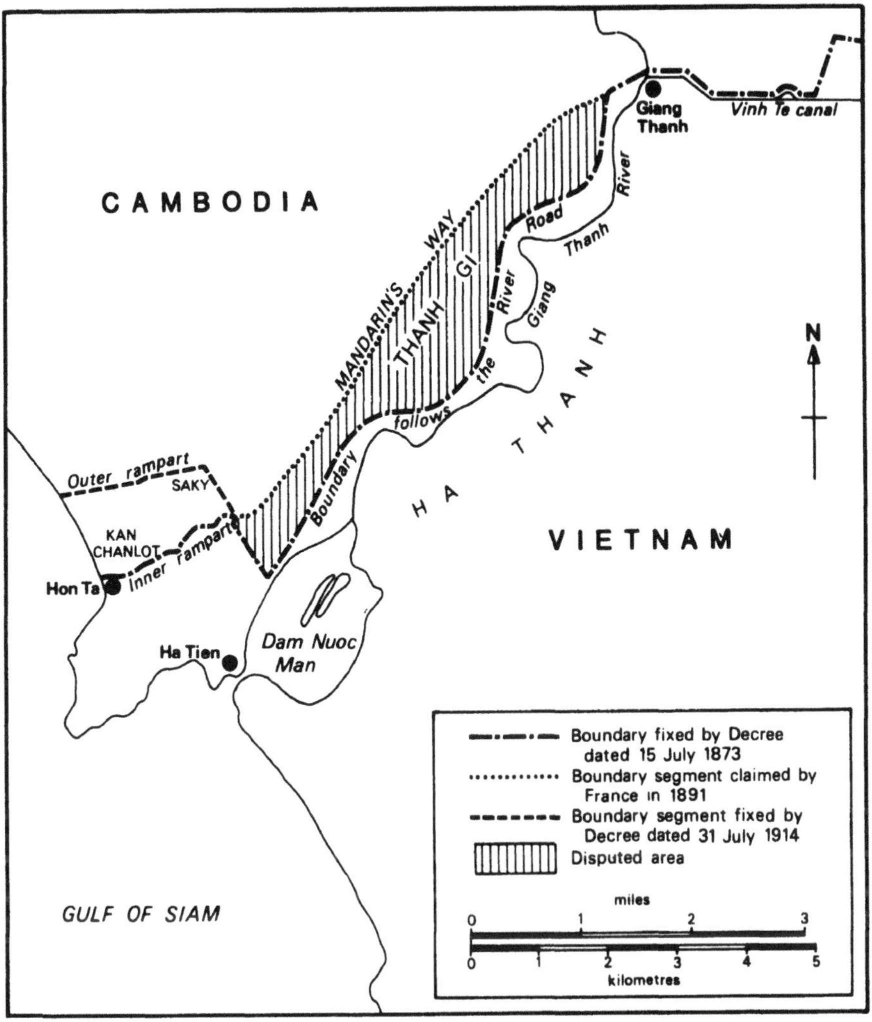

Figure 4.3 The southern sector of the Cambodia–Vietnam boundary.

arbitrator appointed to settle the problem suggested a compromise which allowed the Italian settlements certain grazing rights in France in exchange for the French residents being awarded rights over woodland in Italy. This compromise, which was accepted, was simplified by an earlier convention which permitted the free circulation of people and property within a zone 20 kilometres wide lying astride the boundary.

Another functional dispute which used to cause friction between Iran and Iraq flared into war in September 1980. According to the protocol agreed by Britain, Persia, Turkey, and Russia in 1913, the eastern sector of the boundary was drawn along the eastern bank of the Shatt-el-Arab. In 1937 small modifications of this line gave Iran an anchorage in the river opposite Abadan. However, on occasions Iraq used its sovereignty over the rest of the river to interfere with Iranian shipping and commerce. In 1975 the two countries agreed to move the boundary from the Iranian bank to the thalweg of the river. In September 1980, after sporadic disputes about traffic on the river, Iraq abrogated the 1975 agreement and hostilities erupted.

Functional boundary disputes are comparatively rare when compared with territorial and positional disputes. In recent years functional disputes have usually been associated either with territorial disputes or with other international disputes. By imposing new regulations at borders it is possible for states to show their displeasure with a neighbour, as Pakistan did when it effectively blocked the transhumance movements of the Powindas. African states in the post-colonial period have sometimes resorted to closing the border with a neighbour to record displeasure with a neighbour's policies or to increase national security at a time when civil wars were generating large numbers of refugees. Between 1964 and 1977 there were 24 reported cases of African boundaries being closed, and sometimes this produced serious consequences because Africa has more landlocked states than any other continent and because these states have vulnerable economies. In some cases the closure of the boundary was designed to force an alteration in a neighbour's policy. For example, in February 1976 Kenya refused to allow the passage of Uganda's trade through Mombasa because President Amin had announced that Uganda had legitimate claims to large areas of western Kenya. These claims were abandoned within a few days of the boycott being imposed. In October 1975 Benin, which was formerly called Dahomey, closed its border with Togo on the ground that this country had been involved in an attempted coup against President Kerekou of Benin. Suspicions were voiced in Ghana and Togo that the closure of the boundary was for another reason. Due to congestion in Lagos harbour many vessels were discharging cargoes for Nigeria at Ghanaian and Togolese ports and arranging for them to be transferred by lorry to Nigeria. By closing the boundary with Togo and severing this land connection, Benin was serving notice on Nigeria that it would be wiser to unload cargoes in Benin and avoid the risk of this disruption.

Another interesting case concerned allegations by Lesotho in November 1976 that South Africa had closed three border posts on Lesotho's eastern border to force that country to recognize the independence of the Transkei. Lesotho alleged in the United Nations Security Council that the closure resulted in 250 000 people being held hostage in the most rugged part of that mountainous state and in interference with Lesotho's major exports, which are livestock and migrant workers. There are 36 border crossings along Lesotho's boundary and 3 of them are with the Transkei; they are Qacha's Nek, Tele Bridge, and Ramatsilitshek. The latter is a rough track and mainly used by rustlers. None of the border posts was closed, but at the three posts Transkei officials required purchase of a visa, costing one Rand, which was valid for multiple journeys for a period of three months. In fact several thousand Lesotho citizens were paying this fee and crossing the boundary during the period it was alleged the boundary was closed. The flow of trade between Lesotho and the outside world was unaffected by the border regulations, yet Lesotho was able to persuade the United Nations that it was being severely affected and in consequence a number of countries and the European Economic Community gave aid to Lesotho to offset the alleged damage suffered by its economy! This successful hoax may have been inspired by the actions of Mozambique which closed the boundary with Rhodesia on 3 March 1976 and then appealed to the international community for aid to make up the loss of revenue.

Resource development disputes

The commonest source of such disputes are water bodies which mark or cross any international boundary. Jones (1945) has recorded some disputes connected with minerals, but they are rare. Disputes over maritime resources are considered in the next chapter.

Boundaries were often drawn to coincide with rivers in order to allow easy recognition, and the disadvantages of such features as boundaries have already been considered. In many other cases, however, except when the boundary coincided with a watershed, river basins were divided between adjacent states. When the boundary coincided with a watercourse the agreement usually contained a clause providing equal rights for nationals from both sides. Generally, the clause did not define the position with regard to the tributaries of boundary waters nor make provision for the joint control of rivers which crossed the boundary. Often this was because the border areas were underdeveloped and the use of rivers for hydroelectricity and irrigation had not been envisaged. It was only when the border areas became more closely settled and advances in technology made possible the use of the rivers for purposes other than navigation that disputes about the use of boundary and other waters developed. Many terms are used by different writers in referring to rivers forming and crossing the boundary; it is proposed here to distinguish three types. 'Boundary waters' are those features within which the boundary is drawn; this term is preferred to 'contiguous waters' used by Griffin (1959). 'Tributaries of boundary waters' form the second group. The term is entirely descriptive and it is essential to distinguish the tributaries from the boundary waters. Rivers which cross a boundary are called 'successive rivers'. This is a term suggested by Griffin which seems more satisfactory than any other, such as 'divided rivers', since boundary waters are also divided.

Griffin (1959) has shown that customary international law requires that no action should be taken in respect of the boundary waters which will diminish their value and usability to the other state. The Guadalupe–Hidalgo Treaty of 1848 between America and Mexico was explicit on this point in respect of the Gila and Bravo rivers: 'Neither [state] shall without the consent of the other construct any work that may impede or interrupt, in whole or in part, the exercise of this right [free navigation]: not even for the purpose of favouring new methods of navigation' (Miller 1937, p. 217). In fact an American company did interfere with the course of the river and a Mexican complaint to the American federal courts was successful, so that the company had to make restitution to Mexico (Burns 1912).

A dispute over tributaries of boundary waters developed between Britain and the United States in 1980. The Chicago municipal authority tapped Lake Michigan by means of a canal to augment the flow of the Des Plaines River which flowed eventually into the Mississippi River. This scheme was designed to improve the city's system for disposing of sewage. Britain objected that the diversion was lowering the level of Lake Michigan. By 1926 it was estimated that Lakes Michigan, Huron, Ontario, and Erie had fallen 5 inches to a new mean lake level. Since every inch represented 60–80 tons in the carrying capacity of vessels, Britain argued that Chicago was impairing the navigability of the lakes. Britain was successful in this case and the extraction of water from the lakes was reduced to an agreed volume (Simsarian 1938).

Disputes over boundary waters are less common than disputes over successive rivers. This is due to two factors. First, most treaties governing boundaries along boundary waters contain clauses guaranteeing both sides equal rights and prohibiting any use which adversely affects the other state. Secondly, since both states have access to the same parts of the river a unilateral action by one country can be met by retaliation by the other country.

In considering the utilization of successive rivers, it is clear that the lower riparian may be harmed if the flow of water is diminished, whereas the upper riparian rights may be infringed if the river is dammed downstream to produce flooding beyond the boundary. The classical example of the first situation concerns Egypt's anxiety that the flow of the Nile should not be diminished through irrigation projects in the Sudan. This matter was carefully controlled through the Nile Waters Agreement of 1929 under which Britain undertook not to interfere with the quanitity, level, or date of the river's régime. The Sudan has continued to respect this agreement, although there is not yet any agreement on how additional supplies of water should be apportioned. When Lake Kariba was formed on the Zambezi between Northern and Southern Rhodesia, Portugal demanded and secured guarantees of a certain minimum flow through Mozambique. The flow of 35 000 cubic feet per second is sufficient to allow navigation of the lower Zambezi throughout the year. At the same time the Rhodesian governments gained assurance from the Union of South Africa and Angola that they would not draw additional supplies from the Zambezi above the lake.

Glassner (1970) has provided a very interesting and comprehensive account of a dispute between Chile and Bolivia over the Rio Lauca. This successive river rises in swamps on the high semi-arid plateau, called the Altiplano, 75 miles east of the Chilean port of Arica. The first 45 miles of the river's course lies in Chile and there are then 90 miles of the river in Bolivia; it drains into the landlocked Lago de Coipasa, which is set in a salt flat with the same name. In 1939 the Chilean government announced that it intended to divert water via a tunnel into the valley of the Rio Azapa in order to allow irrigation of the fertile soils in this arid valley. Bolivia protested against this proposal because it was considered to be contrary to the 1933 Declaration of Montevideo, by which member states agreed that the upper riparian state had the right to use the river provided it did not modify the hydrological conditions and the natural régime of the river. At intervals since 1939 the dispute has flared and died down and there has been no final solution to the question, even though Chile is using the scheme. Bolivia argued that the withdrawal of water from the Rio Lauca would increase the rate of salt accumulation in the Bolivian sector of the river and adversely affect the population living there. According to Glassner (1970, p. 198), Bolivia failed to produce evidence of the damage it alleges has been caused.

A serious dispute developed between India and Bangladesh over the Farraka Barrage scheme on the River Ganges. In 1951 India began to plan for the diversion of water from the River Ganges into the Hooghly River in order to flush sediment out of the port of Calcutta and improve conditions for navigation. Pakistan objected to the proposals, but India began construction in 1961 and completed the barrage in 1975. Bangladesh had succeeded East Pakistan by this time and was maintaining objections to the scheme. The scheme itself is very old, and there are references to it during the first half of the 19th century. Bangladesh is fearful that the withdrawal of water from the River Ganges, 11 miles before it enters Bangladesh, will reduce water volume in the Granges Delta, which comprises much of Bangladesh, to levels where navigation, fishing, and irrigation will be adversely affected. Bangladesh is also concerned that the reduced flow will increase the rate of salt accumulation in parts of the delta. Although there is no reference in literature published by the two governments to the possibility of coastal changes in the delta being caused by the new patterns of water flow and sediment discharge, it should be a consideration. There is plenty of evidence in other parts of the world to demonstrate that alterations to the upper parts of river systems might produce changes in the form and stability of deltas at the rivers' mouths. India argued that the amount of water diverted would not adversely affect Bangladesh and asserted that the supply of water from the Brahmaputra River was most critical to Bangladesh and that this river was not being efficiently used. After a period of acrimonious debate the two countries initialled an agreement on 30 September 1977 providing for the implementation of short-term dry season sharing arrangements of the Ganges waters and for the development within five years of a long-term solution for augmenting the flow of the River Ganges. By the terms of the agreement Bangladesh was awarded a larger share than India had been prepared to grant earlier in the discussions, and the amount diverted through the Hooghly River was set at about half the volume which officials in West Bengal claimed was essential.

Upstream flooding resulting from the construction of dams has often occurred. In 1897 the Canadian Dyking Company made a dam on Boundary Creek in British Columbia, which resulted in the flooding of 80 000 acres of Idaho; this of course reduced the rateable value of that state (Simsarian 1938). A contemporary example is provided by the Aswan High Dam in the United Arab Republic, which has flooded part of the Sudan and caused the resettlement of 35 000 Nubians in the Khashm el Girba area astride the Atbara River.

The Franco-Spanish dispute over the waters of Lake Lanoux provides a convenient example of how technological advances trigger disputes of this kind. Soon after World War II France decided to dam the lake, which normally drained towards Spain, and force the water over a drop of 780 metres into the Ariège valley. The water would then be returned by a tunnel to the course of the River Font, which was tributary to the Serge River in Spain. A canal from the French side supplied water to Spanish irrigation schemes. Spain objected to this plan on the grounds that it infringed the Treaty of Bayonne of 1866. The eighth, eleventh, and twelfth Articles of the *Acte Additionnel* provided that both states had sovereignty over water within their boundaries; that the downstream riparian had a right to the 'natural waters which flow from higher levels without the hand of man having contributed thereto'; that the riparian rights of the upstream state should not be harmed; and that there should be consultation on all new works. Spain requested that France should make the dam less than the planned height to increase the natural flow towards Spain and reduce the electricity production by 10 per cent. France refused this request, and in 1957 the International Court of Justice adjudged that France's plan did not infringe the 1866 agreement.

Since 1955 some attention has been directed to air pollution in one country which adversely affects another state. Such problems were considered in Chapter 1 in the section dealing with boundaries in the air.

The rôle of maps in boundary disputes

In the previous chapter the importance of using maps to understand the evolution of the boundary was stressed; it is equally important in examining boundary disputes to use maps to understand the basic nature of the problems. Not only are maps used by scholars concerned with objective research, they are also used with increasing frequency by lawyers trying to prove a particular point and by governments trying to influence the opinions of citizens and other governments (Murty 1964, 1983). This section examines the characteristics which useful maps will possess and the circumstances under which their use will be most appropriate.

Looking first at the characteristics of the map, it is important that the political boundaries under consideration should be marked on them. If the boundaries are not marked, there may be an effort to use the fact as evidence that no boundary existed, but such evidence would not be conclusive. Dorion (1964) and Sinnhuber (1964) have both written useful papers on the way in which boundaries are represented on maps. The scale of the map is also very important, and maps rapidly decline in usefulness as the scale is reduced. Scales of about 1 : 50 000 or larger are the most satisfactory; when the scale is reduced to 1 : 1 million the map is useful only when the boundary follows a pronounced physical feature such as a watershed or a mountain crest. If it is desired to compare the representation of the same boundary on two maps of different scales, it is essential to reduce the large-scale map to the smaller scale since the reverse process cannot be carried out with a sufficient degree of accuracy. For example, the boundary disputed between India and China in Assam is shown on two treaty maps. The first accompanied a secret Anglo-Tibetan treaty and is at a scale of 1 : 500 000; the tripartite agreement has a map at a scale of 1 : 3.8 million. It has always been assumed that the two lines are identical, but the scales of the maps make comparison very difficult, and the tragedy for India is that it is only the larger-scale map, attached to the secret treaty, which is at a scale sufficient for reasonable identification of the boundary in the landscape. There is, of course, no reason why China should accept such a document.

The accuracy of the map is also a critical factor, and historic maps which are so often used in connection with boundary disputes differ much more widely from each other in accuracy than modern maps produced with the benefit of satellite photography and modern photogrammetry. Attention must also be paid to the author of the map because it is a reasonable inference that boundaries will be shown more accurately on official maps than on unofficial maps and that certain government departments, such as those concerned with foreign affairs, will be more careful in boundary representation than other departments, such as mining or meteorology. Because maps represent the region at a particular time it is important to match the date of the map to the period of boundary construction or dispute under consideration. For example, it is difficult to fathom the dispute between the United States and Mexico in the 1850s without reference to Disturnell's map of 1847, which was used by the commissioners.

There are five situations in which maps are likely to contribute to the analysis of boundary disputes. First, it is appropriate to concentrate on the map attached to a particular treaty when the map alone defines the boundary or when the attached map is given precedence over the description in the text. For example, the Sino-British Treaty of 9 June 1898 defined the extension of Hong Kong on a map at a scale of 1 : 314 000. The Anglo-Belgian treaty of 1926 regarding the boundary between what are now Tanzania and Burundi specifically noted that if there was any conflict between the text and the map that the map took precedence.

Secondly, maps are vital when they alone provide the evidence needed to interpret the text. For example, the Anglo-Belgian agreement of 22 November 1934 defined the boundary between what are now Tanzania and Rwanda by reference to a series of pillars built on the headlands of a lake. When the lines between these pillars intersected the shore the line diverged to follow the shore until it intersected the ray joining the next two pillars. Only a map of that period will allow the shoreline of that time to be identified and the boundary to be defined with accuracy. Similarly, in May 1872, General Goldsmid drew a boundary between Afghanistan and Persia and defined it as following the outer edge of the naizar, which is a belt of reeds. Since that feature varies in extent as a result of burning, grazing, and collection for thatch, it would have been necessary to have maps showing the naizar's extent in 1872 to locate the boundary. It was recognition of this problem that contributed to a redefinition of the boundary in 1905.

Thirdly, the map may be useful in a situation when evidence is conflicting. Weissberg (1963), in an excellent article, describes how a map was decisive in settling a dispute between Belgium and Holland. A Belgian enclave in Holland was under dispute and the judges were finally very impressed by the fact that a map in 1841 showed the Belgian enclave very clearly, in colour, so that its existence was obvious to anyone glancing at the map.

Fourthly, maps may be very useful in helping to understand the reasoning by which boundary architects selected a particular line.

Reference was made in the last chapter to the rôle of charts based on Vancouver's explorations in persuading the negotiators to define the Anglo-Russian boundary along the crest of a range parallel to the coast. Finally, maps are very useful in discovering the attitude of governments towards their boundaries. The phrase 'cartographic aggression' is now commonly used to describe the inclusion on a map by one state of territory which is under the control of a neighbouring state. For example, Guatemala regularly produces maps showing Belize as a Guatemalan province, and many Tanzanian maps indicate that it owns part of Lake Tanganyika, whereas according to the relevant treaties the boundary follows the eastern shore of the lake.

There was a time when the value of maps as legal evidence in boundary disputes was heavily discounted as Hyde (1933) and Sandifer (1939) indicated:

> It may be doubted whether a series of maps, however numerous, proves that the boundary which they unite in prescribing is necessarily the correct one, to be accepted as the juridical basis of the proper frontier, especially when they are contradicted by trustworthy evidence of title. (Hyde 1933, p. 316)

> Maps can seldom, if ever, be taken as conclusive evidence in the determination of disputes which may arise concerning the location of the boundary. (Sandifer 1939, p. 157)

That time seems to have passed and Weissberg, after reviewing a number of recent cases, was able to assert that maps sometimes had an important rôle: 'Maps may be regarded as strong evidence of what they purport to portray. They may be termed and treated as admissions, considered as binding, and said to possess a force of their own' (Weissberg 1963, p. 803).

Conclusions

Most governments take the development of boundary disputes very seriously. This is because friction in the borderlands may be likened to a bushfire in south-east Australia during a dry January. If it is not controlled quickly, it might spread widely and do great damage. Governments are aware of the correctness of Siegfried's assertion nearly 50 years ago:

> The study of boundaries is dangerous . . . because it is thoroughly charged with political passions and entirely encumbered with afterthoughts. The people are too interested in the issues when they speak of boundaries to speak with detachment: the failing is permanent! (Siegfried, in Ancel 1938, p. vii)

Except in the comparatively few cases when governments have a political interest in fanning the flames of discord in the borderland, efforts are concentrated on ending the dispute soon.

The success in achieving this aim often varies with the type of boundary dispute. Territorial disputes arise when one government requests a neighbour to cede some territory for one of two reasons. It will argued either that the territory is presently held illegally or that it would be better for all concerned if ownership of the area was transferred. In the second case the arguments will usually be based mainly in history and geography. Such disputes are rarely settled quickly and some last for a very long time and have periods of intense activity separated by intervals of apparent disinterest on the part of the claimant.

Positional disputes arise when the evolution of the boundary is incomplete, and the most common cause is the presence of ambiguous phrases in the delimitation of a boundary which has never been demarcated. Generally, positional disputes involve comparatively small areas and in such cases they can be solved quite quickly. Of course, there are exceptions and one is provided by the long time it took Mexico and the United States to settle the Chamizal dispute, which arose because the Rio Grande changed its course by cutting through the necks of meanders. In a few rare cases positional disputes involve large areas, as was the case between Argentina and Chile in the Andes and between the United States and Britain in the Alaskan Panhandle. In such cases the positional disputes have the stubborn characteristics of territorial disputes.

Functional and resource development disputes are often less difficult to solve because the solution does not require any movement of the boundary with the consequent transfer of territory. The adjustment of regulations by which a government's functions are exercised at border crossings can be easily altered to reduce friction if there are real grounds for complaint by the neighbouring government. The need to consult with neighbours about the ordered exploitation of transboundary resources is now well known, whether the matter concerns rivers, coalfields, or oilfields. That knowledge has been employed in settling some seabed boundaries where it is common for a clause to specify that the parties will confer on the best way to exploit any mineral resource which straddles the marine boundary.

The importance of boundary disputes to the state of relations between neighbouring countries has made them targets for research by scholars from disciplines which include law, history, geography, politics, and economics. Geographers have always played a leading rôle because boundary disputes often require detailed analysis of terrain, drainage patterns and the distribution of population, and reference to many maps of varying scales.

Although some scholars would concentrate on a single aspect of boundary disputes, such as their effect on international relations, a comprehensive examination of the subject would answer the following questions: What is the *cause* of the dispute? Why did it develop at *a particular time*? What are the *aims* of the state making the claim or the complaint? Which *arguments* are used by the rival sides in defence of the position adopted? What *results* followed either from the settlement of the dispute or its continuation?

References

Ancel, J. 1936. *Géopolitique* (Geopolitics). Paris: Delagrave.
Ancel, J. 1938. *Les frontières* (Boundaries). Paris: Armand Colin.

Boggs, S. W. 1940. *International boundaries: a study of boundary functions and problems.* New York: Columbia University Press.
Bowman, I. 1923. *The new world.* New York: World Book Company.
Burns, W. T. 1912. The Horcon Ranch case. *American Journal of International Law,* **6** (1), 478–85.

Caroe, Sir O. 1961. Pathans at the crossroads. *Eastern World* **15**, 12–13.

Dennis, W. J. 1931. *Tacna and Arica.* New Haven, Conn.: Yale University Press.
Dorion, H. 1964. La représentation géographique des frontières litigieuse: le cas du Labrador (The geographical representation of disputed frontiers: the case of Labrador). *Cahiers de Géographie de Quebec* **9**, 77–87.

East, W. G. 1937. The nature of political geography. *Politica* **2**, 259–86.

Fisher, H. A. L. 1940. *The background and issues of the war.* Oxford: Clarendon Press.
Fraser-Tytler, Sir W. K. 1953. *Afghanistan.* London: Oxford University Press.

Glassner, M. I. 1970. The Rio Lauca: dispute over an international river. *Geographical Review* **60**, 192–207.
Griffin, W. I. 1959. The use of international drainage basins under customary law. *American Journal of International Law* **46**, 50–80.

Hartshorne, R. 1936. Suggestions on the terminology of political boundaries. *Annals, Association of American Geographers* **26**, 56–7.
Hasan, K. 1962. Pakistan–Afghanistan relations. *Asian Survey* **11**, 14–19.
Hertslet, Sir E. 1909. *Map of Africa by treaty* (3 volumes plus atlas). London: HMSO.
Hill, N. L. 1976. *Claims to territory in international law and relations.* New York: Greenwood Press.
Hinks, A. R. 1921. Notes on the technique of boundary delimitation. *Geographical Journal* **58**, 417–43.
House, J. W. 1959. The Franco-Italian boundary in the Alpes-Maritimes. *Transactions, Institute of British Geographers* **26**, 107–31.
Hsu, I. C. Y. 1965. *The Ili crisis: a study of Sino-Russian diplomacy 1871–81.* Oxford: Clarendon Press.

Hyde, C. C. 1933. Maps as evidence in boundary disputes. *American Journal of International Law* **27**, 311–16.

India, Ministry of External Affairs 1961. *Report of the officials of the governments of India and the People's Republic of China on the boundary question.* New Delhi: Government Printing Office.

Ireland, G. 1938. *Boundaries, possessions and conflicts in South America.* Cambridge, Mass.: Harvard University Press.

Johnson, R. W. 1966. The Canada – United States controversy over the Columbia River. *University of Washington Law review* **41**, 676–763.

Jones, S. B. 1945. *Boundary making, a handbook for statesmen, treaty editors and boundary commissioners.* Washington, DC: Carnegie Endowment for International Peace.

Keesing's Contemporary Archives 1976. Volume 22. Bath: Keesing's Publications.

Keesing's Contemporary Archives 1982. Volume 28. Bath: Keesing's Publications.

Kirk, W. 1962. The inner Asian frontier of India. *Transactions, Institute of British Geographers* **31**, 131–68.

Miller, H. 1937. *Treaties and other acts of the United States of America*, Vol. 4. Washington, DC: Government Printing Office.

Murty, T. S. 1964. Boundaries and maps. *Indian Journal of International Law* **4**, 367–88.

Murty, T. S. 1983. *Paths of peace: studies on the Sino-Indian border dispute.* New Delhi: ABC Publishing House.

Office of National Assessments 1984. *Maritime boundaries in the southwest Pacific region.* Research Memorandum 1/454. Canberra: Australian Government Printer.

Pounds, N. J. G. 1951. The origin of the idea of natural frontiers in France. *Annals, Association of American Geographers* **41**, 146–57.

Pounds, N. J. G. 1954. France and 'Les limites naturelles' from the seventeenth to twentieth centuries. *Annals, Association of American Geographers* **44**, 51–62.

Prescott, J. R. V. 1975. *Map of mainland Asia by treaty.* Melbourne: Melbourne University Press.

Prescott, J. R. V., J. H. Collier & D. F. Prescott 1977. *The frontiers of Asia and Southeast Asia.* Melbourne: Melbourne University Press.

Qureshi, S. M. M. 1966. Pakhtunistan: the frontier dispute between Afghanistan and Pakistan. *Pacific Affairs* **39**, 99–114.

Ryder, C. H. D. 1926. The demarcation of the Turco-Persian boundary of 1913–14. *Geographical Journal* **66**, 227–42.

Sandifer, D. V. 1939. *Evidence before international tribunals.* Chicago, Ill.: Foundation Press.

Sanguin, A. L. 1983. La bordure franco-italienne des Alpes-Maritimes ou les consequences de la modification d'une frontière internationale (The Franco-Italian border in the Alpes-Maritimes or the consequences of change in an international boundary). *Mediterranée* **47**, 17–25.

Simsarian, J. 1938. The division of waters affecting the United States and Canada. *American Journal of International Law* **32**, 488–518.
Sinnhuber, K. A. 1964. The representation of disputed political boundaries in general atlases. *Cartographic Journal* **1**, 20–8.
Sykes, Sir P. M. 1940. *History of Afghanistan* (2 volumes). London: Macmillan.

Taberd, J. L. 1837. Note on the geography of Cochin-China. *Journal of the Royal Asiatic Society, Bengal* **2**, 737–45.
Taussig, H. C. 1961. Afghanistan's big step. *Eastern World* October, 15.
Temperley, H. W. V. 1920–1. *History of the peace conference of Paris* (5 volumes). London: Hodder & Stoughton.
Touval, S. 1966. The Moroccan–Algerian territorial dispute. *Africa Research Bulletin* **3**, 631–3.
Triggs, G. D. 1983. *Australia's sovereignty in Antarctica: the validity of Australia's claim at international law*. Programme in Antarctic Studies 61, Melbourne University, Melbourne.

Vietnam Foreign Ministry 1975. *White paper on the Hoang Sa [Paracel] and Truong Sa [Spratly] Islands*. Saigon: Government Printer.

Ward, F. K. 1932. Explorations on the Burma–Tibet frontier. *Geographical Journal* **80**, 465–83.
Weissberg, G. 1963. Maps as evidence in international boundary disputes. *American Journal of International Law* **57**, 781–803.

5 *Maritime boundaries*

> The sovereignty of a coastal State extends beyond its land territory and internal waters, and in the case of an archipelagic State, its archipelagic waters, over an adjacent belt of sea described as the territorial sea. (United Nations 1983, p. 3)

Since 1930 there have been four international conferences seeking to establish a comprehensive law of the sea. The conferences held in 1930 and 1960 failed to produce any definite results. The 1958 conference produced four conventions which were endorsed by several states and which still apply. The 1973 conference ended in December 1982 when 119 delegations signed the United Nations Convention on the Law of the Sea. That Convention will come into effect one year after the sixtieth country has ratified it. Although it is not expected that this event will occur for some years, there is plenty of evidence that many important rules of the convention are already being observed by a number of states.

Competition for exclusive control of areas of seas and seabed in the present period has been likened to the scramble for colonies by European countries in the 18th and 19th centuries. Just as those periods witnessed the construction on a grand scale of boundaries which today still form the framework of national sovereignty, so this modern period is recording the proliferation of maritime limits which will determine national control of areas of the seas in the foreseeable future.

This chapter begins by examining the origin of maritime claims by states and then continues with a description of zones which can be claimed. There is then an analysis of the problems associated with drawing the three basic maritime boundaries. These boundaries are the baseline, the edge of the continental margin, and international boundaries between adjacent or opposite states.

The origin of national maritime claims

The earliest cases of sovereignty being exercised over the oceans are shrouded in antiquity, but there are early records of rulers granting favours to subjects in connection with adjacent seas. These favours involved exclusive access to shallow fishing grounds, the right to mine salt from saline marshes, exemption from harbour taxes, and unhindered passage through narrow straits. Davis (1984) has also discovered that Aborigines along the coast of Arnhem Land in northern Australia divide the coastal waters between clans by lines which they insist have existed for the thousands of years that this coast has been settled.

The construction of navies allowed the defence of the state's territory at sea and discouraged unfriendly alien vessels from sailing too close to hostile shores. The development of trade was soon followed by the incidence of smuggling, and this was an activity which could be most easily detected and prevented in coastal waters rather than after the goods had been landed. The growth of coastal fishing industries promoted claims which excluded foreign competition. Some small, weak countries tried to avoid involvement in the squabbles of larger neighbours by proclaiming belts of neutral waters around their shores. For these and other reasons the attention of rulers and administrators was increasingly directed to control over coastal waters.

Fenn (1926), who has written a most detailed account of the origin of the theory of territorial waters, identifies four main contributors during the period before the 17th century. The first was made by the Roman glossators, who agreed that the emperor had the right to punish wrongdoers at sea in exactly the same way that he punished them on land. In the 12th century Azo successfully argued that the emperor had the right to limit the communal nature of the oceans by granting privileges based on long and uninterrupted use. In the 14th century Bartolus, who taught law at Pisa, asserted that any country owned islands which were within 100 miles of the coast and had authority over the intervening sea. Finally, in the 16th century, Gentilis enunciated that coastal waters were a continuation of the territory of the state whose shores they adjoin. It therefore followed that the territorial rights which the sovereign possessed on land extended over the coastal waters. Fenn is certain that 'after Gentilis it is literally correct to speak of territorial waters in international law' (Fenn 1926, p. 478).

There was still the related problem of deciding the areal limits which should apply to these maritime rights. Fenn regarded the delimitation of territorial waters as 'a mere matter of detail' which should be solved by politicians rather than lawyers. In fact this detail was not solved until December 1982, and there are still some countries which would dispute the rules most countries appear prepared to accept.

The first claim to a continuous zone of territorial sea measuring 3 nautical miles was made by Sweden on 9 October 1756, and this represented the convergence of two different concepts. The first determined that the territorial waters were continuous and the second that their proper width was 3 nautical miles. These two concepts had different origins; the first came from Scandinavia and the second from the Mediterranean. From 1598 Denmark, the most powerful Scandinavian state, had been reserving exclusive fishing grounds around Iceland within 2 leagues of the coast. The Scandinavian league measures 4 nautical miles. During the reign of Christian IV (1588–1648) the reserved zone varied from 2 to 8 leagues in width, and in his successor's reign the limit was fixed at 4 leagues. This claim brought Denmark into collision with more powerful countries including Britain, France, Holland, and Russia. For example, in 1743 Russia compelled the Danish governor of Finnmark to allow Russian vessels to operate within 1 league of the coast. If this permission had not been granted, the Russians threatened that the land boundary would be closed, effectively stopping trade and the transhumance movement of Lapps with their herds of reindeers. In 1745 Denmark formally reduced its claims to 1 league, but this failed to satisfy the French authorities, who were involved in further diplomatic exchanges with Denmark over the fate of two English prizes captured by French vessels. France informed Denmark that Danish sovereignty would only be conceded over waters measuring 'Three Miles, the Possible Reach of Cannon Shot from the Land' (Prescott 1975, p. 39). This was the limit generally accepted in areas of the Mediterranean and southern Europe.

In 1610 Holland referred to the cannon shot rule during a dispute with Britain, and Walker (1945, pp. 211–18) has shown that this concept was common amongst countries such as France and Holland in the 17th century. Bynkershoek is usually credited with formulating this generalization into international law in 1703 in an effort to avoid the imprecision associated with claims to areas of water within sight of land:

> Therefore it evidently seems more just that the power of the land [over the sea] be extended to that point where missiles are exploded . . . [the power of the land] is bounded where the strength of arms is bounded; for this as we have said guards possession. (Balch 1912, p. 414)

There was still some debate about the exact application of this rule because it was unclear whether it applied to guns actually in position or to an imaginary line of guns placed around the entire coast. The narrow interpretation operated in favour of the strong, well armed states, but the general interpretation placed all countries on an equal footing. It was probably this consideration which led Galiani, in 1782, to propose that, instead of waiting to see where a neutral country mounted its coastal pieces, each country be awarded a belt of coastal waters 1 league wide. This southern Mediterranean league measured 3 nautical miles, and the new proposition, which had already been successfully employed by Sweden since 1756, received general acceptance. Baty (1928, pp. 517–32) has provided a comprehensive list of cases where countries accepted 3 nautical miles as the limit of territorial waters in the first half of the 19th century. It is impossible to disagree with the American diplomat who informed the French government in 1864 that 'no other rule than the three miles rule was known or recognized as a principle of international law' (Crocker 1919, pp. 659–60). That was certainly the position maintained by most unilateral declarations and bilateral agreements in the period before the 1920s when the most powerful countries refused to recognize any wider claims. Such claims were made by Italy, Spain, and Russia but without much success. However, at the 1930 conference the 47 delegations were unable to agree on the maximum width of claims to territorial seas. Agreement on the maximum widths of zones was eventually reached in December 1982, although some countries still claim more than the distances specified in the new Convention.

National and international maritime zones

There are five maritime zones which coastal states may claim and one zone which will eventually be controlled by an international body (Fig. 5.1).

Proceeding seawards from the shore, the first zone which states can claim consists of *internal waters*. This zone need not be continuous because internal waters lie landwards of straight baselines, and these can only be drawn when coasts have particular characteristics, which are considered later. With a single exception, internal waters are legally indistinguishable from the land territory of a country. This exception occurs in respect of those internal waters created by the construction of new straight baselines. If such waters had not previously been considered internal waters, alien vessels would continue to enjoy the right of innocent passage through them. The right of innocent passage means navigation through territorial waters for the purpose of traversing those seas or proceeding to a port in a manner which is not prejudicial to the peace, good order, and security of the coastal state. Submarines are required to navigate at the surface when exercising the right of innocent passage.

The next national zone consists of *territorial seas*. This zone is measured seawards from a baseline drawn according to rules set out in the 1982 Convention and should not exceed 12 nautical miles in width. By the end of 1985, 85 countries claimed territorial seas 12 nautical miles wide. Claims to narrower seas were made by 22 states and of these 16 claims were to seas 3 nautical miles wide. Most of the states claiming this traditional width were either located in western Europe or were former European colonies which had recently achieved independence. The most powerful state in this group was the United States. Denmark and Norway still claimed the Scandinavian league of 4 nautical miles and Israel, Greece, and Turkey claimed the traditional Mediterranean width of 6 nautical miles. Turkey only claims this width off its Mediterranean shores; in the Baltic Sea it claims 12 nautical miles.

Claims to seas wider than 12 nautical miles were made by 25 states, of which Albania was the only European representative. All the others were countries in Africa or South and Central America, and the claims vary from 20 to 200 nautical miles.

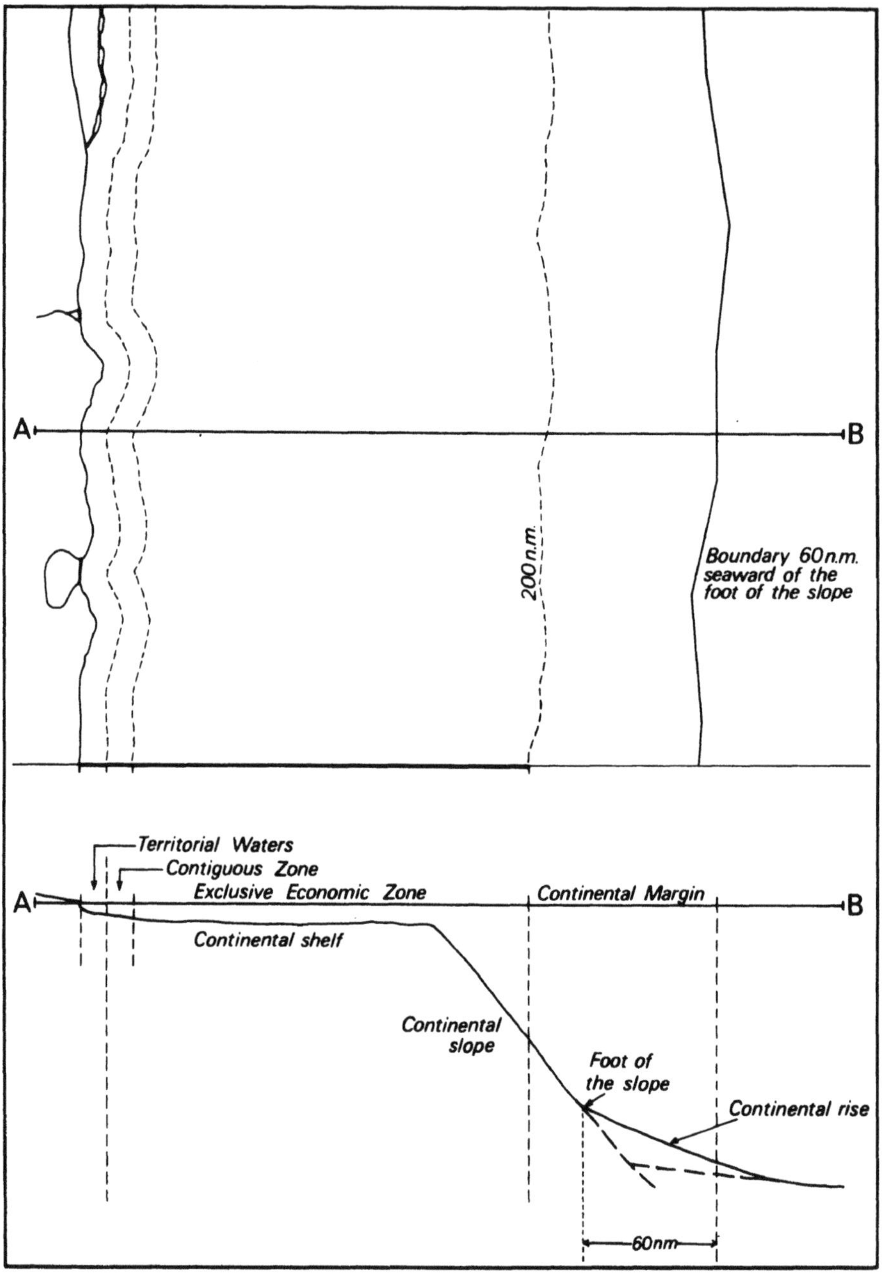

Figure 5.1 National maritime zones.

The right of innocent passage for vessels does not apply to aircraft passing over the territorial waters, and specific agreements must be concluded for such movements.

The third zone which states possess is the *contiguous zone*. This zone does not have to be claimed by states which can automatically exercise the rights associated with these waters. The contiguous zone extends for 12 nautical miles from the outer edge of the territorial waters, and in it the state may exercise controls necessary to prevent or punish infringements of its customs, fiscal, immigration, and sanitary regulations which apply to the territorial waters.

The fourth zone, which may extend for 200 nautical miles beyond the baseline from which the territorial seas are measured, is called the *exclusive economic zone*. In this region the coastal state has sovereign rights for the purpose of exploring and exploiting, conserving, and managing natural resources, whether living or non-living, of the seabed, subsoil, and superjacent waters and of using the seas and winds for the production of energy. This means that no alien can conduct any economic enterprise such as fishing or mining without the permission of the coastal state. However, aliens do have the right to navigate vessels through the exclusive economic zone, to overfly the area, and to lay or repair submarine cables on the seabed. In addition to sovereign rights, the coastal state may also exercise jurisdiction in respect of the establishment and use of artificial islands, marine research, and the preservation of the marine environment. One of the fears of developed countries is that some states will not distinguish between territorial seas and exclusive economic zones in the exercise of authority.

The *continental shelf* is the fifth zone which may be claimed. It is really only necessary to make a separate claim to the continental shelf if it is wider than 200 nautical miles, because a claim to an exclusive economic zone will provide control over the seabed and subsoil out to that distance. The claim to the shelf beyond 200 nautical miles is restricted to the mineral and other non-living resources of the seabed and subsoil together with living organisms belonging to sedentary species. At the harvestable stage these organisms are either immobile on or under the seabed or are incapable of moving, except in contact with the seabed. Thus no state has exclusive rights to fish the waters which overlie the continental shelf more than 200 nautical miles from the baseline. Indeed the waters outside the exclusive economic zone constitute the high seas within which all states have equal rights to navigate, fish, overfly, and conduct scientific research.

The sixth zone, which is considered to be the common heritage of mankind, consists of the *seabed and ocean floor and subsoil thereof beyond the limits of national jurisdiction*. That long descriptive phrase has been shortened in the 1982 Convention to the Area. This means that the Area begins where national claims to the exclusive economic zone or the continental shelf ends. The 1982 Convention provides for an international body known as the Authority to control mining of the Area.

It is now necessary to explore the problems of defining the boundaries of these six zones. The discussion can be divided into three parts. First, the baseline from which the territorial sea is measured must be considered. The construction of this line will determine the extent of any internal waters which lie landwards. The outer limits of territorial waters, the contiguous zone, and the exclusive economic zone will be easily fixed by simply measuring the prescribed width from the baseline. If there is no claim to the continental shelf, the outer limit of the exclusive economic zone will separate national claims from the Area. Secondly, it is important to consider how the outer edge of any continent margin wider than 200 nautical miles is defined. This limit is independent of the baseline except in one minor situation, and where a claim to the continental shelf is made this line will separate national claims from the Area. Thirdly, it is essential to examine the question of how countries draw common boundaries through overlapping claims to maritime zones.

The definition of the baseline

The 1958 Convention on the Territorial Sea and the Contiguous Zone codified the rules for drawing baselines which had been evolving over the previous century. Those rules, with minor modifications, were incorporated in the 1982 Convention and some new rules were added. Because the 1982 Convention has not come into force, many baselines are based on the 1958 Convention, but it is proposed to describe the rules set out in the 1982 Convention since some countries have already started to draw baselines according to the additional rules this document contains.

The normal baseline is a low-water mark around the coast of the state and around the coast of any islands the state might own. There are many low-water marks and each country must decide which it will use. Since most countries seek to secure the largest possible areas of sea and seabed, they normally choose the low-water mark which lies furthest seawards. Such a line would be the lowest astronomic tide. This is the lowest tide which can be predicted to occur under average meteorological conditions and under any combination of astronomic conditions. This level can only be properly obtained by studying tidal records over a full tidal cycle of 18.6 years. If the coast of a particular state has a low tidal range and if the offshore gradient of the seabed is steep, it will not really matter which low-water mark the country selects. In such circumstances it will probably be convenient to select the low-water mark already marked on the country's navigation charts. A survey of 60 declarations about baselines showed that 42 countries simply referred to the low-water mark without providing any further detail. Only Australia used the lowest astronomic tide, although seven countries in either the Middle East or West Africa referred to the lowest low-water mark or the lowest ebb tide. Under freak weather conditions it is possible for sea level to retreat below the lowest astronomic tide, but such occurrences cannot be predicted. Only Ethiopia defined its baseline as the maximum annual high tide. This is not a level shown in tide tables, and it is not clear whether the baseline is recalculated each year or if it is based on the average of a number of years. The tidal range in the southern Red Sea is higher than in the northern section, so Ethiopia is forfeiting some territorial waters by using this baseline.

Straight lines may be substituted for the low-water mark under special circumstances, and it is convenient to divide these straight lines into two classes. First, there are the *local* straight baselines. These are found along short sections of coast and they link longer sections where the low-water line is used. These lines are most commonly found closing the mouths of rivers and bays. Secondly, there are the *regional* straight baselines. These lines are drawn along coasts which are deeply indented or fringed with islands, along coasts which are unstable and retreating rapidly, and around the islands which form the territory of archipelagic states.

Local straight baselines

According to the 1982 Convention, the mouths of rivers can be closed by a local straight baseline if the river flows directly into the sea. The French version of this Convention translates the word directly by the phrase *sans former d'estuaire* (without forming an estuary). There is no evidence that states are resisting the temptation to close estuaries and it is clear that this provision is open to abuse. Indeed, an identical provision in the 1958 Convention has already been specifically cited by Argentina and Uruguay as justification for closing the mouth of the Rio de la Plata with a straight line measuring 120 nautical miles.

Bays owned by a single state may be closed by a straight line if they meet two conditions. First, the mouth of the bay must not be wider than 24 nautical miles. If there are islands in the entrance to the bay so that there are a number of channels, then the combined width of the channels must not exceed 24 nautical miles. If the mouth exceeds 24 nautical miles in width, a closing line may be drawn within the bay where it narrows to that width. The second condition relates to the surface area of the bay, which must be larger than the area of a semicircle with a diameter equal to the measured entrance to the bay. For purposes of measuring the area of the bay, islands in the bay are not distinguished from water (Fig. 5.2).

States may also close bays which are wider than 24 nautical miles if they do so on historical grounds. Historic bays are only mentioned in the 1982 Convention when it is noted that the rules for defining bays 'do not apply

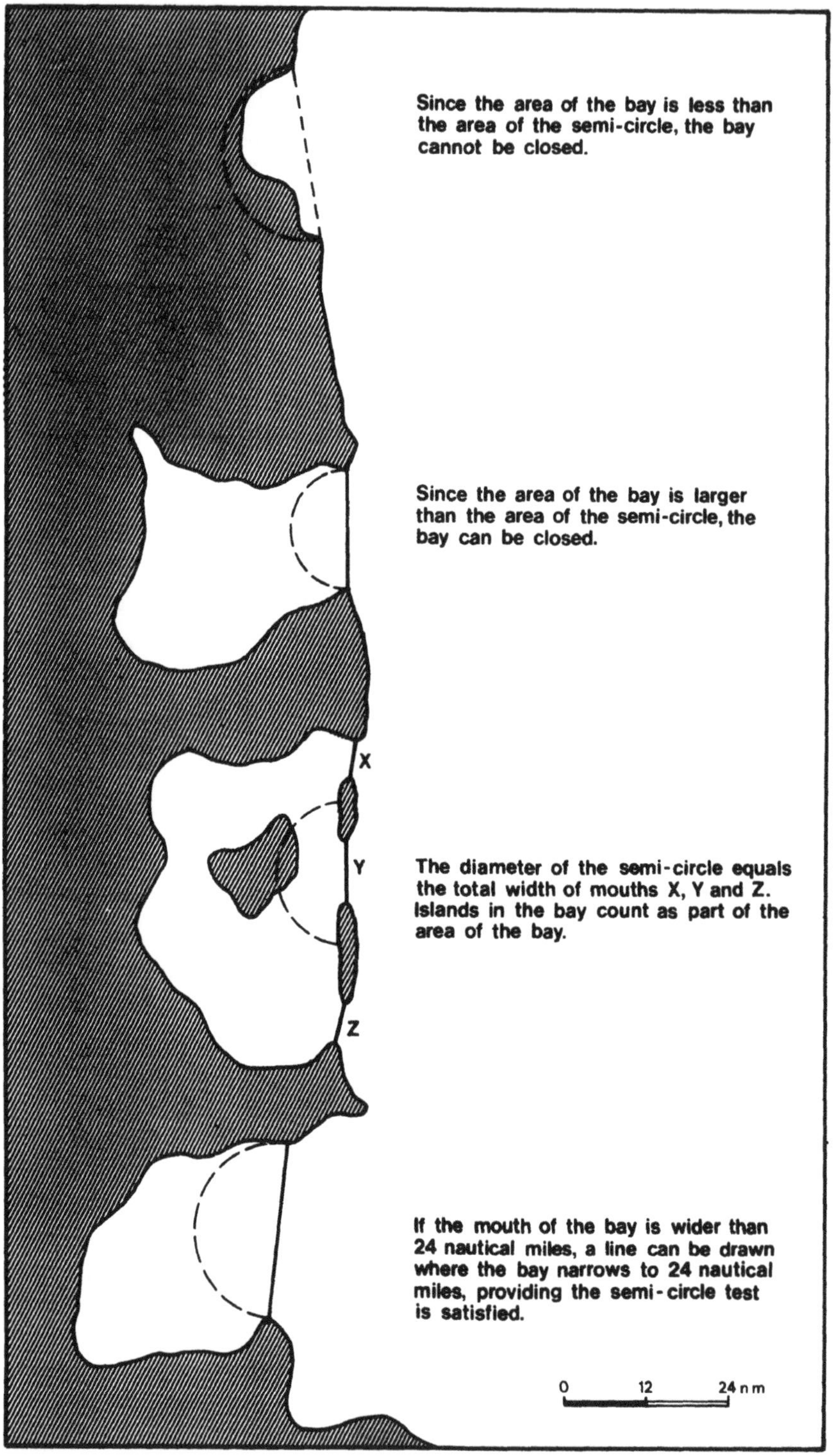

Figure 5.2 Claims to bays.

to so-called 'historic bays' (United Nations 1983, p. 5). It is generally considered that historic bays must satisfy three conditions. They must have been used exclusively for a long time by the claimant; a formal assertion of sovereignty must be made; and that claim must be accepted by other countries. Canada's claim to Hudson Bay as an historic bay is the best example of this type of claim.

States are entitled to include in their territorial seas roadsteads which would otherwise be situated wholly or partly outside the outer limit of the territorial sea. In this case there is no change in the normal baseline, but the outer limit of the territorial sea is deflected to include the zone where ships may anchor to load or discharge cargo. In such cases the width of the territorial sea might be wider than 12 nautical miles.

Before dealing with regional straight baselines it should be noted that territorial seas can also be claimed from low-tide elevations near the coast. A low-tide elevation stands above water at low tide and is submerged at high tide. Figure 5.3 illustrates the use of low-tide elevations as points on the baseline, and three points should be noted. First, low-tide elevations can only be used if they lie within the territorial sea generated from land in the normal way. The features marked B and C in Figure 5.3 can be used to claim additional territorial waters. Secondly, low-tide elevations cannot be used as stepping stones to claim territorial waters distant from the shore. Thus the feature marked D cannot be used as part of the baseline even though it lies within the territorial waters generated from the features labelled B and C. Thirdly, the feature marked A cannot be used to claim territorial waters because it lies more than the width of the territorial waters from land. The fact that it lies within the territorial waters created by the bay closing line is of no account.

Regional straight baselines

Although some countries, including England, had drawn very long closing lines across wide gulfs as early as the 14th century, it was not until 1951 that the concept of regional straight baselines became part of international law. In 1937 Norway announced that it would measure its territorial sea from a series of straight lines drawn around the outer edge of its fjord coast. This coast is fringed with a swarm of tiny islands known as the *skjaegaard*. This rule created problems for the captains of British trawlers who often operated very close to the Norwegian coast. Britain and Norway agreed to test the validity of the Norweigian claim in an action before the International Court of Justice, and Norway won the case when the Court ruled that the method of straight baselines was not contrary to international law (International Court of Justice 1951).

The language of the judgment found its way via an International Law Commission of 1956 into the 1958 Convention and was then transferred

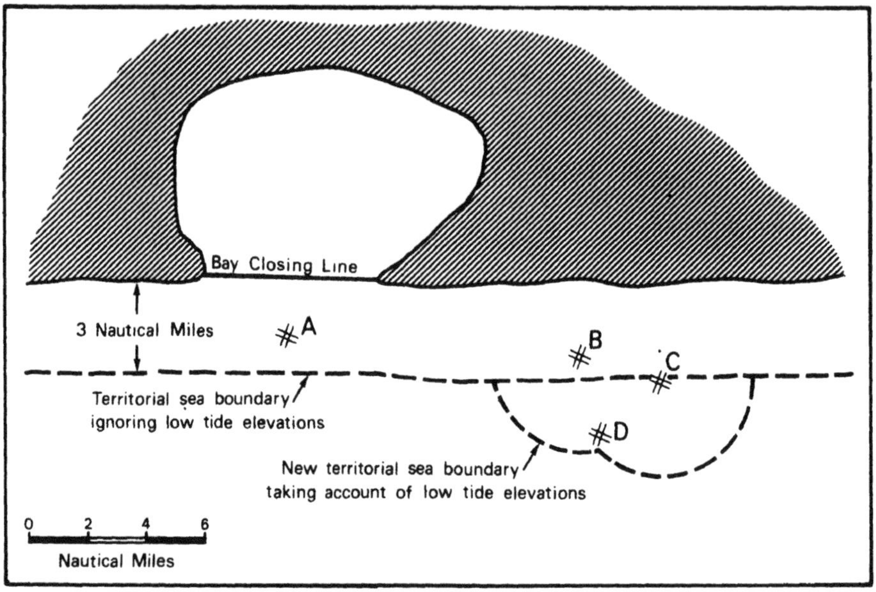

Figure 5.3 Claims from low-tide elevations.

without alteration into the 1982 Convention. But though the thread of ideas and language is continuous from the International Court of Justice in 1951 to the Convention of the Law of the Sea in 1982, it is evident that the interpretation of the ideas and language has undergone a remarkable transformation.

It is the hallmark of a straight baseline drawn according to the exacting standards set by Norway that it consists of many short segments of straight lines connecting segments of low-water mark along the shores of islands or headlands. When islands are involved there are many of them and they are close to the coast. Although such baselines produce a very large increase in the area of internal waters, they do not push the outer limit of the territorial sea away from the coast to any significant extent, and they have absolutely no effect on the outer limit of the exclusive economic zone. Countries such as Canada, Chile, Finland, and Yugoslavia have drawn straight baselines which are models of propriety and beyond criticism. In contrast, by using the straight baselines of Bangladesh, Burma, Ecuador, Italy, Spain, and Vietnam as precedents, it would be possible to justify straight baselines along any coast. There is no evidence that the extravagant claims which these baselines represent have provoked widespread international opposition. The United States has been the most vigorous defender of the original standards for straight baselines. The despatch of naval vessels to Libya's Khalij Surt is probably the most dramatic action in this endeavour. In terms of providing useful information to governments and scholars, The Geographer of the United States' State Department has made a very significant contribution through the factual series entitled *Limits in the seas.*

The 1982 Convention introduced two new rules which enable regional straight baselines to be drawn. The first concerned unstable coasts:

> Where because of the presence of a delta and other natural conditions the coastline is highly unstable, the appropriate points may be selected along the furthest seaward extent of the low-water line and, not withstanding subsequent regression of the low-water line, the straight baselines shall remain effective until changed by the coastal State in accordance with this Convention. (United Nations 1883, p. 4)

Deltas are formed when the sediment deposited at the mouth of a river fills that mouth and then continues the building process so that the newly formed land protrudes into the sea (Samojlov 1956). Bird (1976, p. 209) has classified deltas according to their shape, which is a function of the type of sediment, the rate at which it is supplied, and the rate at which it is removed by the sea. It is clear that the existence of a delta alone is not sufficient to justify a set of straight baselines. It is necessary that the edge of the delta is being eroded because this provision is designed to give relief to countries that are experiencing a retreat of their baseline.

The retreat of the delta could be caused by variations in a number of elements. For example, if a river is dammed, then the reservoirs and lakes will act as settling ponds and the supply of sediment to the downstream area will be reduced. If waves and currents are removing sediment from the shore of the delta faster than it is being delivered, the delta will retreat. Even if no dams are built but the river's discharge is being reduced by diversion to irrigation and industrial uses, the river will be unable to maintain former sediment loads and erosion might occur. Bird and Ongkosongo (1980) have described how the delta associated with the Rambatan River in Indonesia was cut back after the Anyar Canal was built for an irrigation project. The construction of harbour works might deflect waves and currents and cause them to attack a new section of coast and erode it. The removal of sand and gravel for the construction industries might lower the level of parts of the delta and make them vulnerable to invasion by the sea during exceptional tides or storms. Similarly, the removal of vegetation, especially mangroves in the Tropics, will often made the coasts of deltas easier to erode.

Even without human intervention, deltas will change their shape, and growth in one area might be matched by retreat elsewhere. The distributaries of deltas sometimes break their banks during floods and establish new courses. This might mean that the coast at the mouth of the abandoned channel is not nourished with sediment and it might be eroded. The great weight of sediment in some deltas causes subsidence; this characteristic has been noted in some of the older sections of the delta of the Mississippi River.

The phrase 'highly unstable' which is used in the 1982 Convention is not capable of any precise mathematical interpretation, but it has the flavour of rapid, short-term fluctuations. Orlova and Zenkovich (1974) have recorded retreat of the shore near the Damietta mouth of the River Nile at a rate of 40 metres per year. Nor is it clear over what period this instability should be mentioned. For example, is it only permissible to redress the instability which occurred after the 1982 Convention was signed, or can the instability of earlier decades be used to justify straight baselines?

Since the article dealing with unstable coasts begins with the phrase 'Where because of the presence of a delta and other natural conditions', it appears that a delta must be present for the subsequent provisions to be invoked. There are other unstable coasts in retreat which lack deltas. Volcanic action can sometimes create unstable coasts, and along the Siberian and Canadian coasts of the Arctic Ocean tundra cliffs are often subject to spectacular erosion during the onset of spring. It is possible in view of the way that other rules of the convention have been stretched that countries with unstable coasts which lack deltas will seek to gain the relief this rule affords.

Article 47 of the 1982 Convention sets out the rules for drawing archipelagic baselines. They may be constructed around archipelagic states, which are defined as states consisting wholly of one or more archipelagos. It was the plain intention that such baselines should not be drawn around the offshore archipelagos of continental states, but that particular horse has already bolted! Spain has drawn archipelagic baselines around the eastern Canary Islands, Italy has done the same for Sardinia, Ecuador has enclosed the Galapagos Islands, Denmark has surrounded the Faeroes by straight lines, and Australia has drawn straight baselines around the Houtman Abrolhos Group.

Archipelagic states are entitled to draw straight baselines around the outermost points of their outermost islands, providing five tests can be satisfied. The baselines must include the main islands; they must not depart to any appreciable extent from the general configuration of the coast; not more than 3 per cent of the lines can exceed 100 nautical miles in length; no line can be longer than 125 nautical miles in length; and the baselines must enclose an area of water which is at least equal in extent to the area of land but not more than nine times that land area. The first three tests are too imprecise to allow consistent application. The concept of 'main islands' could apply to the largest islands, the most populous islands, the islands of greatest cultural significance, or the most productive islands. Restricting the proportion of segments which can measure between 100 and 125 nautical miles means nothing when the state can determine the number of segments. Since states can draw their archipelagic baselines around the outermost points of the outermost islands and reefs, it is likely that different surveyors would draw different lines to preserve what they perceive to be the general configuration of the group.

The rule that no line can be longer than 125 nautical miles appears to be unambiguous, although there could be disagreement about the nature of reefs which can be used as turning points in the straight baseline system. The requirement that the ratio of water to land within the baselines must fall within the values of 1 : 1 and 9 : 1 is useful. It immediately excludes at the lower end of the scale states such as Australia, Japan, and the United Kingdom which cannot enclose an area of water equal to their territory. At the upper end of the scale, states such as Kiribati are excluded because the area of water which the baselines would enclose is more than nine times the area of the land. In fact Kiribati might be able to enclose a very small part of its territory formed by the islands of Abaiang, Tarawa, Maiana, Kuria, Aranuka, and Abemama. The only problem with this test is that the Convention slightly obscures the distinction between water and land! Waters within fringing reefs may be counted as land, and there is no clear definition of a fringing reef. Further, it is permissible to count as land water which overlies steep-sided oceanic plateaus which are enclosed or nearly enclosed by a peripheral chain of limestone islands and drying reefs. It is generally understood that this option was included for the express benefit of The Bahamas.

There are 12 archipelagic states which can surround their entire territory with straight baselines; they are Antigua, The Bahamas, Cape Verde, Comoros, Grenada, Indonesia, Jamaica, Maldives, Philippines, St Vincent and the Grenadines, São Tomé and Principé, and Vanuatu. If Vanuatu was successful in claiming Matthew and Hunter Islands from France, these islands coulds not be incorporated in the archipelagic baselines. Six states could draw archipelagic baselines about some of their islands; they are Fiji, Kiribati, Papua New Guinea, Solomon Islands, and Tonga.

The outer edge of the continental margin

The 1958 Convention on the Continental Shelf defined the outer limit of national claims as the 200 metre isobath or the depth of water where exploitation of the seabed's natural resources was possible (Prescott 1975, p. 145). This was an unsatisfactory arrangement because it would permit the steady advance of national claims seawards as technical skills and mechanical innovations were improved.

The 1982 Convention adopted a more rigorous approach to this question. Although this particular section of the Convention is headed 'continental shelf', the text makes it clear that it is the entire continental margin which is being treated. The continental margin consists of three elements in most cases. The continental shelf slopes gently seawards and is terminated by a steepening gradient which marks the top of the continental slope. Orlin (1975) has calculated that the average depth at which this junction occurs is 130 metres, although he has found extremes of 20 and 550 metres. The continental slope descends towards the continental rise at an average gradient of 4°17′. The average depth at which the slope intersects the rise is 2500 metres, although Orlin (1975) has recorded extremes of 1500 and 5000 metres. The continental rise consists of an apron of debris, derived mainly from the continental shelf and the continental slope, which masks the junction between the slope and the abyssal plain. The average gradient of the rise is 1°26′ at its landward edge and 0°4′ near the abyssal plain. The relationship of these various features is shown in Figure 5.1.

Unfortunately this average continental margin is only a statistical abstract; these three elements combine in a variety of forms around the world. In some cases the rise has been replaced by a deep trench, as off the eastern Philippines, and in others the rise is 400 nautical miles wide, as off Dakar in West Africa.

The 1982 Convention empowers states to claim the continental margin beyond its territorial sea throughout the natural prolongation of its land territory. The phrase 'natural prolongation' was introduced into the language of lawyers by a judgment of the International Court of Justice in 1969. West Germany had sought relief from the fact that claims from the Netherlands and Denmark had prevented it from claiming the seabed to the middle of the North Sea. The Court found for Germany on the ground that it was entitled to the seabed which was the natural prolongation of its land territory. This fine phrase was eventually incorporated into the 1982 Convention without anyone explaining what it meant or how it could be determined when two countries shared a continental margin. The Court has resisted any temptation it might have felt to remedy this defect in cases involving Britain and France and Libya and Tunisia (Brown 1983).

Fortunately the rest of the section dealing with the continental margin is less ambiguous. States can determine the outer limit of their continental margin, where it extends more than 200 nautical miles from the baselines, by one of two methods. Initially both methods require that the foot of the continental slope is identified and marked on a chart. The foot of the slope is defined as the maximum change in the gradient at the base of the slope. There could be problems in fixing this line, but the distances between the various alternatives are unlikely to be large.

Once the foot of the slope has been surveyed states face two choices. The first allows them to draw the limit of their national claim anywhere within 60 nautical miles of the foot of the slope. The second option is more difficult to apply. When the thickness of sedimentary rocks has been measured, the national claim may be drawn through those points where that thickness is at least 1 per cent of the shortest distance between the site and the foot of the slope. This means that if a country wanted to fix its boundary more than 60 nautical miles from the foot of the slope, it would have to discover sedimentary rocks with a thickness of 1111.2 metres. This figure is calculated by taking 1 per cent of 60 nautical miles.

To simplify the outline of the final boundary, states may construct the line out of straight segments measuring not more than 60 nautical miles. Countries may use whichever method gives them the largest claim on any section of coast, but the lines must not exceed one of two absolute distances. The first is 350 nautical miles seawards of the baseline from which the territorial sea is measured; the second is 100 nautical miles seawards of the 2500 metre isobath. There are also general provisions to ensure that states, such as Iceland, do not make enormous claims to the oceanic ridges which extend for thousands of miles across the seabed.

Once a state has selected the boundary of its continental margin it must be considered by an international commission established under the terms of the 1982 Convention. This commission will consist of 21 experts, who will operate in subcommittees of seven members. They will consider the proposed boundaries and make recommendations on the boundary's correctness. If the boundary is deemed to conflict with the convention's rules, the coastal state is required to submit a revised proposal in a reasonable time. This international commission is involved to ensure that there is no infringement of national claims into the Area.

It is likely that many countries will delay any submissions to this commission for as long as possible. So long as the authority of states to the outlying parts of their continental margin is not challenged, there is no reason to take any action other than the collection of data which can be used to refute any challenge in the future. Since the only possible challenge could come from the Authority, set up by the Convention to regulate mining on the deep seabed, the matter might be delayed for a very long time.

International maritime boundaries

Where two adjacent countries share the same coastline and where two countries are separated by a comparatively narrow belt of sea, some of their maritime claims will overlap. The 1982 Convention offers assistance to countries faced with the need to draw a common boundary at sea. Article 15 deals with boundaries between territorial seas:

> Where the coasts of two States are opposite or adjacent to each other, neither of the two States is entitled, failing agreement between them to

> the contrary, to extend its territorial sea beyond the median line every point of which is equidistant from the nearest points on the baselines from which the territorial seas of each of the two States is measured. The above provision does not apply, however, where it is necessary by reason of historic title or other special circumstances to delimit the territorial sea of the two States in a way which is at variance therewith. (United Nations 1983, pp. 5–6)

This is a long-winded way of saying 'States should agree on their common territorial sea boundaries'. The introduction of the concept of historic title and other special circumstances allows any country to introduce any argument it wishes in an effort to gain the lion's share of the overlapping waters.

There is no advice on how boundaries between contiguous zones should be determined, and Articles 74 and 83 dealing with boundaries between exclusive economic zones and continental margins respectively use the same language:

> The delimitation of the exclusive economic zone [continental shelf] between States with opposite or adjacent coasts shall be effected by agreement on the basis of international law, as referred to in Article 38 of the Statute of the International Court of Justice, in order to achieve an equitable solution. (United Nations 1983, pp. 26, 29–30).

The optimism which such language generates is quickly dissipated when Article 38 is read. It enjoins the International Court of Justice to reach decisions by applying international conventions expressly recognized by the contesting states; international custom; the general principles of law recognized by civilized nations; and judicial decisions and the teachings of the most highly qualified publicists. I am assured by international lawyers that there is nothing in these requirements which would prevent any country from introducing any matter which it considered to be relevant to its case.

Median lines

Despite the unlimited scope which these three Articles give to states to introduce special arguments in favour of some line other than a median line, most maritime boundary agreements concluded to the end of 1984 contained elements of equidistance. An examination of distinct boundary sectors shows that opposite states drew 44 sectors which had an element of equidistance whereas six sectors contained no such characteristic. When adjacent states were considered, it was found that 20 sectors possessed an element of equidistance and 9 were without it. The term 'sector' is used to refer to each discrete boundary. For example, the 1969 agreement between

Indonesia and Malaysia defined three continental shelf boundaries. They separated the claims of the two states in Malacca Strait, in the South China Sea north-east of Singapore, and in the same sea north of the coast of Borneo. The first two sectors were based on the principle of equidistance, but the third was not constructed on that basis. So this single agreement provides information about three boundary sectors. In contrast, the 1974 boundary drawn between the claims of India and Sri Lanka was extended in 1976; in this case the original boundary and its extension count only as one sector.

Because many maritime boundary agreements do not state how the boundary was drawn, it is necessary to resort to a series of tests to discover whether the principle of equidistance played any rôle in the boundary's construction.

The first step is to mark the boundary on a chart of convenient scale on which the baselines claimed by the two countries are also depicted. For long boundaries it may be necessary to use charts with a scale of 1 : 1 million or 1 : 2 million; for short boundaries it will be possible to use scales of 1 : 500 000. If the agreement is illustrated by an official map, it will often be convenient to use it.

Once this chart has been prepared or acquired, it is then necessary to test each turning point on the boundary to discover the relationship it bears to the nearest points on each coast. In using charts drawn on the Mercator projection, it is important to remember that the scale will vary and will increase towards the poles. The variations in scale will be least on large-scale charts near the Equator and greatest on small-scale charts near the poles. If, after making the measurements with due care, it is found that the points are equidistant, then the boundary is established as a median line.

However, even if the points are found to lie away from the median line, it would be wrong to conclude immediately that the principle of equidistance played no part without considering the following possibilities.

First, it is possible that the boundary was constructed on very large scale charts which had all the important baseline points marked in the positions revealed by the latest surveys and verified by each country. Such charts would usually be kept secret and would certainly be more accurate than the charts generally available for the public. So it is possible that the charts on which the measurements are made do not have all the correct and essential information marked, or, where such information is shown, it might be in the wrong position.

It is fairly safe to assume that if turning points and termini are defined by co-ordinates which include degrees, minutes, and seconds that large-scale accurate charts have been used, since 1 second represents about 31 metres. To produce such precise locations detailed surveys must be available.

Secondly, it is possible that the boundary appears to follow a course other than the median line in some sections because the two countries have agreed to disregard some points on the baseline. Conversely, the two countries may have agreed to use one or more features which would not normally be considered part of the baseline. For example, in negotiating a boundary through the Coral Sea between Australia and France's New Caledonia, France permitted Australia to use Middleton Reef as a basepoint. Middleton Reef is a low-tide elevation 125 nautical miles from the nearest Australian territory.

Countries may also use secret baselines. Although Malaysia has never proclaimed any straight baselines, it is evident from Malaysian maps showing the limits of territorial waters that they have been derived from straight baselines. Such baselines may have been drawn to counter Indonesia's archipelagic baselines.

When some median lines are constructed they follow an irregular course with short segments and frequent changes in direction. Such lines make inconvenient lines for the departments charged with surveillance. For this reason some of the points on the median line might be omitted to create a smoother boundary which navigators can identify more easily than the median line. Such a line can be drawn so that each party gains and loses equal areas of sea and seabed when the smoothed line is compared with the median line.

In some cases countries have agreed to discount some feature which gives a considerable advantage to one of the countries. For example, the boundary drawn by Iran and Saudi Arabia in 1968 gave only half-effect to Iran's Kharg Island. This means that median lines were drawn giving full effect to Kharg Island and then ignoring the island entirely. The area bounded by the two resulting lines was then divided between the two states. In its judgment on the Anglo-French boundary in the English Channel, a Court of Arbitration determined the final segment of the boundary by giving half-effect to Britain's Scilly Isles. Although the discounting of basepoints appears to be by half in all known cases, there is no reason why the discount should not be one-third or 12 per cent.

In a few rare cases some of the points on a boundary between two states are equidistant between the baseline of one state and a point on the baseline of a third country. This situation occurs in the boundary agreements reached by Britain and West Germany, the United States and Venezuela, and the Dominican Republic and Venezuela. In the first case the boundary is equidistant between the United Kingdom and Denmark and the Netherlands; in the second and third cases Venezuela has been allowed to take advantage of basepoints located in the Netherlands Antilles.

There were two striking features about the maritime boundary agreements which had been reached by the end of 1984. First, 73 per cent of the agreements dealt with the continental margin. This suggests that there has been a preoccupation amongst many states with securing control over distinct areas of seabed, presumably in the hope that these areas might contain fields of petroleum or natural gas. It is also true that continental shelf boundaries are easier to draw because their surveillance is a comparatively simple matter compared with the surveillance of boundaries limiting territorial waters.

The second obvious feature about those agreements is that 70 per cent are located in enclosed or semi-enclosed seas, including the North Sea, the Baltic Sea, the Mediterranean Sea, the Black Sea, the Caribbean Sea, and the Persian Gulf. Another 6 per cent of the agreements are located in the south-west Pacific Ocean where there are swarms of islands divided amongst a number of countries and colonial territories.

The negotiation of maritime boundaries can follow the same range of paths as the international land boundaries set out in Chapter 3. In common with land boundaries, each maritime limit occupies a unique position and resulted from bilateral negotiations which took into account a distinct set of circumstances. These circumstances, which might occur in a variety of combinations, are mainly related to political, geographical and economic issues.

Factors which complicate boundary negotiations

Dealing first with the political circumstances which might complicate boundary negotiations, it is clear that an absense of formal relations between countries or serious tensions where formal relations exist will make it hard to negotiate a maritime boundary. It seems entirely improbable that Jordan and Israel could discuss maritime boundaries, just as it seems certain that any agreement on a maritime boundary between Iraq and Iran will only occur after the war in the Persian Gulf has ended.

Negotiations can be made more difficult if one of the parties is wedded to a particular historic viewpoint which the other cannot accept. The resolution of the Beagle Channel dispute was delayed by the traditional Argentinian view that Chile could never be an Atlantic power. Vietnam's conviction that the Sino-French boundary agreement of 1887 divided the waters of the Gulf of Tongking will have to be set aside before there is any chance of China agreeing to a maritime boundary in these seas.

Difficulties can arise if part of the boundary is not determined and one of the governments changes its attitude to the question. The continental margin boundaries drawn by Australia and Indonesia in 1971 and 1972 terminated before the western section had been divided. The Indonesian attitude to this question has hardened and now argues that Australian islands should be discounted and a median line drawn between Indonesia and the Australian mainland. In the earlier agreements all islands had been given at least some effect, including Ashmore Reef, which is Australia's most northerly outpost in the Timor Sea.

Islands seem to provide the geographical circumstances which create most difficulties for boundary negotiations, and there are two aspects to consider. First, there is the problem of deciding whether a feature which stands above the high-water mark is a rock or an island. Rocks can only be used to claim territorial seas if they cannot sustain human habitation or an economic life of their own. Any island, however small and barren, can be used to claim all maritime zones. The Convention apparently believes that the criteria set out in the context of human habitation and economic life will enable the distinction to be made between rocks which can only be used to claim territorial seas and rocks and islands from which all types of maritime zones can be claimed. In practice the matter is very difficult to determine because of uncertainty about the terms 'human habitation' and 'economic life' (Hodgson 1973).

The location of islands provides the second main difficulty. When the islands of one country are located close to the shore of a neighbour the islands restrict the claim of the neighbour. This situation was the root cause of the dispute over boundaries around the British Channel Islands. France argued successfully before a Court of Arbitration that Britain's ownership of the Channel Islands should not prevent France from claiming some of the continental shelf near the middle of the English Channel. The poor relations between Greece and Turkey in the Aegean Sea relate most directly to the close proximity of Greek islands to the Turkish mainland. If these islands were given full effect in drawing a median line between the two countries to allocate the seabed, Turkey would receive only a very narrow strip of the continental margin adjacent to that mainland.

The character of the seabed shared by two states can complicate boundary negotiations. This is particularly the case when a trench or deep submarine valley divides the shelf into two unequal parts. In such cases the country which owns the shore farthest from the declivity will often argue that its axis marks the limit of the natural prolongation of both countries. Such a situation has been responsible for disputes between Britain and Norway, Australia and Indonesia, Libya and Malta, and the Unites States of America and Canada.

There are two economic circumstances which might complicate boundary negotiations. First, if there is a marked disparity in the wealth and resources of the two parties, the poorer state might argue that it should be favourably treated so that the disparity might be reduced if the seas and seabed turn out to be productive. This is an argument which Papua New Guinea exploited with skill during its negotiations with Australia over the boundary through Torres Strait.

Secondly, if it is known that the overlapping area contains resources of petroleum and natural gas, the parties may be reluctant to compromise until they have exhausted every effort to secure the entire resource.

Conclusions

There are three sets of boundaries which coastal states might have to draw. The first is the baseline from which the territorial sea, the contiguous zone, and the exclusive economic zone are measured. Rules governing the construction of straight lines which deviate from the low-water mark are included in the 1982 Convention, but they are generally ambiguous. In any case, so many baselines have been drawn and maintained in defiance of these rules that their force has seriously eroded.

The second boundary limits the outer edge of the continental margin and must be drawn by those states which possess margins wider than 200 nautical miles. The rules in the 1982 Convention are an improvement on those contained in the 1958 Convention, but they are still untried. Problems of strict interpretation will certainly arise in the future, but it is unlikely that countries will rush to test these new rules before the commission created to approve claims to wide margins.

The third set of boundaries must be drawn by states which have claims overlapping with those of neighbours. There are no effective rules for drawing such lines and they will be the product of hard bargaining between states similar to the negotiations which settle trade and defence arrangements.

References

Balch, T. W. 1912. Is Hudson Bay a closed or an open sea? *American Journal of International Law* **6**, 402–15.

Baty, T. 1928. The three-mile limit. *American Journal of International Law* **22**, 503–37.

Bird, E. C. F. 1976. *Coasts*, 2nd edn. Canberra: Australian National University Press.

Bird, E. C. F. & O. S. R. Ongkosongo 1980. *Environmental changes on the coasts of Indonesia*. Tokyo: United Nations University.

Brown, E. D. 1983. The Tunisian–Libyan continental shelf case. *Marine Policy* **7**, 142–62.

Davis, S. 1984. *Aboriginal tenure and the use of the coast and sea in northern Arnhem Land*. Unpublished MA thesis, University of Melbourne.

Fenn, P. T. 1926. Origins of the theory of territorial waters. *American Journal of International Law* **20**, 465–82.

Hodgson, R. 1973. *Islands: normal and special circumstances*. Washington, DC: US Department of State.

International Court of Justice 1951. *Reports of judgements, advisory opinions and orders*. The Hague: International Court of Justice.

Orlin, J. 1975. Offshire boundaries: engineering and economic aspects. *Ocean Development and International Law* **3**, 73–88.
Orlova, C. & V. P. Zenkovich 1974. Erosion of the shores of the Nile Delta. *Geoforum* **18**, 68–72.

Prescott, J. R. V. 1975. *The political geography of the oceans.* Plymouth: David & Charles.

Samojlov, I. V. 1956. *Die Flussmündungen* [River mouths]. Gotha: Haack.

United Nations 1983. *The law of the sea: United Nations Convention on the Law of the Sea.* New York: United Nations Secretariat.

Walker, W. L. 1945. Territorial waters: the cannon shot rule. *British Yearbook of International Law* **22**, 210–31.

6 *Border landscapes*

In the first chapter, it was stated that one of the principal interests in boundaries of any political geographer relates to the way in which a boundary or frontier influences both the landscape of which it is a part and the development of the policies of the states on either side. This view resulted from acceptance of the dictum, repeated by nearly all the authors reviewed, that it was meaningless to consider the boundary outside the context of the flanking state areas. Lapradelle termed this zone *le voisinage*, and 'border landscape' is suggested as an equivalent term. Political geographers are interested in boundaries because they mark the limits of political organization, which varies over the Earth's surface. Variations in political systems are often accompanied by variations in regulations concerning economic activity and the movement of people, goods, and ideas. The results of these variations are likely to be most clearly seen in the neighbourhood of the boundary, whether state functions are rigidly applied at the boundary or whether the states combine to minimize the adverse effects of the boundary upon the border inhabitants. Few workers have selected this subject as the focus of their study, but many have included important references to it. Minghi (1963) did not suggest a separate group of such studies in his proposed classification of case studies but included important papers by House (1959), Ullman (1939), and Nelson (1952) in categories dealing with the effects of boundary change and the characteristics of internal boundaries.

House (1982) has identified the need for much more detailed research into the nature of borders:

> There is an urgent need both for empirical and comparative studies of a dynamic nature for frontier [border] situations, whether these involve confrontational or co-operative relationships, and for a more coherent set of theoretical frames within which to study such situations. (House 1982, p. 264)

In an effort to answer his own call, House produced an important study of the borderland between Mexico and the United States along the Rio Grande. After describing the evolution of the boundary and the characteristics of the people living on both sides, he analysed the rich variety of transactions within the borderland, which included water management, migration, smuggling, and tourism.

He usefully stressed the set of spatial relationships which can be profitably considered. They are portrayed in Figure 6.1, which is adapted from his work (House 1982, p. 10). This figure shows the relationships between administrations, commercial enterprises, and individuals in the border, in provincial centres, and in the countries' capitals.

At the conclusion of his study House drew attention to some of the problems which face geographers undertaking research in borderlands. He emphasized the difficulties of differentiating the transactions, perceptions, and contributions of actors at the provincial and capital levels, and he also warned of the problems associated with data bases for the neighbouring countries which were not strictly comparable and which would not allow multivariate analysis.

This seminal study by House also answers a much older call by Minghi (1963) for more attention to the normal situation of boundaries and borderlands. There is a temptation for scholars to concentrate on dramatic events associated with boundaries, but they are dramatic because they are exceptional, and for every border incident which is reported in newspapers and broadcasts there are thousands of routine transactions occurring within the world's borderlands.

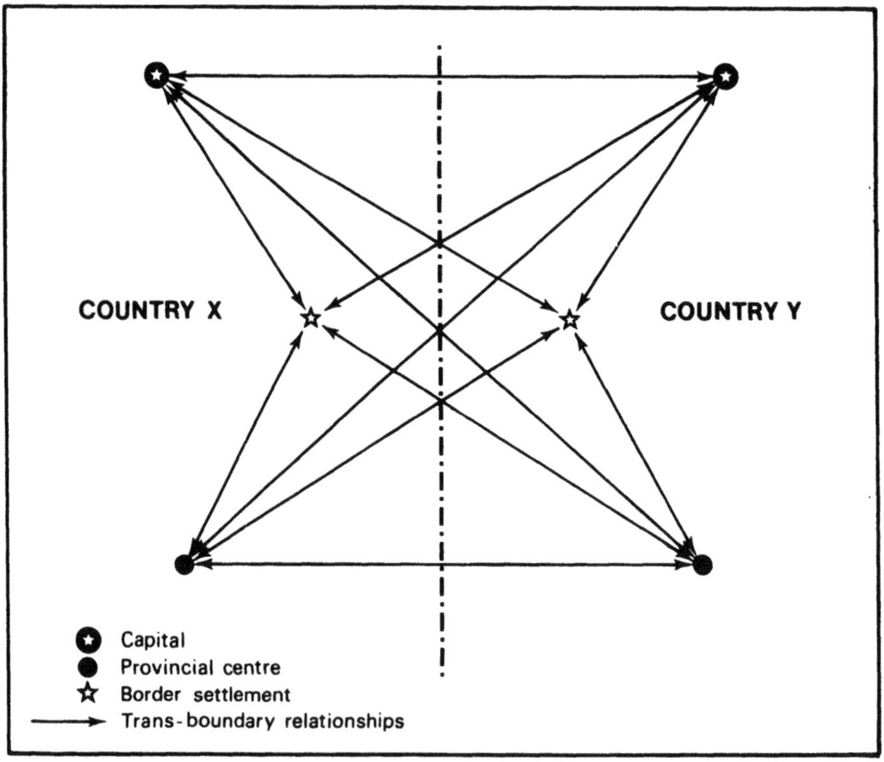

Figure 6.1 Border relationships (after House 1982, p. 10).

Aspects of border landscapes

Research by geographers into border landscapes covers four broad topics. First, there is the consideration of the boundary as an element of the cultural landscape. A boundary's physical existence results from the demarcation of the boundary and the construction of buildings, defences, and systems of communications to give effect to state functions applied at it. The physical difference between an internal and international boundary is usually outstanding, but there are also important differences between different international boundaries. The suggestion has already been made that the appearance of any boundary in the landscape is a guide to the functions applied there, and the stringency with which they are applied.

Secondly, a geographer may wish to examine the extent to which variations in landscape and land use on either side of a boundary can be explained by the proximity of two different political systems and the regulations which they have developed. In this connection it is important to distinguish such cases from those where variations in landscape and land use result from the coincidence of the boundary with some linear physical feature such as a watershed which is also a climatic divide. Population distribution is one phenomenon related to land use which may be partially explained by the nearness of the boundary. There are also cases where the existence of a boundary results in the duplication of transport, administrative, and retail services.

In addition to variations in the physical appearance of the landscape, geographers are also concerned to understand the variations in the population and economic structures of the borderland, which can only be uncovered by analysis of statistics. For example, House (1982, pp. 81–2) shows how the Rio Grande marks a clear division between the high level of agricultural employment in Mexico and of service industries in the United States.

The remaining two aspects are not directly concerned with the cultural landscape, but they may be conveniently considered here, since they may be the medium through which the boundary influences the cultural landscape. Thirdly, there is the influence of the boundary's presence and operation upon the attitudes of the border inhabitants. Fourthly, there is the influence which the boundary has upon the policies of the state, and in this connection it may often be difficult to separate the effects of the nature of the boundary from those of the nature of the state beyond the boundary. Each of these aspects is considered in greater detail.

Boundaries as features of the landscape

A boundary is usually demarcated only when the separated authorities believe this to be necessary. Some of the most important internal boundaries, which decide where persons may vote, at what level rates will be levied, and the schools children must attend, are never marked on the ground but are shown on maps hanging in municipal and local government offices. In some cases, for reasons of pride local authorities in major cities will indicate when travellers along main routes are entering or leaving them. On the other hand, nearly all international boundaries are marked on the ground in some way simply because most states feel it is desirable that their limits should be understood by their neighbours.

The boundary may be identified in the landscape by two sets of features. First, there is the indication of the boundary by means of markers, cut lines, fences, and notices. Secondly, there are various constructions designed to allow the smooth application of state functions at or near the boundary. Many travellers will be familiar with customs posts located near barriers across main roads and stations built on the boundary to allow passengers to be subjected to customs and immigration regulations. It should also be noted that international airports and seaports and coastal defences are types of border landscape.

Since nearly all international boundaries are demarcated in some way, it is possible to draw certain inferences from the nature of the demarcation about the nature of state functions and the relationship between the separated states. The following paragraphs indicate the type of conclusion which may be reached in the three most common situations, although the reader is reminded that each boundary tends to be distinct.

If an international boundary is not demarcated, there are three probable explanations. First, the states concerned may not feel that demarcation is necessary or of high financial priority. Many of the international colonial boundaries were not demarcated because of the expense involved and the improved relations between the colonial powers. Secondly, boundaries may not be demarcated because the exact position of the boundary is disputed due to some ambiguity in the definition of the boundary. This cause retarded the demarcation of many South American boundaries and currently continue to prevent any demarcation of the boundary between Somalia and Ethiopia. When a boundary is disputed, however, it will often be found that military and police installations are located in the border area in order to preserve rival claims. One obvious reason for not demarcating an international boundary may be found in the unfavourable nature of some environments. Where international boundaries lie within hot deserts or high mountain ranges, demarcation is often regarded as unnecessary on the grounds of security and impossible on the grounds of finance. However, such views may change as mineral exploration raises the possibility of discovering worthwhile deposits in deserts, and improved military techniques and changed military power reduce the defensive value of mountain ranges. It seems likely that India will insist on boundary demarcation when its dispute with China is settled. Where a boundary is drawn within an unfavourable environment, it will often be found that both the states establish their police and customs posts on the edge of the area. This means that the intervening area is not under direct and continuous control.

If the boundary is demarcated, two general situations can be distinguished. The first occurs when the demarcation is maintained and the second when it is neglected.

Looking first at the case where the demarcation is maintained, two extreme conditions can be described which will define a spectrum of variations and combinations.

At one end of the scale there are the boundaries between allies, such as Canada and the United States of America. Here the boundary vistas are carefully cut and the boundary monuments kept in good repair, even on the more remote western borders. This is largely for reasons of administrative convenience, not to restrict circulation. Structures to allow the application of state functions are located at the important recognized crossing points. Along such boundaries there is often an absence of permanent fortifications. At the opposite end of the scale there are those boundaries between unfriendly states where the boundary demarcation is maintained in order to prevent circulation and to simplify defence. The boundary in such cases is often marked by an obstacle such as a fence or a wall, and guard posts are located at regular intervals along the entire length. A strip of land adjacent to the boundary may be cleared to make observation easier and illegal boundary crossing more difficult. The civilian population is often evacuated from such a border. The crossing points on these boundaries are few and heavily guarded. These features may all be seen on some sections of the boundary between East and West Germany especially in Berlin, and the land boundaries of Israel with its Arab neighbours.

Between these two extremes may be found a wide variety of boundary forms, and there may be significant variations along any boundary. Boundary form is also likely to vary with changing circumstances. For example, Morocco built defensive walls along part of its boundary with Algeria and part of the boundary drawn in 1900 between Mauritania and Spanish Sahara. These walls are designed to make it more difficult for forces belonging to the Polisario Front from invading the territory of former Spanish Sahara now occupied by Morocco. The Algerian section of boundary had not been demarcated, but the Mauritanian section had been marked by a few pillars.

When a boundary has been demarcated and that demarcation has not been maintained, the political geographer may find this a useful pointer to changed political situations. In some cases colonial powers at the height of their competition carefully marked their boundaries, but once the period of competition gave way to a period of development, with its attendant concentration on internal affairs, the border areas were often neglected either for reasons of priority or to avoid incidents. Boundary vistas became overgrown and people, animals, and vegetation destroyed boundary pillars. The reestablishment of these boundaries by the independent successors to the colonial states often causes friction and gives rise to territorial or positional boundary disputes.

Contrasting landscapes on opposite sides of the boundary

Both international and internal boundaries may mark changes in the landscape and the economic activity which results from the separate areas being subject to different regulations. It follows, however, that the greatest variations occur between different countries and they will usually be more striking than the differences associated with internal boundaries.

Platt & Bucking-Spitta (1958) published an interesting descriptive account of the Dutch–German boundary, which passes through areas of agriculture, industry, and mining. In examining agricultural landscapes and production in three different areas, Platt and Bucking-Spitta found no significant differences on opposite sides of the boundary. The three areas were the polder country near the North Sea, the moor-edge settlements of Bellingwolde and Wymeer, and the flood plain and diluvial terraces of the River Rhine. Not only were the areas used by farms of similar size, divided into similar field patterns producing approximately the same proportions of various crops, but the building styles were also similar on both sides. The textile industry north of the Rhine, including Enschede and Gronau, was originally established as an 'international industry' and its operation was unaffected before World War I by the presence of the boundary. Since then the two areas have tended to become more national in character, but there is still some movement of workers across the boundary. In 1953, when the study was made, conditions were more satisfactory in Holland and there was accordingly a greater volume of movement of Germans into Holland than Dutch citizens into Germany. In the small coalfield around Kerkrade, Platt and Bucking-Spitta found no variations in distribution, method of production, and output which could not be explained by the differing physical circumstances of the coal reserves on each side.

Platt and Bucking-Spitta then examined the political and economic organization of the border areas and again found that there were similar forms on both sides of the boundary. Their conclusion was interesting, and it seems to provide an important principle:

> The preceding chapters have dealt with the characteristics of the border areas. Landscape and occupance have been found similar on opposite sides. The boundary is not natural and does not separate different uniform regions. It could be pushed east or west without changing the appearance of things in general on opposite sides . . . although the forms of areal organization may be similar on opposite sides of the boundary the organizations themselves, the units of organization, political, economic and social, as they have developed through years of human activity, are generally separate. A shift of the boundary in either direction disturbs the organization in units small and large on both sides, and generally does damage which can only slowly if ever be repaired. (Platt & Bucking-Spitta 1958, p. 85)

Platt and Bucking-Spitta's conclusions indicate quite clearly that although the boundary separates areas with almost identical landscapes and systems of organization there remain vital intangible differences of political attitude and social custom. These are aspects which a geographer finds difficult to measure and about which only general ideas could be formulated, even after a long residence in the borderlands.

This descriptive study was principally concerned with the present, unlike that made by Daveau (1959), which considered the Franco-Swiss borderland in the Jura over several centuries. This admirable study, which has not attracted the attention it deserves, shows clearly how the influence of the boundary on the borderlands will vary as political and geographical circumstances change. It would be surprising if this was not the case, but it is valuable to have such a clear demonstration. Daveau shows how the presence of the boundary has influenced the development of agriculture, forestry, and industry since it was established.

The area considered lies west of Lake Neuchâtel, and the principal agricultural activity was the raising of stock and the preparation of dairy products, especially cheese. During the early 18th century there was a considerable measure of interpenetration of pasture lands, and Daveau calculated that Swiss farmers owned 400 hectares of French land whereas French farmers owned 1000 hectares in Switzerland. The areas of greatest French colonization were near les Verrieres and la Brevine, which carried a low density of Swiss. After the Reformation, many Swiss and French farmers sold their land across the boundary because of the rise of national consciousness, the complex regulations covering alien landownership, and the problem of maintaining the property when the countries were at war. During the 19th century further changes took place: there was a withdrawal of French population from the upper pastures of the French border and a concentration in the valleys. This coincided with a considerable increase in the Swiss herds, many of which were accommodated on leased French pastures. At the same time there was a greater concentration of beef and veal production at the expense of dairy products, which were now manufactured mainly in the valleys on both sides of the border.

The extension of Swiss control over French border areas was increased by the fall of the French franc compared with Swiss currency during World War I. However, the threats of further falls in the value of the French franc encouraged Swiss farmers to lease land rather than buy it outright. They were in the fortunate position of being able to fatten the cattle cheaply in France and sell the meat at the higher Swiss rates in Switzerland. The French villages in the valleys do not benefit from the Swiss occupation of the uplands since the herdsmen generally bring all their provisions from Switzerland. The Conventions of 1882 and 1938 allow free circulation of agricultural products within a zone 10 kilometres wide on each side of the boundary. The actual wording of the relevant sections allows a number of interpretations, and variations are found between the arrangements in the Department of Doubs and the Canton of Vaud. Some landscape differences are noted where the boundary crosses a valley. One air photograph clearly shows the contrast at the boundary between the small strip fields of Amont in France and the summer pastures of Carroz in Switzerland, even though the physical character of the landscape on both sides is the same.

> In many forests of the mountains, the boundary between France and Switzerland is visibly marked by the misuse of our trees, and the traveller who remembers this fact will be able, without guide and without difficulty, to recognize the boundary in any place, and touch with certainty a Swiss fir tree with one hand and French fir tree with the other. (Le Quinio, quoted in Daveau 1959, p. 386)

This quotation shows that even in the 18th century landscape differences could be noticed in respect of the exploitation of forests. Daveau shows that this situation exists today, although for different reasons. At first the forest was regarded as a useful buffer zone between the Swiss and French, and restrictions were imposed upon the cutting of wood. In the period before 1750 there were many cases of wood in French territory being illegally cut by Swiss citizens. In the post-1750 period the situation was reversed: French depredations in Swiss forests were the general rule and there were minor wars as the Swiss attempted to protect their territory. The conflicts eventually stopped as the areas became more densely populated and customs patrols became more active on each side. In 1882, under a boundary Convention, customs officers were given authority to pursue illegal woodcutters across the boundary, and the duty-free transport of 15 000 tons of firewood was allowed within a zone 10 kilometres wide on each side of the boundary. In 1938 this figure was reduced to about 9000 tons.

Once again, as in the case of pasturage, the decline of the French franc in relation to the Swiss franc allowed Swiss citizens to gain an advantage. More and more forest in the French border has been bought by Swiss, who have then realized quick profits by wholesale cutting and clearing without any plans for reafforestation. The denuded lands are converted to pasture, which is then sold or leased to Swiss cattle farmers. This practice produces striking landscape differences on opposite sides of the boundary. One air photograph included by Daveau shows how the forests on the French side of the boundary have been almost completely cleared whereas those on the Swiss side remain in a well kept condition (Daveau 1959, Plate 11).

The analysis of farming and forestry in the borderland led Daveau to conclude that the economic boundary lies in Switzerland's favour to the west of the political boundary. The reciprocity written into the border Conventions is meaningless as long as the French franc stands at an unfavourable rate to the Swiss franc. This is a clear warning to geographers not to accept written guarantees in boundary treaties at face value; their application must be tested.

Daveau also examined the influence exerted by the boundary upon the watchmaking industry of the Franco-Swiss Jura. The industry began in Switzerland and the first French factory was at Besançon. Before 1834 the French government tried to protect the French industry from Swiss competition by high tariffs and the suppression of smuggling, which was a profitable occupation for French and Swiss citizens alike. In 1826 it was decreed that all French watchmakers must have their premises at least 7 kilometres from the boundary. The duty on Swiss watches was reduced after 1834 to 4 per cent *ad valorem* in 1836. This low rate did not make smuggling worth while and the activity declined rapidly, but by 1842 the smugglers realized that it was worth while to take French watches to Switzerland and then reimport them as Swiss watches. This flourishing trade continued for some time.

In the last years of the 19th century a tariff war between France and Switzerland resulted in a distinct rupture between the two industries. This helped certain sections of both industries. The Swiss apparently captured more of the market in small watches. In France certain workers who had specialized in making escapements could no longer export their product to Switzerland and they turned to the manufacture of complete watches. Further, France began to deal directly with overseas markets which had previously been supplied by Swiss merchants who bought from French producers. We can see then that the regulations laid down by the governments affected the location of the industry on the French side and the traffic in watches and caused variations in the type of production.

Kibulya (1967) studied the border landscape along the Zaire–Uganda boundary between the Ruwenzori Mountains and Lake Albert. He described sharp contrasts between the peripheral areas of the two countries which had their origin in colonial policies and which have been maintained because the post-colonial governments have not significantly changed those colonial policies. The Ugandan side was characterized by a dispersed settlement pattern serviced by many adequate roads. There was a high population density resulting from a declining death rate consequent on improving medical services. Most of the land was farmed and coffee was the main cash crop. The Congolese borderland was characterized by a nucleated settlement pattern poorly serviced by badly made roads. There was a comparatively low density of population resulting from a high rate of infant mortality offsetting the high birth rate. The Congolese farmers produced only subsistence crops from a small proportion of the area available. These differences are very interesting in view of the fact that the colonial boundary divided the Amba tribe when it was drawn through a uniform landscape. Kibulya found that Nyahuka in Uganda was the main trading centre for the whole borderland and that, although Ugandan currency circulated freely in the Congolese border, the Congolese franc had no currency in the Ugandan areas.

Reitsma (1971, 1972) has shown how the agricultural policies of the United States and Canadian governments have encouraged the growth of distinct agricultural regions on opposite sides of the boundary as it passes through the plains west of the Great Lakes.

These studies by Platt and Bucking-Spitta, Daveau, Kibulya, and Reitsma have shown that international boundaries may lie through identical cultural landscapes or mark significant changes in land use and economic activity. However, they would all agree that, even though the border landscape may be uniform, the two sides have a human distinctiveness which is difficult to measure but which is nevertheless real to people living in the borderland. Daveau and Kibulya have shown how important it is for scholars to be aware of currency variations in explaining changes in the significance of the boundary over a long period. This view is confirmed by the experience of Sevrin (1949), who studied transboundary population movements on the Franco-Belgian borderland. In 1929 there were 10 219 Belgians working in the French borderland, and this figure declined to 2757 in 1946 as a result of the war and the attendant decline in the value of the franc. By 1947 the number had increased to 8810, as there was an upsurge in textile production and as measures were implemented to maintain the exchange value of the French franc.

Ullman (1939), Nelson (1952), Rose (1955), Logan (1968), and Best (1971) have all written studies on the significance of internal boundaries as factors influencing the development of landscapes within a country.

Ullman's study of the eastern Rhode Island – Massachusetts boundary revealed that the boundary, which was accordant in situation but discordant in site, did influence the establishment of industries south of the Fall River. The industries were located in Rhode Island to gain tax concessions, but the tenements for the workers were located in Massachusetts so that the workers could continue to enjoy the superior social and cultural amenities offered by that state. It was noticed that the boundary became zonal in some respects because water, gas, and electricity services were common to sections of both states and because some private properties spanned the boundary.

Nelson (1952) examined the boundaries of the Vernon area of California to assess their contribution to an understanding 'of the areal distribution and functioning association' of various elements of the urban landscape. He plotted the distribution of residential, commercial, public, industrial, and transport land use and found that the boundary of Vernon coincided with significant landscape differences. There was a remarkable concentration of land devoted to industry and transport within Vernon almost to the complete exclusion of the other three categories. It also emerged that other boundaries in the Vernon area did not coincide with similar distinctions in land use. This suggests that Nelson was fortunate in his selection of Vernon as a case study and that the technique he used will be valuable only in a few instances. Although it may be true that a simple land-use analysis will rarely show significant correlations with political boundaries, if Nelson's technique is carried a stage further and quality of land use is studied, it should have much wider application. For example, instead of distinguishing residential and other types of land use, it will be necessary to examine the residential category in greater detail, collecting information about house types and rateable values. In a cosmopolitan city it may be of significance to record the nationality of house occupants or owners.

A further technique useful in such studies has been suggested by Mackay (1958), who has applied Dodd's interactance hypothesis to boundaries. By comparing the value of actual and computed interactance between cities or areas separated by a political boundary, Mackay suggests that it will be possible to obtain a measure of boundary interference. This would be useful in assessing the significance of individual boundaries and in comparing the significance of two or more boundaries. These studies would be limited to areas for which detailed statistics are available, and Mackay warns that factors other than the boundary may produce differences between computed and actual interactance. It would also seem worth while to investigate the extent to which the model prepared by Dodd is applicable to boundary studies in areas dissimilar to North America in political, economic, and social development.

Minghi (1963) has correctly drawn attention to the comparative neglect of Losch's work (1967) on the distortion which boundaries might produce in economic patterns. In view of the laudable desire of many geographers to introduce a greater measure of mathematical precision into political geography, it is surprising that this subject has not attracted more attention.

The studies reviewed in this section suggest that internal boundaries do influence the development of the cultural landscapes in many ways and that the analysis of these relationships is a worth while, though neglected, aspect of political geography, which could provide valuable information for those interested in economic and urban geography. The influence of the internal boundaries is usually less spectacular than that of international boundaries and will be revealed only by careful research. There are exceptions to this generalization and one is provided by the only county in northern California to allow casino gambling. The construction of hotels and motels transformed the urban landscape, which now contrasts strongly with the adjacent rural areas. Before leaving this subject it is necessary to counsel care in the use of statistics in demonstrating the influence of the boundary upon the cultural landscape. Federal boundaries always form the basic framework for statistical divisions, which may also coincide with some sections of internal boundaries. It is therefore not surprising that in many cases the political boundary will appear to separate areas with different population densities and per capita agricultural outputs. Such differences may only be apparent; field examination may show that the differences in population density and intensity of activity do not coincide with the political boundary.

It was suggested in the introduction to this chapter that a boundary may exert some influence upon the attitudes of persons living in the borderland and on the policies of states separated by the boundary. Although House (1982) makes some interesting comments about the perceptions of Americans and Mexicans to the opportunities and problems created by the Rio Grande as a boundary, not enough studies have been made to justify or reject this concept. Accordingly, the following views must be regarded as tentative.

The attitudes of border dwellers

If the influence of the boundary upon the attitudes of border dwellers is examined first, three points can be made. First, no one can doubt that 'frontiersman' denotes a person with a particular kind of philosophy and character; this point was made most convincingly by Turner and has been repeated by others such as Kristoff (1959). Since most international borderlands bear some relationship to a frontier, we can expect the boundary to have some measure of influence. The influence is likely to be exerted through the opportunities the presence of the boundary offers for economic gain, the inconveniences presented by the boundary to everyday living, and a greater awareness of the security needs of the state. One would expect the Belgians living on the German border to have a different awareness of the need for military preparedness from the Belgians living in the interior of the country and along the French border. The differences between the security viewpoints of the border and core dwellers is likely to be greater in countries larger than Belgium. The great difficulty is to measure the extent to which the attitude of borderland dwellers is distinctively influenced by the presence of the boundary. In some cases the borderland may give support to a specific political party, which will express clear views on questions of tariff and boundary and security arrangements. In other cases referenda relating to the boundary may be informative. This is obviously true when the referenda concern the movement of a boundary, as some did in Europe after World War I, but is also true when they concern altering the functions of the boundary, as in the Australian referendum on the formation of a federation.

Logan (1968) examined the influence of the boundary on the attitude of border dwellers to the proposed federation. Two views are held by historians on this matter. Parker (1949) holds the opinion that most voters judged federation in terms of regional economic interest:

> Federation by establishing a customs union, and perhaps by setting limits to railway competition and ensuring unhampered transport on the Murray system which flows through three colonies, is inevitably attractive to border residents and repulsive to the urban interests which rely on discriminatory legislation in order to participate in an otherwise uneconomic commerce. (Parker 1949, p. 1)

Blainey (1949), on the other hand, believes that there is a considerable variation within the voting patterns of the borderlands, suggesting that this influence is not uniform. When the referendum statistics were plotted on a map the result tended to support Parker's concept. In southwestern Victoria the border constituencies of Lowan, Normanby, and Portland recorded a higher proportion of affirmative votes than the state average (82 per cent) and the metropolitan vote (77 per cent). In the Victoria and Albert constituencies of southeastern South Australia, 78 per cent of the voters were in favour of federation compared with the state average of 67 per cent and the metropolitan figure of 55 per cent. Although proximity to the boundary is not the only factor involved, the figures are suggestive. Neglect of the more remote state areas by the metropolitan administrations may also have been responsible for the larger vote; during the 1860s there was an unsuccessful attempt to form a new state in the borderland of South Australia and Victoria, which was to be called Princeland. There were variations within the border, however, and Blainey's position can be understood. Some electorates which lay close to the boundary but which had no transboundary contacts returned lower percentages of affirmative votes. Port Macdonnell, which lies only 15 miles within South Australia, is an example of this type of electorate.

Borders and national policies

It is difficult to establish beyond doubt that the nature of a particular boundary has influenced state policies because it is neither easy, nor necessarily profitable, to distinguish the influence of the boundary from the influence of the state with which the boundary is drawn. For example, since 1967, farmers along Israel's borders have been discouraged from growing grain crops which can be burned, and fruit crops which will spoil if not harvested at exactly the right time, because of the damage and interference which enemy action might cause. Instead they cultivate crops, such as vegetables, which are not so easily destroyed by enemy action and which can be harvested over a period of weeks. This policy is directly related to the hostility of some neighbouring states, but there are sections of the boundary where it is easier for attacks to be launched by the enemy. It would not be very useful to attempt to separate these two intertwined strands in any analysis of the border.

Hasson and Gosenfeld (1980) have analysed Israel's policy of establishing new agricultural settlements in borderlands. They conclude that since 1967 such settlements have been used to 'project new boundaries for Israel'. In effect, these settlements are an insurance against any future political boundaries drawn by third parties or external agencies, because the Israeli authorities believe that such lines would conform to the distribution of recent settlements (Hasson & Gosenfeld 1980, p. 321).

There are two possible cases where policies may have been directly related to the nature of the boundary. In the second half of 1963 the Australian government began to demarcate more clearly the boundary separating their New Guinea territories from West Irian. This policy probably resulted partly from the unsatisfactory condition of this longitudinal boundary and also from the territorial friction between Indonesia and Malaysia. Had the Dutch remained in control of West Irian, it is unlikely that it would have been considered worth while spending funds on boundary demarcation. It proved to be a wise policy in view of the friction along this boundary caused by the flight of some dissidents from West Irian into Papua New Guinea.

Secondly, there is the example of India, which in the years following independence neglected her apparently secure Himalayan border in order to concentrate upon the development of the remainder of the country and the security of her Pakistan border. Even the reestablishment of Chinese influence in Tibet did not alter this policy, which was only ended when the Chinese allegedly invaded Indian territory in the Himalayas. Apart from examples of inferences of this kind, it seems likely that more concrete examples will have to be derived from historical studies in political geography using first-hand material in archives.

Conclusions

The range of examples provided in this chapter indicates that geographers are aware of the influence which boundaries may exert upon the development of cultural landscapes and that this is a field which does not seem to attract the attention of other scholars, although economists may wish to pursue the ideas of Losch and it would be understandable if historians devoted more attention to this subject. In most cases the studies of border landscapes were incidental to larger studies dealing with the evolution of the boundary and aimed at understanding its problem. This is to be commended rather than criticized because students of boundary evolution and problems should always be aware of, and record, the changes which the presence of the boundary causes in the landscape.

References

Best, A. C. G. 1971. South Africa's border industries: the Tswana example. *Annals, Association of American Geographers* **61**, 329–43.

Blainey, G. 1949. The rôle of economic interests in Australian federation: a reply to Professor Parker. *Historical Studies of Australia and New Zealand* **13**, 224–37.

Daveau, S. 1959. *Les régions frontalieres de la montagne Jurassienne* (Frontier regions of the Jura Mountains). Paris: Trevoux.

Hasson, S. & N. Gosenfeld 1980. Israeli frontier settlements: a cross-temporal analysis. *Geoforum* **11**, 315–35.

House, J. W. 1959. The Franco-Italian boundary in the Alpes-Maritimes. *Transactions, Institute of British Geographers* **26**, 107–31.

House, J. W. 1982. *Frontier on the Rio Grande: a political geography of development and social deprivation*. Oxford: Clarendon Press.

Kibulya, H. M. 1967. Geographic contrasts on the Bwamba–Congo border. In *The political geography of the Uganda–Congo boundary*, H. M. Kibulya and B. W. Langlands (eds), 52–93. Occasional Paper No. 6. Kampala: Makere University Press.

Kristoff, L. A. D. 1959. The nature of frontiers and boundaries. *Annals, Association of American Geographers* **49**, 269–82.

Logan, W. S. 1968. The changing landscape significance of the Victoria – South Australia boundary. *Annals, Association of American Geographers* **58**, 128–54.

Losch, A. 1967. *The economics of location* (translated from 2nd revised edition by W. B. Woglom assisted by W. F. Stolpter). New York: Wiley.

Mackay, J. R. 1958. The interactance hypothesis and boundaries in Canada: a preliminary study. *Canadian Geographer* **11**, 1–8.

Minghi, J. V. 1963. Boundary studies in political geography. *Annals, Association of American Geographers* **53**, 407–28.

Nelson, H. J. 1952. The Vernon area of California: a study of the political factor in urban geography. *Annals, Association of American Geographers* **42**, 177–91.

Parker, R. S. 1949. Australian federation: the influence of economic interests and political pressures. *Historical Studies of Australia and New Zealand* **13**, 1–24.

Platt, R. S. & P. Bucking-Spitta 1958. *A geographical study of the Dutch–German border*. Munster: Geographical Commission for Westfalen.

Reitsma, H. J. 1971. Crop and livestock production in the vicinity of the United States – Canada border. *Professional Geographer* **23**, 216–23.

Reitsma, H. J. 1972. Areal differentiation along the United States – Canada border. *Tijdschrift voor Economische en Sociale Geografie* **63**, 2–10.

Rose, A. J. 1955. The border zone between Queensland and New South Wales. *Australian Geographer* **6**, 3–18.

Sevrin, R. 1949. Les échanges de population a la frontière entre la France et la Tournaisis (Exchanges of population between France and Tournaisis). *Annales de Géographie* **58**, 237–44.

Ullman, E. L. 1939. The eastern Rhode Island – Massachusetts boundary zone. *Annals, Association of American Geographers* **29**, 291–302.

7 *Europe*

> Alsace-Lorraine was acquired by the French through the methods which have led to the consolidation of most modern States, namely, conquest, trickery and cession. (Temperley 1920, p. 159)

The brutal honesty of this statement would be preserved if names such as Tacna, Bornu, and Primorskiy Kray were substitutes for Alsace-Lorraine and the charge was made against Chile, Britain, and Russia respectively. However, it would be an error to assume that because European states had similar territorial ambitions in Europe and the other continents and used similar policies to achieve them that there are close similarities between the evolution of boundaries in Europe and in the other continents. There are more important differences in respect of boundary evolution between Europe and the rest of the world than there are between any other two continents.

The distinctive nature of boundary evolution in Europe

Boundary evolution in Europe was entirely an indigenous process. The changing pattern of major and minor political divisions of considerable areas of the continent have been recorded without a break, certainly since the Roman period. Useful perspectives can be found in general atlases produced by Shepherd (1922), Engel (1957), and Treharne and Fullard (1965). More detailed representations of classical Europe have been provided by Menke (1865) and Smith and Grove (1874). There are also regional atlases, including the remarkably detailed *Geschiedkundige Atlas van Nederland* (1913–32) and the monumental atlas of southeastern France by Baratier *et al.* (1969).

In the other continents, apart from Antarctica, the indigenous process of boundary evolution was overlain and generally halted by colonial activities of the imperial powers, such as Spain and Portugal in South America, Germany and France in Africa, and Britain, Russia, and China in Asia. The advent of imperial power in these lands may be likened to a political unconformity. The processes by which the Incas, Amandebeles, Sioux, Turkomans, Karens, and Aborigines determined boundaries effectively ended when the imperial powers acquired authority in their regions. In some cases approximations of the indigenous boundaries which existed at that time were preserved in comparatively short sections by the imperial powers. Thus the British drew the boundary between Northern and

Southern Nigeria in the vicinity of the most southern advance of the Hausa and Fulani against the Yoruba Kingdom of Oyo.

Even a superficial acquaintance with the evolution of boundaries in continents outside Europe makes it clear that many problems resulted from genuine uncertainties about the distribution of geographical features and the patterns of political authority. These problems may have existed in very early times in Europe, but for the past 2000 years there is no evidence that such uncertainties caused significant problems. The high densities of population, the considerable interregional trade, conquests, migrations, the clearing of forests, the draining of swamps, and the absence of entirely inhospitable terrain ensured that the physical geography of Europe was generally known to authorities at an early time.

Regional geographical knowledge necessary for boundary construction was certainly completed during the feudal age in Europe. Strayer (1965, p. 17) has pointed out that effective feudal government is local since it requires the performance of political functions based on personal agreements between small numbers of people. He and other writers also show that in any particular parcel of land different lords might exercise different authorities related to justice, taxes, the various uses of forests, and forced labour. Genicot (1970) has provided a most illuminating account of this type of situation. Thus it follows that the areas within which these important rights were available were closely defined. Genicot (1970, p. 32) is of the view that by the end of the 11th century the growth of population and the extension of authority had resulted in the remaining frontiers of Europe being refined to fairly precise lines. There were still disputed zones, but their limits were clearly understood.

Careful study of the treaties contained in the collection of European boundary treaties by Hertslet (1875–91) shows that boundary construction was simplified because the continent was divided into a hierarchy of local administrative units with precisely defined limits. The reader is constantly aware that the international boundaries established by the various treaties coincide with the limits of *cantons*, principalities, bailiwicks, *cercles*, counties, Commanderies of the Teutonic Order, circuits, *arrondissements*, bishoprics, duchies, *landgraviates*, *communes*, and parishes. Thus, the boundary negotiations in Europe were not bedevilled by problems such as trying to determine the extent of the territory of the Paramount Chief of Barotseland or the limit of lands occupied by the Saryks on the border between Russia and Afghanistan.

The situation in Europe also gave a positive advantage which was denied to negotiators in other regions. In most regions of Europe the land was divided into small parcels, which were sometimes aggregated into larger administrative units. These local divisions provided a series of building blocks from which national territories could be fashioned. Thus the international bargaining centred on pieces of territory, and once their disposition had been decided there was a ready-made boundary, which might at different sections coincide with known boundaries between parishes, bishoprics, and bailiwicks.

This advantage was not available to the same extent in the Balkans where feudalism as known and practised in western Europe did not exist (Pounds 1947, p. 129). Nor did it apply after World War I, when strategic considerations were a powerful factor in determining the alignment of boundaries.

Just as the physical and political geography of Europe was known clearly to negotiators, so the economic geography of disputed regions was understood. Examples of states pursuing clear economic objectives are the determination of German representatives to secure as many of the iron-ore fields of Alsace-Lorraine as possible in 1871; the competition for coalfields in Silesia between Poland, Germany, and Czechoslovakia in 1919; Yugoslavia's enthusiasm for securing the port of Rijeka by the Treaty of Rapallo; and Rumania's insistence on control over the railway from Timisoara to Arad. Although these are comparatively recent examples, there is no reason to suppose that nobles in earlier periods were not entirely aware of the economic advantages of securing the transfer of specific feudal rights from a defeated neighbour.

In a broad swathe of central Europe stretching from the coasts of East and West Germany and Poland to the shores of the Adriatic, Aegean, and Black seas in the south-east, present national boundaries were fashioned in a series of major wars. To a much greater extent than on any other continent, boundaries in this region were created during the peace conferences which followed widespread conflict. Judged by the speed with which agreement was reached at the Congress of Vienna in 1815, the Congress of Berlin in 1878, the London Conference in 1913, and the peace conferences in Paris which ended World Wars I and II, it is possible to conclude that wars simplify the process of drawing boundaries. The successful powers usually have clear ideas of the territorial arrangements they will demand before victory is achieved, and the defeated states are rarely able to resist most of these territorial adjustments. In most cases there is probably an overwhelming desire on the part of all parties involved to secure an agreement which will allow the abnormality of war to be ended. Successful governments want to demonstrate to the tired populace the gains bought by military sacrifices; and they wish to avoid the need to maintain large armies and navies and to reduce to a minimum the funds which have to be expended on administering the territory of the defeated states. Defeated countries generally want to regain a level of independence necessary for the reconstruction of national morale and wealth.

The Congress of Vienna 1815

The Congress of Vienna signed on 9 June 1815 is a convenient event to start a brief survey of the main events in the evolution of Europe's boundaries. The aim of the victorious powers in these negotiations was to restore the political situations which existed before the Napoleonic Wars (Albrecht-Carrié 1958, pp. 9, 15). France was confined again within the boundaries of 1792, and King Louis XVIII was placed on the French throne. Denmark and Saxony, which had unwisely persisted with support for Napoleon too long, were punished by territorial losses. Denmark lost Norway, which joined Sweden, and significant parts of Saxony were attached to Prussia. There were other comparatively small alterations. The Austrian Netherlands went to Holland and Austria was recompensed by parts of northern Italy.

Thus in 1815 there was a clear tripartite division of Europe. In the west lay the established nations of Portugal, Spain, France, England, Denmark, and Sweden. Mitteis (1975, p. 4) notes that it was in western Europe, including the Iberian and Scandinavian peninsulas, that the earliest and most vigorous growth of rudimentary political units occurred. In the period after 1815, these countries experienced only minor changes in their limits. Eastern Europe was dominated by the Russian, Austrian, and Ottoman empires. Between these two areas lay the unconsolidated zone of Europe comprising the German and Italian states. These congeries of states were separated from each other by Switzerland, which was guaranteed perpetual neutrality by the victorious powers. Strayer (1965, pp. 24–5) has explained the fragmented nature of Germany and Italy at this time. The German kings never used feudalism as the chief support of their government. This meant that the kings continually interfered in the attempts of local lords to make political decisions and thus blocked the growth of powerful feudal principalities. However, the kings were never powerful enough to exert a unified authority over the region, and power fell to regional princes outside the feudal system. In the case of Italy, the German emperors intervened in the affairs of Italian states without being able to control them effectively. In addition, the principal cities were too strong to be dominated by feudal lords but too weak to be able to construct kingdoms by annexing adjacent rural areas.

After 1815 much of Europe's territorial history was concerned with the unification of Germany and Italy and the disintegration of the Ottoman, Austrian, and Russian empires. It is this latter process which is most important for the evolution of European boundaries and it began in southeastern Europe on 21 July 1832 and spread northwestwards.

In 1832 a treaty signed in Constantinople by Britain, Russia, Turkey, and France established the limits of continental Greece (Hertslet 1875, pp. 903–8). The successful Greek war of independence occurred about the same time that Belgium achieved its independence from the Netherlands.

The Congress of Berlin 1878

The next major division of the Ottoman territories in eastern Europe occurred in 1878. Turkey had been defeated in war by Russia, which sought to impose its own terms in the Treaty of San Stefano dated 3 March 1878 (Hertslet 1891, pp. 2672–6). The centrepiece of this treaty was the creation of a large state called Bulgaria which would be subject to Russian domination for two years. This transparent attempt by Russia to dominate the Balkans was resisted by other major interested powers, and new territorial arrangements were made at the Congress of Berlin in a treaty signed on 13 July 1878. The main results of this treaty are shown in Figure 7.1. The previous nominal independence of Romania, Serbia, and Montenegro was made effective within enlarged territories. Russia regained Bessarabia, which it had held from 1812 to the Crimean War, and the Austro-Hungarian Empire secured Bosnia-Herzogovina and Sandjak. Bulgaria was created and Eastern Roumelia was established as a distinct part of the Ottoman Empire in which special provisions would protect the peculiar interests of its inhabitants; for example, the Governor-General had to be a Christian. The remaining Ottoman corridor from the Black Sea to the Adriatic Sea was narrowed by the transfer of Thessaly to Greece in 1881 and the transfer of Eastern Roumelia to Bulgaria in 1885.

The boundaries created by the Treaty of Berlin were described in varying degrees of detail and it was left to European commissions to determine the precise delimitation. They consisted of representatives of Britain, Austro-Hungary, France, Germany, Italy, and Russia, together with delegates from the countries separated by the particular line. Hertslet (1891) has recorded much useful information about the work of these commissions, and it is evident that no efforts were spared to produce boundaries which would reduce international discord by negotiating variations in the lines defined in the Treaty of Berlin. The commissions did not simply enforce boundaries on Turkey without regard to any potential problems. Instead, negotiations to moderate the adverse effects of particular lines continued. This process is revealed clearly in the delimitation of the boundary between Turkey and Montenegro.

The generous additions to Montenegro by the Treaty of San Stefano were reduced by the Treaty of Berlin (Hertslet 1891, pp. 2674–5). The new boundary meant that the main new areas gained by Montenegro were in the upper valley of the Lim River around the settlements of Gusinje and Plava (Fig. 7.2). In the Treaty of Berlin the boundary between Montenegro and Turkey was defined as follows:

Figure 7.1 The Balkans after the Treaty of Berlin 1878.

From there [Secular] the new frontier passes along the crests of the Mokra Planina, the village of Mokra remaining to Montenegro; it then reaches point 2166 on the Austrian Staff Map, following the principal chain and the line of the watershed between the Lim on one side, and the Drin as well as the Cievna (Zem) on the other.

It then coincides with the existing boundaries between the tribe of the Kuci-Drekalovici on one side and the Kucka-Krajna, as well as the tribes of the Klementi and Grudi on the other, to the plain of Podgorica

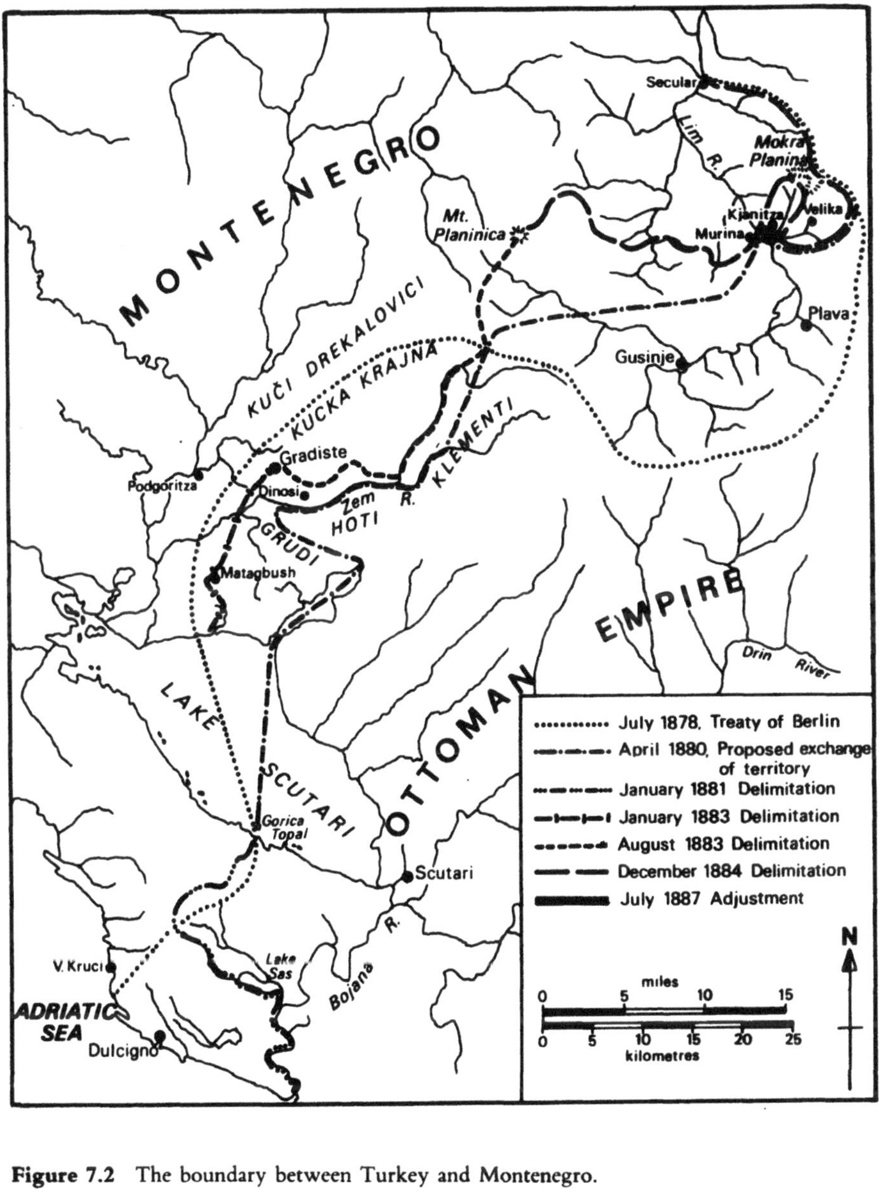

Figure 7.2 The boundary between Turkey and Montenegro.

> [Podgoritza], from whence it proceeds towards Plavnica, leaving the Klementi, Grudi and Hoti tribes to Albania.
>
> Thence the frontier crosses the lake [Scutari] near the Islet of Gorica-Topal, and, from Gorica-Topal, takes a straight line to the top of the crest, whence it follows the watershed between Megured and Kalimed, leaving Mrkovic to Montenegro and reaching the Adriatic at V.[Vieux] Kruci. (Hertslet 1891, p. 2782)

There is plenty of scope for disagreement about some phrases in this description to employ a boundary commission for decades. Does the crest coincide with the watershed separating the basins of the Lim and Drin rivers? How will the ethnic boundary which leaves the Klementi, Grudi, and Hoti tribes to Turkey be fixed? To which country does the Islet of Gorica-Topal belong? Which of the infinite number of straight lines will be used to reach the crest of the watershed south of Lake Scutari?

In fact the meetings of the European Commission and the various agreements concerning this boundary make it clear that there were scarcely any problems of interpretation. The real problem which delayed settlement of this line until 21 December 1884 was the realization by Montenegro that insistence on possession of the salient provided by the upper valley of the Lim River was not in its best interests. Turkey was very happy to retain the areas around Gusinje and Plava, but it became necessary to decide which additional areas should be ceded to compensate Montenegro for the loss of this region.

Before this issue came to a head, the European Commission met from 30 April 1879 to 8 September 1879 and settled the boundary from Vieux Kruci to Lake Scutari (Hertslet 1891, pp. 2890–6). In April 1880, Montenegro and Turkey agreed to an exchange of territory. In return for the areas around Gusinje and Plava, Turkey would cede to Montenegro Kucka-Krajna and a strip of territory occupied by the Grudi, Hoti, and Klementi tribes (Hertslet 1891, pp. 2952–4). This variation from the line defined in the Treaty of Berlin was quickly approved by the major powers and arrangements were made to transfer the two zones. By accident or design the Turkish withdrawal from some areas occurred before the Montenegran forces were ready to move, and the military vacuum was filled by Albanian irregulars, who established themselves in force in fortified positions astride the road between Podgoritza and Scutari.

Earl Granville had described the situation in this region in the following terms:

> Your Excellency [G. J. Goschen] is aware from the reports which have been received that the state of the country in North-east Albania is little short of anarchy. The Turkish officials are powerless to execute justice; murder, violence and forced exactions are prevalent, and the peaceable population is at the mercy of the armed Committees, who, under the

name of the Albanian League, have been allowed to assume absolute authority. (Hertslet 1891, p. 3000)

Turkey was either unwilling or unable to dislodge the Albanians from the eastern part of the area ceded to Montenegro. Negotiations then began to discover if there was any additional territory which could be ceded to Montenegro in exchange for the areas around Dinosi and Metagbush occupied by the Albanians.

The British and Austrian consuls in Scutari, Messrs Green and Lippich, proposed that Turkey should cede the coast between Vieux Kruci and the Bojana River, including the town of Dulcigno. Turkey eventually accepted this arrangement, providing Dinosi was excluded from the area occupied by Montenegro. This was the last serious difficulty, and the way was now open for the delimitation and demarcation of the entire boundary.

The line from the coast at the mouth of the Bojana River to Lake Scutari was fixed by the European Commission meeting in January and February 1881. There was some debate about whether the boundary should follow the west bank of the Bojana River, preferred by Turkey, or its thalweg, preferred by Montenegro, and the matter was resolved in favour of the latter country (Hertslet 1891, p. 3029).

The remainder of the boundary from the northern shore of lake Scutari to Secular was delimited in three sections by a joint Turkish–Montenegran Commission between January 1883 and December 1884 (Fig. 7.2). The reports of these commissions (Hertslet 1891, p. 3193) indicate that in the vicinity of the Lim River the commissioners were able to use old village boundaries to define the international limit. In July 1887 a decision was made by both countries to eliminate the Turkish pedicle about Kjanitza, and this area was transferred to Montenegro.

The Turkish corridor from the Black and Aegean seas to the Adriatic Sea was eliminated west of the Maritsa River during the Balkan Wars of 1912–13. Albania emerged as a separate country and Greece, Serbia, and Bulgaria annexed the remaining areas (Carrié-Albrecht 1958, p. 281). The scene was now set for the impact of disintegration to be experienced by the Austro-Hungarian Empire.

The Treaty of Versailles 1919

At the conclusion of World War I the victorious powers signed separate peace treaties with their five opponents. The Treaty of Versailles was signed with Germany on 28 June 1919, and this was quickly followed by treaties with Austria and Bulgaria signed respectively at St Germain-en-Laye on 10 September 1919 and at Neuilly on 27 November 1919. The treaty with Hungary was signed on 4 June 1920 at Trianon, and finally on 10 August 1920 the Treaty of Sevres was signed with Turkey. The Carnegie Endowment for International Peace (1924) has published a very useful collection of these treaties, together with a commentary and maps prepared by Martin.

Toynbee (1935) distinguished two main aims of the territorial provisions of these treaties:

> In the territorial chapter of the peace treaties, as in the Disarmament chapter, one of the things which the winners of the war were aiming at was to strengthen themselves, and weaken their late enemies comparatively, as much as possible. The second and more ideal and constructive purpose in the territorial resettlement of Europe was to redraw the political map in such a way that as many Europeans as possible would be living under national governments of their own choosing. This was called the principle of national self-determination. (Toynbee 1935, p. 72)

Germany was weakened by the loss of Alsace Lorraine to France, of northern Schleswig to Denmark, and of West Prussia and part of Silesia to Poland. The loss of West Prussia weakened Germany in two ways. First, it forfeited the population and resources of the region; secondly, East Prussia was isolated from the rest of the country. Germany also lost to Belgium four small territories called Eupen, Malmedy, Prussian Moresnet, and Neutral Moresnet.

Bulgaria was weakened by the loss of its coastal lands on the Aegean Sea, which were incorporated into Greece, and by the loss of three salients on its western boundary from which it had attacked Serbian communications (Temperley 1921, p. 455). Major Turkish losses have been considered in the chapter dealing with the boundaries of the Middle East.

The treaties with Austria and Hungary dismembered their former large, united territory. Some territory went to existing states. Romania secured Transylvania and Italy acquired the south Tyrol and Istra, a peninsula in the northern Adriatic Sea. Hungary and Austria were left in possession of small central areas and the northern lands were divided between the reconstituted state of Poland and the new country called Czechoslovakia; the southern area became Yugoslavia (Fig. 7.3).

In an effort to find boundaries which accorded with national settlements, plebiscites were organized in the following areas: northern Schleswig; Eupen and Malmedy; upper Silesia; the Klagenfurt Basin on the borders of Austria and Yugoslavia; Odenburg in the borderland of Austria and Hungary; and the southern and western parts of East Prussia.

Russia started the war on the side of the Allies, but after its transformation into the Soviet Union it was defeated by Germany and forced to sign the Treaty of Brest-Litovsk in 1918. This treaty detached from the Soviet Union a strip of territory occupied mainly by minorities

Figure 7.3 New and recreated states after World War I.

who sought national self-determination. The German plan to dominate this area for itself was ruined by the Allied victory, but it was effectively carried out by the Allies who presided over the emergence of Finland, Estonia, Latvia, Lithuania, and Poland. The Ukraine, which had been a German target, was left with the Soviet Union. However, it could not retain Bessarabia, which once again changed hands and became part of

Romania. Toynbee (1935, pp. 75–7) has perceptively noted that the principle of national self-determination would not have been carried out to such a marked extent in Europe if the Soviet Union had remained attached to the Allies throughout the war.

The Supreme Council established by the Allies to oversee the postwar territorial arrangements played a major rôle in allocating territory to the various new countries. The detailed work of this Council has been described by Baker (1923), Hankey (1963), and Mantoux (1955).

Ogilvie (1922) has provided an excellent survey of the processes by which boundaries were established after World War I. He identified four main stages, which began when the Supreme Council instructed one of the several territorial commissions to begin work. These commissions dealt with the boundaries of Czechoslovakia, Yugoslavia and Romania, and Greece and Albania. The territorial commission consisted of persons expert in the demography, economics, and strategic problems of the region under consideration. Through consultation, they produced a report recommending in general terms the description of a suitable boundary on a map of a scale of 1 : 1 million. The second stage involved a subcommission, which worked on maps at a scale of 1 : 200 000 to produce a more precise line which would be suitable for demarcation:

> In the sub-Commission stage such questions as the military value of hill ranges and the military and economic importance of railway junctions were considered, as were also the possibilities of maintaining intact geographical units, such as intermontane basins and valleys. (Ogilvie 1922, p. 6)

This refined boundary was then submitted to the Central Geographical Committee, which produced the final description of the boundary and defined the latitude which should be granted to the demarcation commissioners. This committee worked on cadastral maps and tried to ensure that the new international boundaries did not partition individual properties. The use of survey points to simplify the process of demarcation and specific details about the boundary to apply to rivers were also the responsibility of the Central Geographical Committee. When the boundary's definition had passed through these three stages it was then incorporated into the relevant treaty, and the final stage involved the demarcation of the boundary.

Although there were many helpful articles and notes by Hinks (1919a, b, c, 1920a) on the subject of the new boundaries in Europe, members of the demarcation teams did not publish accounts at the same frequency as surveyors operating in Africa and Asia. However, Cree (1925) provided an article on the work of the Yugoslav–Hungarian boundary commission. He describes how the commission had a covering letter and detailed instructions, presumably provided by the Central Geographical Committee, as well as the description in the treaty. Before the fieldwork commenced, the delegates representing Hungary and Yugoslavia were required to provide their interpretation of the treaty's description and any sections where an appeal to the League of Nations might be considered. Then, after a preliminary reconnaissance, meetings were held in Hungary and Yugoslavia on the same day to hear the views of interested communities in the borderland. Mayors, presidents of chambers of commerce, large landowners, factory owners, and farmers would attend these meetings, and present their case for a particular line, and answer questions which the members of the commission might ask. The President of the Commission then consulted with the Hungarian and Yugoslav delegates in an effort to secure total agreement. Once agreement had been reached one member of the team would be asked to define the boundary. Then, in turn, other members would be asked to propose any modifications of this line, and each proposed change would be debated until agreement was reached. If no agreement was possible, a vote was taken. The final line was then marked on a map at a scale of 1 : 75 000, and this was signed by the members. The Hungarian and Yugoslav assistant-commissioners then had the responsibility of marking the boundary on the ground. In a few cases involving the railway station at Gola and the water supply for the railway station at Gyekenyes, the commission prepared protocols to govern particular transboundary functions.

Boundary evolution after 1945

After World War II the main boundary changes were in northern and eastern Europe. The Soviet Union regained most of the territory lost by the Treaty of Brest-Litovsk. Although Finland remained independent, the boundary was adjusted in three regions in favour of the Soviet Union and Finland lost its coastlands on the Arctic Ocean. Poland continued its westwards march across the Baltic plains and secured those areas of Germany east of the Oder and Neisse rivers in exchange for eastern areas of Poland which were lost to the Soviet Union. The Soviet Union acquired sub-Carpathian Ruthenia, which formerly had been part of Czechoslovakia, and for the fifth time since 1812 Bessarabia was transferred from one country to another. There were two other minor alterations: Romania, having lost Bessarabia to the Soviet Union, also forfeited southern Dobrogea to Bulgaria; most of Italy's gains along the Adriatic coast in 1920 were lost to Yugoslavia.

Albrecht-Carrié (1958) has pointed to the marked difference of principle between the construction of boundaries in 1919 and 1945:

> . . . at all events the peacemakers of 1919 had adhered to the humane principle that the land belongs to the people and that frontiers therefore

> must be adapted to their existing distribution. Things were now [1945] otherwise and the ethnic map of Europe was brought into much closer agreement with its political divisions through the simple, if crude and inhumane, device of making the people fit the frontiers. (Albrecht-Carrié 1958, p. 605)

The Polish annexation of parts of eastern Germany resulted in the migration of more than 2 million Germans westwards across the Oder–Neisse Line. Further, the fragmentation of Europe which began in the Balkans and proceeded northwestwards in Austria and Hungary continued into Germany, which was divided into two parts. This division caused additional shifts of population westwards.

Judged by the published studies, it was the boundary between Italy and Yugoslavia which provoked most interest amongst political geographers. Moodie (1945), Kiss (1947), and Le Lannou (1947) provided perceptive and detailed studies of the area which Moodie calls the Julian March.

Alexander (1953) has provided a very useful account of the small boundary adjustments made in 1949 along the boundaries of Germany with the Netherlands, Belgium, and Luxembourg.

The Danish–German boundary as an example of boundary evolution in Europe

The territories of Schleswig and Holstein in the south of the Jylland Peninsula marked the fluctuating frontier of contact between people of Danish and German origin for many centuries. In the first half of the 10th century Queen Thyra played a leading rôle in supervising the completion of a wall known as the *Dannewerk*, which extended from the Schley Inlet to the Treene River and marked the southern limit of the Danish Kingdom (Larsdon 1912, p. 5)

In subsequent periods Danish authority extended south of this feature and by the middle of the 19th century Denmark consisted of the kingdom which lay generally north of Ribe and the Duchies of Schleswig, Holstein, and Lauenburg to the south of that town. The king's feudal rights in the Duchies were maintained so long as the succession was always from father to son, but the death of Frederick VI broke that sequence and the crown passed to his cousin Christian VIII (Naval Intelligence Division 1944, p. 130). This started an intense period of competition between Denmark, Prussia, and Austria for control of the Duchies. Palmerston is reported to have said that the Schleswig-Holstein question was so complicated that it was only understood by three people. They were the Prince Consort, who was dead, a German jurist who was in an asylum, and Palmerston, who had forgotten (Collocott & Thorne 1962, p. 603).

The matter was settled by force in 1864 when Denmark was defeated by Prussia and Austria. Denmark had earlier refused to cede Holstein and that part of Schleswig whose inhabitants decided at a plebiscite to join Prussia. The peace treaty of 30 October 1864 cost Denmark the Duchies of Schleswig, Holstein, and Lauenburg (Hertslet 1875, pp. 1630–3). Embedded in the territory of Schleswig were Jutland territories which remained with the King of Denmark (Fig. 7.4). To avoid this complication it was agreed that these territories would also be ceded to Austria and Prussia in exchange for equivalent zones from northern Schleswig which were adjacent to the Danish Kingdom.

The boundary defined in Article V of the 1864 treaty mainly coincides with the thalweg of the Königs-Au and the existing limits of parishes such as Odis, Vandrup, Hejls, and Taps. The boundary was also defined off the west coast of the peninsula; it passed 'at equal distances between the islands of Manoe and Roemoe' (Hertslet 1875, p. 1633). The Austrian gains in Schleswig and Holstein were lost to Prussia through the peace treaty dated 23 August 1866 (Hertslet 1875, p. 1722). However, Article V of this treaty stipulated that the northern districts of Schleswig should be ceded to Denmark if the population decided in a free vote that this was a desirable course. This test of public opinion was never held and Germany was released from this condition by a treaty with Austria dated 11 October 1878 (Parry 1977, 293–4).

The plebiscite which Prussia and Austria had proposed in 1864 and for which provision was made in the Austrian–Prussian treaty of 1866 was eventually held in 1920. After World War I Denmark submitted the Schleswig issue to the Peace Conference and proposed that a plebiscite should be held so that those areas of Schleswig occupied by citizens who spoke Danish and who wished to join Denmark could be attached to that country. The Danish government wisely did not want to include a large German minority which might create future irredentist problems. It was therefore arranged that two plebiscites would be held. The first would be held in Zone 1, which lay between the Danish–German boundary of 1864 and a line drawn just north of Flensburg (Fig. 7.4). This zone would vote as a single unit, and the simple majority of votes would decide whether the area was transferred to Denmark or remained with Germany. Within five weeks of the voting in Zone 1, the plebiscite would be held in Zone 2 and votes would be counted within parishes. Depending upon the results, the boundary would then be drawn in accordance with geographical and economic conditions. The commission which conducted the plebiscites would propose a boundary to the Conference of Ambassadors in Paris, which would then instruct a boundary commission to demarcate the line (Hinks 1920b).

There was a move by some Danish chauvinists to call for a third zone which would extend as far south as the Eider River or at least the

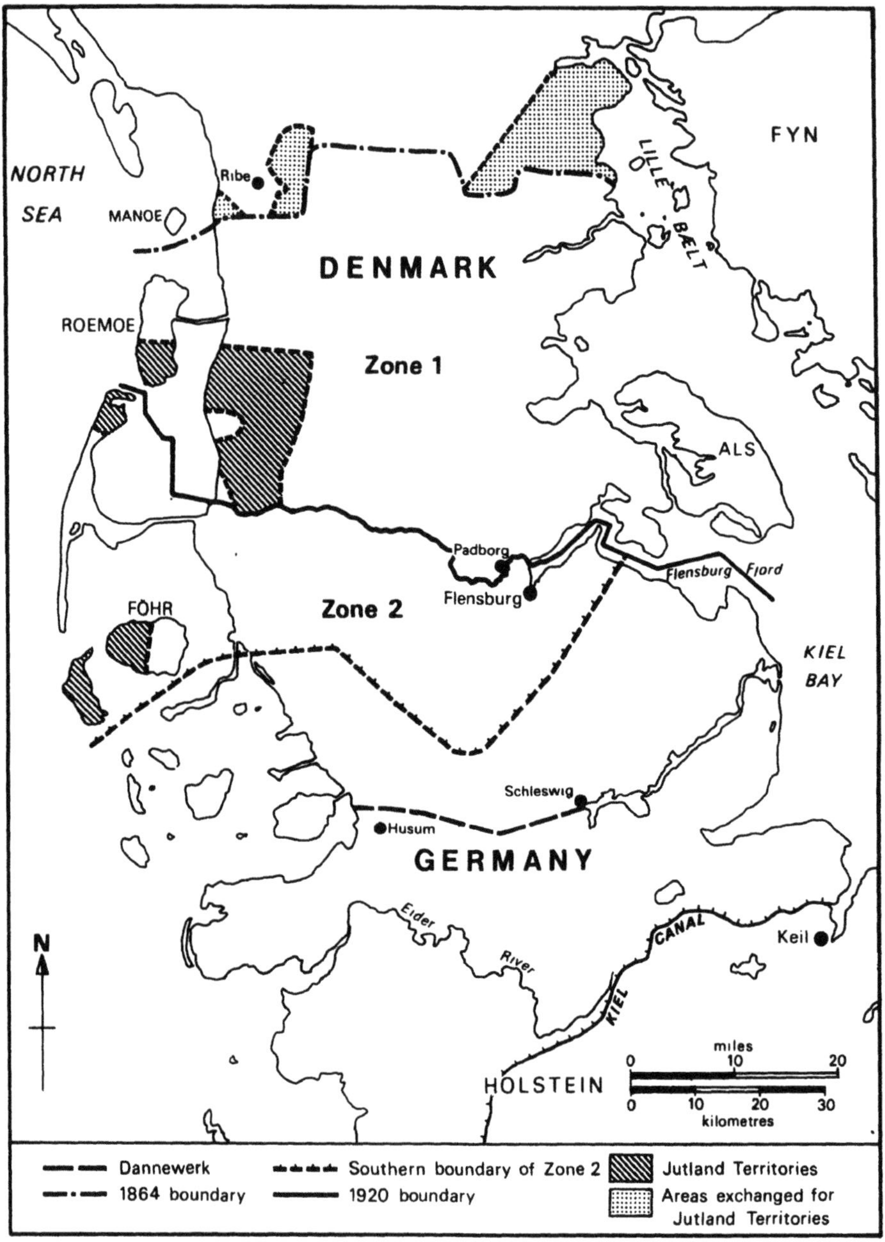

Figure 7.4 The Danish–German border.

Dannewerk, but this proposal was successfully resisted by the Danish authorities, who were fearful that some Germans might vote for union with Denmark because they perceived the economic prospects of that country to be superior to those of Germany.

The plebiscite in Zone 1 was held on 10 February 1920, and 74.2 per cent of the votes favoured union with Denmark. In Zone 2, on 14 March 1920, 79.7 per cent of the votes favoured continued association with Germany (Naval Intelligence Division 1944, p. 519). The only majorities recorded in favour of union with Denmark were found in three tiny parishes on Fohr. The Plebiscite Commission quickly decided that Zone 1 should go to Denmark and Zone 2 should remain with Germany. However, four small alterations were proposed to the boundary between the two zones. First, the boundary was deflected from the median line in Flensburg Fjord to give the port of Flensburg a larger anchorage. Secondly, the boundary was moved northwards so that the entire factory of Kupfermuhle remained with Flensburg. Thirdly, the line was shifted south so that Denmark secured the railway station as well as the junction at Padborg. Finally, the boundary was moved in Denmark's favour so that it controlled the sluices on the Wied Au, which were essential for flood control of some Danish districts.

These proposals were accepted by the ambassadors in Paris and a demarcation commission began work on 5 July 1920. No serious difficulties were encountered, and the description of the boundary shows that it follows streams, municipal limits, and the boundaries of private farms (The Geographer 1968). Hinks (1920b, p. 488) records that median lines on watercourses subject to flooding were calculated between the levels where grass begins to grow, since these levels mark the summer height of the streams. The final boundary was recorded on maps with a scale of 1 : 5000.

Conclusions

> The boundary readjustments of the past have been worked out through a series of gruesome wars; those of the future contain elements of still greater difficulty. Altogether the problem of just and proper boundaries is hopeless and growing worse on the basis of the old statecraft . . . (GilFillan 1924, p. 484)

In his very interesting but gloomy paper, GilFillan rightly stressed the rôles of wars in producing new boundaries in Europe and of territorial disputes in promoting wars. This characteristic and the fact that there was no colonial discontinuity distinguish boundary evolution in Europe from the formation of boundaries in other continents.

It is unlikely that GilFillan was surprised by the onset of World War II, since the boundary adjustments of 1920 had sown the seeds of conflict which were harvested in 1939. But he might have been gratified that developments after World War II produced two trends of which he expressed approval. First, some small states such as Latvia, Lithuania, and Estonia disappeared. Secondly, the organization of the European Economic Community has reduced the divisive nature of some European boundaries and made it easier for people, goods, and ideas to circulate. This international organization fits well with GilFillan's idea that 'the progress of civilisation involves a growth, not a shrinkage, in the size of cooperating groups' (GilFillan 1924, p. 484).

GilFillan published a map of Europe showing boundaries since 1500 according to the length of time they existed. The settled nature of boundaries in western Europe, including the Iberian Peninsula, France, the Netherlands, and Scandinavia, stand out, whereas central and southeastern Europe are covered in a maze of spidery lines. Having experienced another 40 or 60 years of life since that map was drawn, these boundaries in central and southeastern Europe would be more prominent on a modern map. Probably for the first time in European history there is no apparent serious boundary dispute on land. For the present the cycle of territorial dispute, war, boundary adjustment, and new territorial dispute appears to have been broken.

References

Albrecht-Carrié R. 1958. *A diplomatic history of Europe since the Congress of Vienna.* New York: Harper & Row.

Alexander, L. M. 1953. Recent changes in the Benelux–German boundary. *Geographical Review* **43**, 69–76.

Baker, R. S. 1923. *Woodrow Wilson and the Peace Settlement* (3 volumes). New York: Heinemann.

Baratier, E., G. Duby & E. Hildesheimer 1969. *Atlas historique: Provence, Comtat Venaissin, Principauté d'Orange, Comte de Nice, Principauté de Monaco* (Historical atlas: Provence, County of Venaissin, Principality of Orange, County of Nice, Principality of Monaco). Paris: Armand Colin.

Carnegie Endowment for International Peace 1924. *The treaties of peace* (2 volumes). New York: Carnegie Endowment for International Peace.

Collocott, T. C. & J. O. Thorne 1962. *Chambers's world gazetteer and geographical dictionary*. Edinburgh: Chambers.

Cree, D. 1925. Yugoslav–Hungarian boundary commission. *Geographical Journal* **65**, 89–112.

Engel, J. 1957. *Grösser historischer Weltatlas* (Large historical world atlas). Vol. 2: *Mittelalter* (Middle ages). Munchen: Bayerischer Schulbuch.

Genicot, L. 1970. Ligne et zone: la frontière des principautés médiévales (Line and zone: the boundary of medieval principalities). *Academie Royale des Sciences Brussels. Classe des lettres et des sciences morales et politiques*. Bulletin, 5th series, **56**, 29–42.

Geschiedkundige Atlas van Nederland (Historical atlas of the Netherlands) 1913–1932 (3 volumes). 's. Gravenhage: Martinus Nijhoff.

GilFillan, S. C. 1924. European political boundaries. *Political Science Quarterly* **39**, 458–84.

Hankey, Lord 1963. *The supreme control at the Paris Peace Conference: a commentary*. London: Allen & Unwin.

Hertslet, E. 1875–91. *The map of Europe by treaty* (4 volumes with continuous pagination). London: Butterworth.

Hinks, A. R. 1919a. Boundary delimitation in the Treaty of Versailles. *Geographical Journal* **54**, 103–13.

Hinks, A. R. 1919b. The boundaries of Czechoslovakia. *Geographical Journal* **54**, 185–8.

Hinks, A. R. 1919c. The new boundaries of Austria. *Geographical Journal* **54**, 288–96.

Hinks, A. R. 1920a. The new boundaries of Bulgaria. *Geographical Journal* **55**, 127–38.

Hinks, A. R. 1920b. The Slesvig plebiscite and the Danish–German boundary. *Geographical Journal* **55**, 484–91.

Kiss, G. 1947. Italian boundary problems: a review. *Geographical Review* **37**, 137–41.

Larson, L. M. 1912. *King Canute*. New York: Knickerbocker Press.

Le Lannou, M. 1947. Le Venetie Julienne: étude de géographie politique (The Venetzia–Guilia: a study in political geography). *Annales de Géographie* **56**, 13–35.

Mantoux, P. 1955. *Les délibérations du Conseil des Quatre, 24 Mars – 28 Juin 1919* (The deliberations of the Council of Four, 24 March – 28 June 1919). (2 volumes). Paris: Centre National de la Recherche Scientifique.

Menke, T. 1865. *Spruner-Menke Atlas Antiques* (Atlas of the ancient world). Gotha: Justus Perthes.

Mitteis, H. 1975. *The state in the middle ages* (translated by H. F. Orton). New York: Elsevier.

Moodie, A. E. 1945. *The Italo-Yugoslav boundary: a study in political geography*. London: George Philip.

Naval Intelligence Division 1944. *Denmark*. London: HMSO.

Ogilvie, A. G. 1922. Some aspects of boundary settlement at the Peace Conference. *Help for Students of History*, No. 49. Society for promoting Christian knowledge. London: Macmillan.

Parry, C. 1977. *The consolidated treaty series*, Vol. 153. New York: Oceana.

Pounds, N. J. G. 1947. *An historical and political geography of Europe*. London: Harrap.

Shepherd, W. R. 1922. *Historical atlas*. London: London University Press.

Smith, W. & G. Grove 1874. *An atlas of ancient geography: biblical and classical*. London: Murray.

Strayer, J. R. 1965. Feudalism in western Europe. In *Feudalism in history*, R. Coulborn (ed.), 26–48. Hamden, Conn.: Archon Books.

Temperley, H. W. V. 1920 and 1921. *A history of the Peace Conference*, Vols 2 and 4. London: Oxford University Press.

The Geographer 1968. Denmark–Germany boundary. International Boundary Study No. 81. Department of State, Washington, DC.

Toynbee, A. J. 1935. Territorial arrangement. In *The Treaty of Versailles and after*, Lord Riddell (ed.), 70–93. London: Allen & Unwin.

Treharne, R. F. & H. Fullard 1965. *Muir's historical atlas: ancient, medieval and modern*. London: George Philip.

8 *The Americas*

There is an obvious contrast between the three long international boundary segments in North America and the 35 long and short boundary segments in Central and South America. It is proposed to examine boundary evolution in these two continents separately, since they were produced by distinct processes.

North America

The boundary between Mexico and the United States of America has been considered in Chapter 3. This section focuses on the boundaries between Canada and the United States of America.

Nicholson (1954, p. 5) briefly described some of the boundaries between indigenous groups in the period before European colonization of North America, but he concluded that these pre-European boundaries had little or no effect on the evolution of the major boundaries of Canada. The boundary between Alaska and Canada was first negotiated between the British and Russian empires in February 1825 and is generically similar to the Portuguese–Spanish boundary of 1777 in South America, the Sino-Russian limit of 1864 in central Asia, and the 1890 Anglo-French line in West Africa. In contrast, there is no comparable case to the evolution of Canada's southern boundary. In this situation European colonists successfully seceded from the metropolitan power and then negotiated an international boundary with that power as an equal.

In 1763 France was eliminated from North America and restricted to the islands of St Pierre and Miquelon, except for rights in Louisiana later ceded to the United States of America by special treaty. But hegemony in north-east North America, for which Britain had strived so hard, lasted less than 20 years. Peace negotiations between the American colonies and Britain were completed in 1783. Nicholson has recorded that these negotiations placed Britain in the position occupied by France in 1697, with the American colonies playing the British rôle of that period:

> It was natural then that the United States should press for the same boundaries between themselves and Canada as Britain had claimed against the French, and the ultimate result was that Great Britain retained the Quebec, Nova Scotia and Newfoundland of 1763, but lost the Illinois country and the lands south and west of the Great Lakes that had been included in Quebec by the Act of 1774. (Nicholson 1954, p. 18)

Paullin (1932, pp. 52–5) has provided a well illustrated account of the negotiations which led to the Definitive Treaty of Peace on 3 September 1783. On 19 March 1779 the American Congress defined the boundary to be sought at the end of the war. From the Bay of Fundy the proposed boundary lay close to meridian 67° west as far as the junction of the watershed south of the St Lawrence River. This watershed was then followed southwestwards to the Connecticut River, which was followed to parallel 45° north. This parallel was then followed westwards to the St Lawrence and from this junction the line passed directly to the south end of Lake Nipissing. From this point the boundary went directly to the source of the Mississippi River (Paullin 1932, p. 52).

By 8 October 1782, after many proposals and counterproposals by both sides and some contributions from French and Spanish diplomats, the British and American representatives agreed on a line very close to the boundary defined by Congress three years earlier. The British government declined to accept, and one month later a new agreement was reached. The new boundary followed the course of the earlier line as far as the St Lawrence River. It then followed that river and passed through the middle of Lakes Ontario, Erie, Huron, and Superior and the waterways that connect them. West of Lake Superior, the boundary passed westwards to the Lake of the Woods and on to the source of the Mississippi River. An alternative line was offered to Britain at the same time; it simply continued parallel 45° northwestwards to the Mississippi River. Britain declined the offer of territory in what is now Wisconsin and North Dakota for riparian rights in all the lakes except Lake Michigan.

The problem of drawing the boundary from the Lake of the Woods to the Mississippi River was avoided by the sale of Louisiana by France to the United States in 1803. The United States regarded parallel 49° north as the boundary between this territory and British possessions (Paullin 1932, p. 60). Paullin (1923) has described the origins of parallel 49° north as a boundary, and Nicholson (1954, p. 26) observed that it had been so frequently marked on maps as a boundary before 1803 that its acceptance was not surprising. The boundary from Lake of the Woods to the Stony (Rocky) Mountains was defined in an agreement dated 20 October 1818 (Jones 1932).

The American authorities urged that this parallel be used as the boundary to the Pacific Ocean, but Britain hoped to obtain a boundary which coincided with the Columbia River, which enters the sea near parallel 46° north. This hope proved vain, and after refusing offers of parallel 49° north four times Britain proposed its use in 1846 (Nicholson 1954, p. 29). A treaty enshrined this line on 15 June 1846 and the boundary was continued from the junction of the parallel and the coast through Juan de Fuca Strait (Jones 1937).

A number of positional disputes arose when these boundary segments

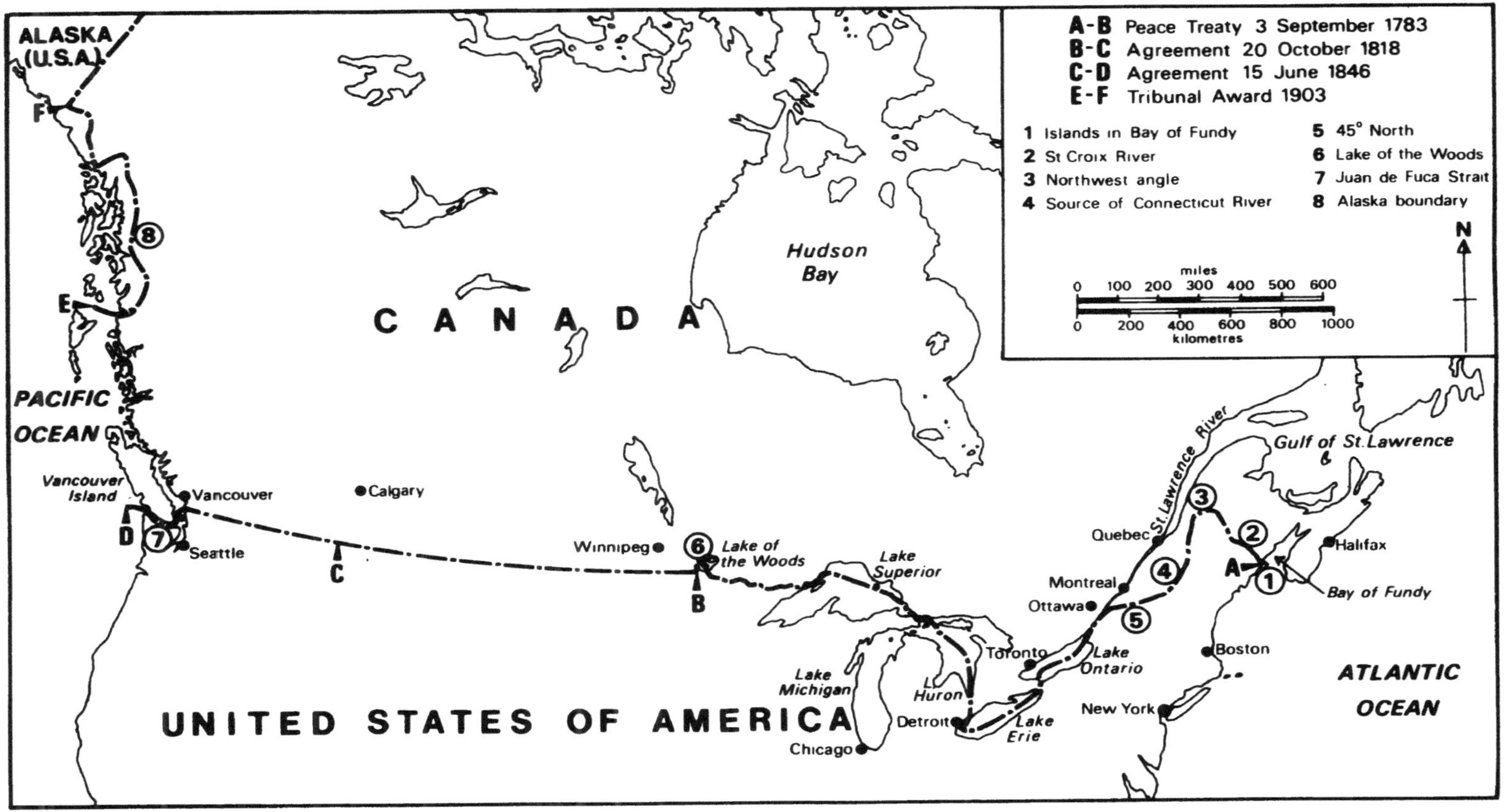

Figure 8.1 Canada's southern boundary.

were demarcated, and their location is shown in Figure 8.1. The disagreement over the ownership of islands at the mouth of the St Croix River was solved on 24 November 1817 by a compromise which saw Canada retain Deer and Campobello Islands. There were three candidates for the title of the St Croix River, and in 1798 it was agreed that the central river, then called the Schoodic, was the intended St Croix River. Nicholson (1954, p. 19) explains that excavation of the remains of Sieur de Mont's winter camp of 1604 at the mouth of the Schoodic convinced the parties that this was the correct river. Once the source of the St Croix River had been determined it was possible to proceed to the definition of the Northwest Angle, which is the point at which the meridian through this source intersects the highlands which mark the southern limit of the drainage basin of the St Lawrence River. This proved a difficult problem and was taken to the King of the Netherlands for arbitration in 1830. His decision was not accepted by the United States, and further bilateral negotiations finally resulted in a division of the disputed zone by the Webster–Ashburton Treaty of 9 August 1842. This treaty also solved the disagreements over the northwesternmost source of the Connecticut River, the correct location of parallel 45° north, and the boundary between Lake Superior and the Lake of the Woods. Parallel 45° north had been marked in the period 1771–4 by the chief surveyors of Quebec and New York, who were respectively called Thomas Vallentine and John Collins. This line was found in 1802 to be as much as three-quarters of a mile north of the true parallel. The King of the Netherlands had ruled that the correct boundary was the true parallel; however, as a contribution to compromise, Britain agreed to retain the Vallentine–Collins Line (Paullin 1932, p. 62). The other two problems arose because the boundary definition of 1783 left disputed zones at the head of the Connecticut River and between the Lake of the Woods and Lake Superior. These disputed zones were divided by the agreement of August 1842.

The remaining problem developed in Juan de Fuca Strait because the two sides could not agree on the location of the 'middle of the channel' in relation to the various islands lying between Vancouver Island and the mainland. This difficulty was resolved by the arbitration of the Emperor of Germany in 1872; he accepted the American view that the boundary should follow the Canal de Haro.

When the United States secured Alaska from Russia on 30 March 1867, it inherited the boundary which Russia had agreed with Britain on 28 February 1825 (Paullin 1932, p. 69). There were three uncertain points about the description of this boundary. First, what was the correct course through the various channels at the mouth of the Portland Canal? Secondly, how did the boundary proceed from the head of the Portland Canal to parallel 56° north? Thirdly, where was the 'summit of the mountains situated parallel to the coast'? The irregular strip of territory between the boundaries claimed by Britain and the United States measured about 500 miles in length and 35 miles in width. Negotiations failed to produce a compromise and the matter was placed before an arbitral tribunal in 1903. This tribunal defined a boundary dividing this disputed strip, except between the Taku and Stikine rivers, and this gap was filled by a boundary defined in an exchange of notes in 1905 (Paullin 1932, p. 71).

Since this time the boundaries between Canada and the United States have been maintained well (Lambert 1965). Problems which have occurred have mainly involved transboundary rivers and the Great Lakes (Bloomfield & Fitzgerald 1958, Piper 1967).

Central and South America

The processes by which international boundaries in Central and South America evolved from colonial origins were prototypes of those followed in Africa and Asia. Four centuries before Britain and France drew a straight line separating their spheres of influence in the unknown areas west of Lake Chad, Portugal and Spain had used a meridian 370 leagues west of the Cape Verde Islands to divide a continent which had not been delineated on reliable maps. More than a century before Russia and China divided Central Asia by a line which was often defined imprecisely, Spain and Portugal had agreed on the approximate limits of modern Brazil. Finally, the division of French West Africa into its constituent elements surrounded by French administrative limits in the 1960s had been preceded 150 years before by the fracturing of Spanish America into states such as Peru and Argentina.

The much earlier establishment and demise of the Portuguese and the Spanish empires in Central and South America compared with other empires in Africa and Asia meant that the independent successor states, such as Chile and Venezuela, inherited boundaries which were not fixed as precisely as the boundaries of countries such as Nigeria and Vietnam. There are a number of other features that make the boundaries of Central and South America distinctive.

The evolution of boundaries in the Spanish zone was partly based on the doctrine of *uti possidetis*, which has already been considered in Chapter 3. This doctrine was defined by the Swiss Federal Council in an arbitral award concerning Colombia and Venezuela in 1922:

> When the Spanish Colonies of Central and South America proclaimed themselves independent in the second decade of the nineteenth century, they adopted a principle of constitutional and international law to which they gave the name *uti possidetis juris* of 1810, with the effect of laying

> down the rule that the bounds of the newly created Republics should be the frontiers of the Spanish Provinces for which they were substituted. This general principle offered the advantage of establishing an absolute rule that there was not in law in the old Spanish America any territory without a master . . . (Scott 1922, p. 428)

When attention is directed to the problems which bedevilled boundary construction in this region, it will become clear that this apparently simple doctrine was often difficult to apply.

There has been more frequent recourse to arbitration to solve territorial disputes in this region than in any other continent. At least 17 sections of international boundaries have been settled by the monarchs of England, Germany, Italy, and Spain, by the presidents of Argentina and the United States, or by tribunals of eminent, impartial jurists. In a number of other cases arbitration was attempted without success.

In common with North America but in contrast with Europe, Africa, and Asia the cultural characteristics of the indigenous population did not play any significant rôle in boundary evolution. A careful study of the work by Ireland (1938), who did for South America what Hertslet and Brownlie did for Europe and Africa respectively, reveals that issues of tribe, language, or religion were not important issues when boundaries were being negotiated.

It is now necessary to consider the evolution of the international and internal colonial boundaries inherited by the independent states of Central and South America before considering the problems they faced in fashioning this inheritance into stable and peaceful limits.

The international colonial boundaries

Before 1810, when independence movements began to end Spanish and Portuguese rule in Central and South America, the colonial powers had agreed on three international boundaries.

The longest boundary was drawn by Spain and Portugal in 1750 and amended in 1777. According to the Treaty of Tordesillas, signed on 7 June 1494, the possessions of Portugal and Spain were separated by the meridian which lay 370 leagues west of the Cape Verde Islands. Because maps at that time were inaccurate and the language of the treaty imprecise, this meridian was identified in a number of locations. Harrisse (1897) has provided information about six determinations of this meridian between 1495 and 1545. The limiting estimates are shown in Figure 8.2. Ribeiro's estimate at 49°45′ west was the best that Portugal could hope for, and even this line left the mouth of the River Amazon to Spain. For 60 years from 1580 the Spanish king also ruled Portugal and therefore all of South America (Herring 1965, pp. 222–3). Burns (1980, p. 70) has described this

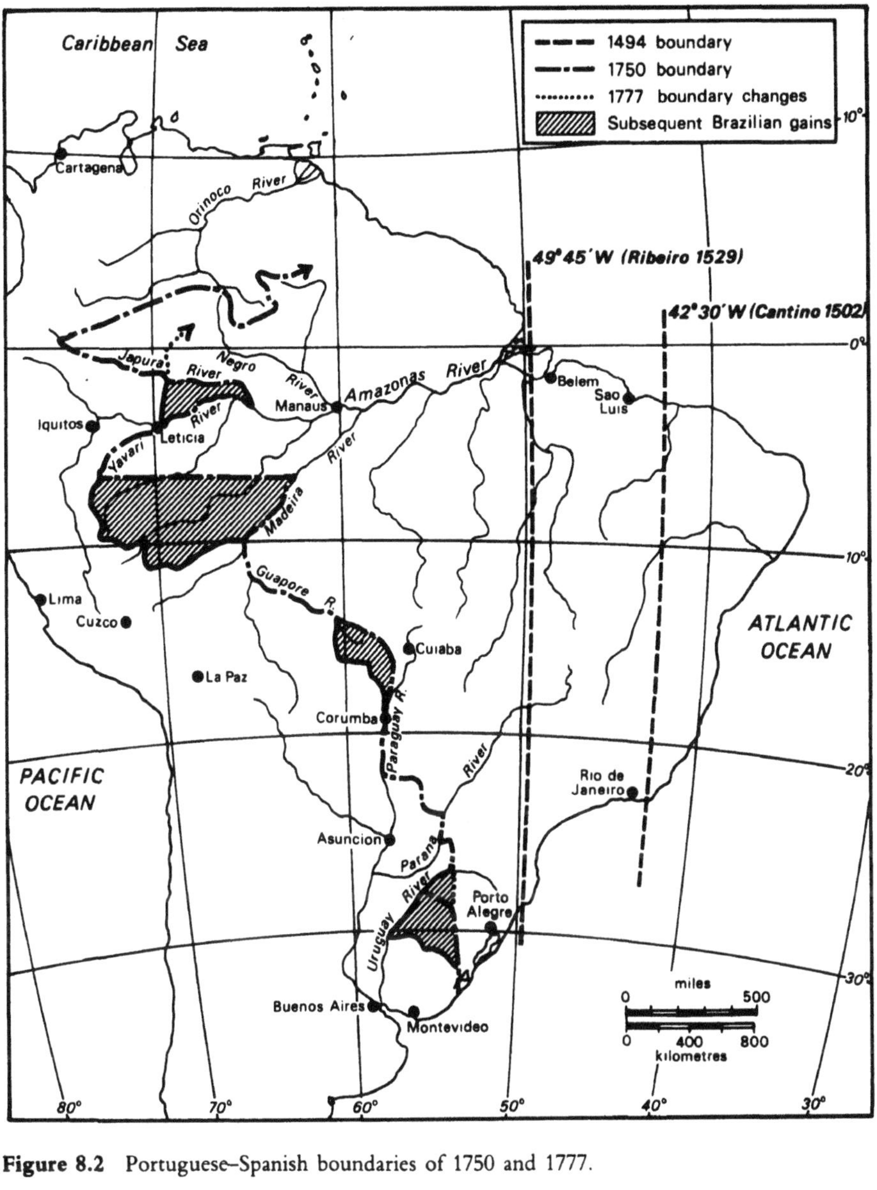

Figure 8.2 Portuguese–Spanish boundaries of 1750 and 1777.

period as Portugal's 'Babylonian captivity', and it produced one decided advantage when Portugal resumed its independence. The Spanish rulers failed to prevent Portuguese migration west of the line defined in 1494, and Philip IV granted much of the Amazon Basin west of that line in hereditary captaincies to two Portuguese nobles in 1637.

Portugal's domination in the Amazon Basin was confirmed by the Treaty of Madrid, which was signed on 13 January 1750. Burns (1980, pp. 70–1) has described how intermarriage within the two royal households at this time made it possible for the boundary to be fixed on the basis of *Realpolitik*. The line was drawn along rivers and other natural features to separate the areas occupied and controlled by each side (Fig. 8.2). The boundary description in the Treaty of Madrid (Parry 1969a, pp. 468–91) indicates that rivers such as the Uruguay, Parana, Verde, Guapore, Mamore, Madeira, Javari, and Japura were reasonably well known. Only in two sections did the line apparently pass through territory which was quite unknown. Between the Madeira and Javari rivers the boundary followed the parallel which passed through a point on the Madeira midway between its confluences with the Mamore and the Amazon. From the headwaters of the Japura the boundary followed 'the summit of the chain of mountains' which separated the Orinoco and Amazon rivers (Parry 1969a, p. 480). Neither the Treaty of Madrid nor the Treaty of San Ildefonso, which amended it in 1777 (Parry 1969b, pp. 361–74), established a northern terminus for this boundary. Spain and Portugal were not prepared to abandon the hope that the Dutch and French settlements on the coast of Guiana might eventually be withdrawn.

There were two main amendments to the line in 1777. First, Spain acquired the area known as the Seven Missions on the east bank of the River Uruguay (Wilgus 1943, p. 108). Secondly, instead of proceeding to the headwaters of the Japura River as far as the mountains separating the Orinoco and Amazon Rivers, the line in that area was drawn to leave to Portugal those Portuguese settlements on the Japura and Negro rivers (Fig. 8.2). It is not possible to be sure of the course of this section of line which eventually rejoined the mountains separating the Orinoco and Amazon rivers.

The boundary of 1777 proved to be the high-water mark of Spanish territory in South America. When Brazil inherited this line in December 1822 it set about establishing definite boundaries with its neighbours and, as Figure 8.2 shows, Brazil gained four considerable areas beyond the line of 1777. In fact, the only retreat from that line occurred on 30 October 1909 when Brazil ceded to Uruguay that part of Lagoa Mirim adjoining the coast between a stream called San Miguel and the Yaguaron River which forms the boundary between the two countries.

Although the Portuguese and Spanish authorities might have hoped that the Spanish and Dutch settlements might be withdrawn from Guiana, Portugal had signed a number of agreements with the French before 1750. The Treaty of Utrecht, dated 11 April 1713, fixed the Japoc or Vicente Pinzon River as the limit between Portuguese and French possessions (Ireland 1938, p. 147). In the second half of the 19th century a dispute developed between France and Brazil over the correct identity of this river. France favoured the Araguari River whereas Brazil selected the Oyapock River. On 1 December 1900 the Swiss Council, which had been asked to arbitrate this issue, declared in favour of Brazil (Ireland 1938, p. 151).

Spain also agreed on a boundary with France in 1777 to separate their possessions on Hispaniola. This was a remarkably detailed boundary for that period and was marked by 221 pairs of stone pillars along its length of about 170 miles (Parry 1969b, pp. 279–312). Some sections of this boundary were retained when a new definition was agreed by Haiti and the Dominican Republic on 21 January 1929 (The Geographer 1961).

The tiny island of Saint Martin at the northern end of the Lesser Antilles was divided between France and the Netherlands in 1648. France secured the northern 20 square miles and the Netherlands gained the remaining 14 square miles when Spanish rule was withdrawn after a period of eight years.

The limits of British Honduras and Guiana and of Dutch Surinam were not drawn until Spanish and Portuguese authority had been ended in this region.

The internal colonial boundaries

Since the Portuguese colony remained intact when Brazil was created, this section is only concerned with the administrative boundaries of the Spanish zone in Central and South America.

When French Indochina and French West Africa were replaced by a number of independent states, the limits of those countries were fairly well known. There were some problems such as that between Burkino Faso and Mali, but generally there were reliable maps on which lines had been clearly drawn and these lines had often been published in a gazette, even if they had not been demarcated. In addition, independence came only to the highest rank of administrative divisions, which was the colony. *Cercles* and districts into which most colonies were divided did not become independent states. The situation in Spanish America when colonial rule ended was quite different.

First, some of the primary administrative limits of the viceroyalties were not known. This situation occurred in some areas because the high ranges or the unhealthy jungles were inhospitable. In some other more favourable areas the low density of population made issues of jurisdiction unimportant. Secondly, it was not only viceroyalties which aspired to independent nationhood. *Audiencias*, captaincies-general, and *intendencias* in different areas and in different combinations also sought to become sovereign states. Unfortunately, the limits of these subordinate administrative units were often defined no more closely than those of the viceroyalties. Thirdly, the limits of some of these territories were altered by Spanish edicts shortly before Spanish power was broken and sometimes for reasons which had little to do with the good government of the people concerned. Fourthly, in some cases, in a manner copied 200 years later by British colonial officers in Nigeria and East Africa, administrators governed beyond their strict limits with the agreement of neighbouring officers because it was more convenient for both parties. These general impressions of the Spanish administration were derived from a variety of studies, including Chapman (1938), Delper (1974), Herring (1965), and Thomas (1956). Those who wish to pursue the subject in more detail should read the studies by Cunningham (1919) and Fisher (1926, 1929).

The overthrow of Spanish authority in Central and South America has been described in many books (Thomas 1956, Herring 1965, Veliz 1968); it is only necessary here to describe the correspondence between the new independent states and the former Spanish administrative areas. Figure 8.3 shows the situation in 1828, which was 20 years after France's intervention in metropolitan Spain had given fresh impetus to nationalist movements in the Spanish colonies. The Viceroyalty of New Spain had split into two parts. The Captaincy-General of the Kingdom of Guatemala, apart from the territory of Chiapas, formed the United Provinces of Central America, and the remainder of the Viceroyalty constituted the independent state of Mexico.

The Viceroyalty of New Grenada remained intact as the new state of Gran Colombia. The northern and southern parts of the Viceroyalty of Peru separated to form Peru and Chile respectively, and the coastal zone between them became part of Bolivia. The Viceroyalty of La Plata broke into four states. Most of the territory formed modern Argentina. Paraguay formed about the core of the Intendencia of Paraguay and the Audiencia of Charcas formed much of the new state of Bolivia, which also acquired the Intendencia of Potosi from the Viceroyalty of Peru and so secured an outlet to the Pacific Ocean. Uruguay was created in 1828 from much of the Banda Oriental of the Viceroyalty after negotiations between Argentina and Brazil.

Further political cleavages occurred along the former administrative limits of the Spanish colonies. In 1830 the state of Venezuela was fashioned out of the Captaincy-General of Caracas, which had been part of Gran Colombia. In the same year Ecuador was created in much of the territory which had formed the Audiencia of Quito. The reduction of Gran Colombia was completed in 1903 when Panama, resentful over decades of neglect, seceded. In 1838 the United Provinces of Central America split into Guatamela, Honduras, El Salvador, Nicaragua, and Costa Rica, which had previously been subordinate units of the Captaincy-General of Guatamela. Thus by 1838 the present states, with the exception of Panama, had been created. However, although the heartlands of these states were clearly identified the search for precise territorial limits occupied some for another century.

Figure 8.3 Post-colonial states in 1828.

Boundary negotiations amongst independent states

In a very interesting paper dealing with the boundary between Ecuador and Peru, Wright (1941, p. 265) concludes that South American boundary negotiations usually started with an examination of strict legal rights and ended with some sort of bargaining. That appears to be true in all cases, except Brazil's boundaries with British and Dutch Guiana and the boundaries between British, Dutch, and French Guiana.

Discussion between Brazil and its neighbours about strict rights usually began with an examination of the treaty of 1777. Figure 8.2 shows that Brazil gained four areas beyond that line. The area east of the Uruguay River was gained during a war with Argentina, which ended with the creation of Uruguay in 1828 (Burns 1980, pp. 166–7).

The other three areas were gained because pioneers from Brazil moved more easily along the tributaries of the Amazon, such as the Guapore, Madeira, Purus, Jurua, Javari, and Japura, than Andean plateau dwellers who had to face the prospect of overcoming a marked change in environment. Fifer (1966) has provided an excellent account of the economic and political forces that shaped the boundary with Brazil and Bolivia, and it is clear that similar factors encouraged Brazil's advance into areas which should have belonged to Peru or Colombia according to the treaty of 1777.

Throughout the detailed study by Ireland (1938) there is frequent reference to long preliminary debates about the former limits of Spanish areas when boundaries were being negotiated. For example, such debates occurred between Costa Rica and Panama, Guatemala and Mexico, Colombia and Ecuador, Colombia and Venezuela, Argentina and Paraguay, and Bolivia and Peru. Wright (1941, p. 258) shows the difficulties associated with such debates. In 1830 Ecuador inherited the boundary dispute between Peru and Gran Colombia. A war between those two states in 1829 had persuaded Peru to surrender the main contested zone between the Maranon and Caqueta rivers east of the Andes (The Geographer 1980, p. 2). Since Ecuador was much weaker than Gran Colombia, Peru decided to maintain its claim to this area and to the smaller areas called Northern Tumbes and Jaen west of the Andes.

Ecuador's claim to the three disputed areas was based on an edict of the Spanish king creating the Viceroyship of Santa Fe de Bogota in 1717 (Wright 1941, p. 257). Peru's claim was based on a *cedula* dated 15 July 1802 by which the Spanish king redefined the northern boundary of the Viceroyship of Peru. Although Peru argued that this last change was decisive in settling the matter of rights, Ecuador asserted that, in contrast with normal procedures in such cases, the *cedula* said nothing about sovereignty and that this ruling was never put into effect because by 1802 the Spanish administration was ineffective in this area (Wright 1941, p. 258).

It was noted earlier that arbitration was frequently used as a means of resolving boundary disagreements in Central and South America. On some occasions arbiters were specifically asked to interpret Spanish orders relating to former territories. For example, on 14 September 1881 Colombia and Venezuela asked Alfonso XII, the Spanish king, to decide which territory lay within the Captaincy-General of Caracas by royal acts of the former sovereign to 1810 (Ireland 1938, p. 209). The award by the king's successor recited the uncertainties of the various proclamations and then described an approximate compromise line.

On 30 December 1902 Bolivia and Peru requested the government of Argentina to determine the limits of the Audiencia of Charcas in 1810. That area would then belong to Bolivia. The arbiter was requested to take into account laws of the Indies, *cedulas*, royal orders, Intendente Ordinances, maps, and 'all documents which, having an official character, might have been used to give the true significance and effect to such royal provisions' (Ireland 1938, p. 103). The arbiter was then asked to resolve the question equitably if the various acts did not define a clear line! This was the course adopted by the arbiter in the award of 9 July 1909 when Peru received 60 per cent of the disputed area.

Although there have been many skirmishes between both regular and irregular forces along disputed borders in Central and South America since 1810 full-scale war has only been decisive on two occasions in producing major changes in the alignments of boundaries. By the War of the Pacific from 1879 to 1883 Chile moved its northern boundary from parallel 23° south to the vicinity of 18° south. Dennis (1931) has given an excellent account of the territorial arrangements produced by this war. Chile's territorial gains were at the expense of Peru and Bolivia, and it was the latter country which again suffered after a war with Paraguay which lasted from 1932 to 1935.

There is an extensive area bounded by the Rio Pilcomayo on the south and the Rio Paraguay on the east called Chaco Boreal. Bolivia laid claim to this region in 1878 after the southern portion had been awarded to Paraguay by the arbitration of President Hayes of the United States on 12 November 1878 after he had been requested to act by Argentina and Paraguay. Five agreements between Bolivia and Paraguay in 1879, 1887, 1894, 1907, and 1927 failed to secure ratification by the governments of one or both countries, and the matter was finally settled in Paraguay's favour in a treaty dated 21 July 1938 after the mediation of Argentina, Brazil, Peru, Uruguay, and the United States (Ireland 1938, pp. 66–95, The Geographer 1978).

Apart from disagreements about the interpretation of historical rights, the greatest difficulties associated with boundary negotiations concerned rivers. This is not surprising in a continent which uses rivers frequently as boundaries. The differences between Brazil and France over the identification of the Japoc River which were mentioned earlier were repeated in negotiations between Argentina and Brazil, Bolivia and Brazil, Brazil and Paraguay, Dutch Surinam and French Guiana, and Dutch Surinam and British Guiana (Ireland 1938, pp. 16, 53, 119, 243, 245). Although President Hayes made the Rio Pilcomayo the boundary between Argentina and Paraguay in 1878, it was not until 1945 that the two countries were able to agree on the co-ordinates of points which marked the boundary through the poorly drained section of its central course (The Geographer 1979).

The use of mountain ranges to define boundaries created problems for Bolivia, Argentina, and Chile. The last two countries included a boundary description in their treaty dated 23 July 1881 which offered a choice of two lines. The first was the line of highest crests and the second was the watershed. This situation, which arose because the treaty draughtsmen thought the two lines coincided, was solved by British arbitration (Ireland 1938, pp. 22–7). Bolivia and Argentina faced the problem of deciding which of two *cordillera* in the Andes was intended by an agreement in 1891 (Ireland 1938, p. 10).

Contemporary boundary disputes

In North America the only boundary disputes are located offshore. In addition to some offshore disputes in Central and South America, there are also disputes over land boundaries which have defied solution (Fig. 8.4). There are eight territorial disputes, three positional disputes, and two concerned with resource development.

Territorial disputes

The 1982 war in the Falkland Islands and the protracted arbitration of ownership of islands in the Beagle Channel which ended in May 1985 make it clear that territorial issues still arouse strong political passions in Central and South America. Recent studies of such questions indicate that these disputes wax and wane in importance (Veliz 1968, pp. 403–12, Day 1982.).

The dispute over ownership of the Falkland Islands and Dependencies has a long history. Discovered in 1592 and named in 1690 by British captains, the islands passed under French control when settlers from that country arrived in 1764. These voyagers from St Malo named the islands *Les Malouines*, which provided the basis for the Argentinian name of Malvinas.

French rights were ceded to Spain, which then recognized a British settlement in 1771 (Day 1982, p. 341). The British settlers withdrew in 1774 and a Spanish garrison then occupied the islands until 1811. The islands became a base for American and British sealers in the second and third decades of the 19th century. These sealers complained to their respective governments after authorities in Buenos Aires claimed the islands in 1820 and established a governor there in 1829. The expulsion of the Argentinians was started in 1831 by American naval forces and completed by the British navy the following year. Since then, British

Figure 8.4 Boundary disputes in Central and South America.

sovereignty has been maintained. Christie (1951, p. 265) has described Argentina's protests against British claims in 1834, 1841–9, and after 1884. Goebel (1927) provides a useful summary of Argentina's case for ownership of the islands. A good outline of the British case is contained in the *Polar Record* (1956).

Efforts by Britain in 1955 to have the matter settled by the International Court of Justice failed when Argentina refused to submit to the Court's authority. Nor has the British victory in 1982 solved the issue in a permanent way. The new government which replaced the discredited administration of President Galtieri after the war reaffirmed Argentina's claims to the islands (Barston & Birnie 1983, Perl 1983).

At the other end of the continent there is the unresolved claim by Guatemala to Belize. Menon (1979) has provided a useful survey of the legal basis on which both sides contest this matter. Guatemala's strictly legal and historic case, which has some merit, appears to be effectively blocked today by Belize's arguments about the modern concept of self-determination. It may be that tension over this issue will be reduced by a formula which guarantees Guatemala unimpeded access through its territorial waters in the Caribbean Sea. Claims of 12 nautical miles by Belize and Honduras would surround Guatamela's territorial waters, and that would be regarded as unsatisfactory. There is some evidence that Belize is prepared to compromise on this particular problem (Prescott 1985, pp. 348–9).

In January 1983 the Guatemalan authorities announced that the claim to all the territory of Belize would be reduced to the southern district of Toledo, including the town of Punta Garda. However, when this administration was replaced in August 1983 Guatamela's claim to all Belize was reasserted.

In the western Caribbean Sea off the coast of Nicaragua there is a group of islands which form the San Andres y Providencia Intendancy of Colombia. The islands are Roncador Cay, Bajo Nuevo, Quita Sueno, Serrana, Serranilla, Cayos de Alberquerque, Courtown Cays, Isla de San Andres, and Isla de Providencia. Nicaragua claims all but the last four islands in this group.

The origins of this dispute can be traced back to 1803 when Spain delegated to the Viceroyalty of Santa Fé de Bogotá, of which Colombia was then a part, responsibility for the defence of Isla de San Andres, Isla de Providencia, and the Mosquito Coast in the north of modern Nicaragua. According to Nicaragua, this order was rescinded in 1806 and defence responsibilities were returned to the Captaincy-General of Guatamela, of which Nicaragua was then a part (Day 1982, p. 359).

In 1928, during intervention by American forces in Nicaragua, that country was persuaded to sign a treaty with Colombia on 24 March. Under the terms of this treaty Colombia's sovereignty was recognized over Isla de San Andres, Isla de Providencia, Santa Catalina, and other islets and reefs forming part of the San Andres Archipelago. Nicaragua's sovereignty was recognized over the Mosquito Coast and the Great and Little Corn Islands. The treaty specifically excluded any concern with Roncador Cay, Quita Sueno, and Serrana because these islands were in dispute between the United States and Colombia. The clear implication of the reason for excluding these islands is that Nicaragua had no claim to them. In a treaty between Colombia and the United States in 1972 the

American claim to these three islands was withdrawn, although Colombian sovereignty was not specifically recognized.

On 4 February 1980 Nicaragua abrogated the 1928 treaty with Colombia. If Nicaragua ever succeeded in securing these islands, it would also gain claims to the sea and seabed covering 70 000 square nautical miles (Prescott 1985, p. 347).

The United States military base of Guantanamo occupies an area of 117 square miles on the southeastern coast of Cuba. This base was acquired in 1903 when the United States recognized the independence of Cuba and the lease was confirmed in 1934. During President Castro's rule, Cuba has regarded the American presence as illegal and has refused all payments of rent (Day 1982, p. 363). Since 1960 the United States has insisted that the 1934 agreement can only be altered or terminated with the consent of both parties, and there is no current prospect of any change in the present position.

After Britain gained Berbice, Demerara, and Essequibo from the Netherlands in 1815 it joined them to form British Guiana in 1831. Venezuela had seceded from Gran Colombia in 1830, and these new neighbours continued a boundary disagreement which had occupied the Netherlands and Spain for the previous 70 years. Venezuela claimed all the territory west of the Essequibo River and Britain opted for a line traced by Sir Robert Schomburgk in 1841 (Braveboy-Wagner 1984, pp. 95–104). The problem appeared to be settled by an arbitral award on 3 October 1899 which gave Britain most of the disputed area. Alas, in 1949 a memorandum dictated by the Venezuelan representative at the tribunal was published after his death. This memorandum alleged that the decision of the tribunal was reached by an illegal compromise and that political pressure had been used to achieve it. Not surprisingly, Venezuela reopened the question and extended its claims to the offshore waters bordering the disputed zone.

According to Day (1982, pp. 381–2), the major issue in dispute is the extent of Dutch control before 1814. The Guyanan authorities assert that the Dutch controlled the Essequibo River from Kykoveral, which was established in 1616. It is also asserted that the Dutch exercised control over trade and the indigenous population from posts on the Pomeroon, Barima, and Cuyuni rivers. Arguments advanced by Venezuela rely on the geographical unity of the land between the Essequibo and Orinoco rivers, on the discovery and exploration of the disputed area by Spaniards, and on the activities of Spanish missionaries. Wilgus (1943, p. 298) has produced a detailed map showing the various boundaries proposed by the two sides between 1768 and 1899, when the arbitral boundary was established.

On 17 February 1966, three months before Guyana became independent, Britain and Venezuela signed an agreement which estabished procedures for settling this boundary dispute. A mixed commission was formed and given four years in which to find a solution. In 1970, when the final report of the commission was presented, it was clear that the commission had failed in its task. Instead of proceeding to select a means for peaceful settlement provided by Article 33 of the United Nations Charter as the 1966 agreement required, Britain, Guyana, and Venezuela signed the Protocol of Port of Spain on 18 June 1970 (Day 1982, p. 383). This agreement placed a moratorium on assertions of territorial claims by the two sides for 12 years and required the two governments to explore all possible avenues for improving relations between the governments and peoples of both countries. When the protocol was due for renewal in 1982, Venezuela declined to extend its duration and noted that the 1966 agreement would be studied to discover alternative ways to reach a solution to the problem. Unfortunately there is no reason to expect an early end to this dispute.

West of Haiti lies Navassa Island which was claimed for the United States on 1 July 1857 by Peter Duncan. This small limestone island consists of an undulating plateau surrounded by white cliffs up to 50 feet high. The island was once the site of a phospate mine, but today it is only used by visitors on holiday. Haiti also claims Navassa Island and the authorities arranged for a church to be built there in the 1950s. When Cuba and Haiti delimited their maritime boundary on 10 October 1977 they ignored the existence of Navassa Island. If the United States insisted on its full legal claim to waters and seabed around the island the enclosed area would measure about 4100 square nautical miles (Prescott 1985, p. 353).

This issue briefly attracted some attention in June 1981 when a group of amateur radio operators from Haiti landed to transmit messages and raise the Haitian flag. They were met by a detachment of American marines who did not interfere with the activities of the Haitians after recording their names and addresses (Day 1982, p. 385)!

As explained earlier in this chapter, Peru and Ecuador have contested the ownership of three areas since 1830, when Ecuador was formed. Before that the dispute had concerned Peru and Gran Colombia. Fighting between the forces of Ecuador and Peru in January 1981 continued a tradition which began in 1859. The history of this dispute has been recounted by Wright (1941), Zook (1964), and The Geographer (1980), and it centres on the interpretation of the northern limit of the Viceroyalty of Peru in 1810. Mediation by other American states has limited serious conflict since 1942, but there can be no absolute guarantee that this border will not cause prolonged fighting in the future.

Ever since Bolivia lost its coastlands along the Pacific Ocean to Chile in 1879, successive Bolivian administrations have aimed to regain at least a corridor to the sea. Day (1982, pp. 356–8) and Ireland (1938, pp. 63–6) have described various attempts by the Bolivian authorities to achieve this aim. In 1975 Chile offered a corridor under the following conditions. Bolivia should cede to Chile an area equal to the extent of the corridor; Day (1982, p. 357) records that the area sought by Chile is probably in the Potosi Department, which is potentially rich in minerals. Bolivia should buy the Arica-La Paz railway and pay compensation for the port installations. Chile should receive full rights to use the waters of the Lauca River and Bolivia should undertake to keep the area demilitarized. For a time it seemed possible that there was a basis for discussion, but in 1978 Bolivia broke off relations with Chile because it was not showing any flexibility.

Any cession of a corridor through former Peruvian territory would have to be sanctioned by Peru according to the terms of the supplementary protocol signed by Chile and Peru on 3 June 1929. The Treaty of Ancon, signed on the same day, arranged for the return of Tacna to Peru. There are reports from Chile that Peru refuses to allow such a corridor to be ceded to Bolivia and insists that if cession occurs it must be through former Bolivian territory. Such a corridor would isolate the northern part of Chile and it has been rejected for that reason.

Positional disputes

A dispute over the location of their common boundary commenced between El Salvador and Honduras in 1838 when the Federation of Central American States was dissolved. Honduras claimed territory corresponding to the Bishopric of Honduras and El Salvador claimed a collection of administrative areas established during the period of Spanish rule. These two claims overlapped. Prior to 1969, when serious fighting developed between the two countries the only section of boundary which had been firmly settled lay along the Goascoran River (Day 1982, p. 372).

In 1969 the government of Honduras introduced regulations which meant that thousands of migrants from El Salvador had to return home. These migrants or their forebears had been crossing into uninhabited areas of Honduras since the 1920s and farming small areas. This conflict was soon ended by the Organization of American States and in June 1970 the two sides agreed to establish a demilitarized zone 3 kilometres wide on either side of the boundary. A decade later the two sides signed a general peace treaty on 30 October 1980, and Section IV of this treaty delimited the boundary for 225 kilometres (Day 1982, p. 374). The sectors of boundary remaining to be delimited are located on the edges of Chaletanango and La Union Provinces of El Salvador and in the delta of the Goascoran River. A mixed commission was appointed to define these sectors and it was instructed to consider documents issued by the Spanish civil and ecclesiastical authorities and any other legal, historical, or human factors appropriate to the application of international law. The sides have agreed that if this commission is unable to reconcile the different views of each country the matter will be sent to the International Court of Justice.

There is a positional dispute between Guyana and Suriname which dates from 1794. In that year the authorities in Amsterdam ruled that the boundary between the two Dutch possessions, called Surinam and Berbice, was the west bank of the Courantyne River (Corantijn in Dutch) (Day 1982, p. 378). The governors of the two Dutch provinces had been arguing over the ownership of rich cotton lands along the coast in the vicinity of this river.

In 1799 Britain acquired these two areas and retained the services of the Dutch governors. These men confirmed the earlier boundary in an Act signed on 20 January and published on 7 February 1800:

> That the west sea coast of the River Corentyne up to the Devil's Creek, besides the west bank of the said River, hitherto considered belonging to the Government of the Colony of Surinam, be declared and acknowledged henceforth to belong to the Government of the Colony of Berbice. (*Revised laws of Guiana* 1949, Vol. 3, p. 6)

The reference to the sea coast was probably made bacause the mouth of the river is nearly 5 nautical miles wide and open to rough seas on occasions. The Act also stipulated that all islands in the river belonged to Surinam together with an unidentified 'Post established on the west bank of the River Corentyne'. The Dutch authorities regained control of both colonies in 1802 and then lost Berbice again to Britain in 1803. British Guiana was formed from the territories of Berbice, Essiquibo, and Demerara in 1831.

When the boundary was defined in 1794 and confirmed in 1800 the extent of the Courantyne River was not known. In 1841 Sir Robert Hermann Schomburgk explored the hinterland of British Guiana and reported that the river continued southwards along its main tributaries, the Coeroeni and Kutari rivers. These rivers were then shown as the boundary between British Guiana and Dutch Surinam until 1899. In that year an arbitral commission was working on the boundary between British Guiana and Venezuela. The Dutch authorities now argued that the New River, discovered in 1871 by Barrington Brown, was the proper continuation of the Courantyne River (Fig. 8.5). The area in dispute measures 13 000 square kilometres.

The dispute soon waned as a Dutch geographer called Yzerman announced in 1924 that the Kutari and Coeroeni rivers occupied a larger catchment area than the New River (Day 1982, p. 379). The Dutch government used this argument to dismiss claims from its colony for action against Britain. An offer was then made to Britain by the Netherlands to sign a treaty confirming the Courantyne, Coeroeni, and

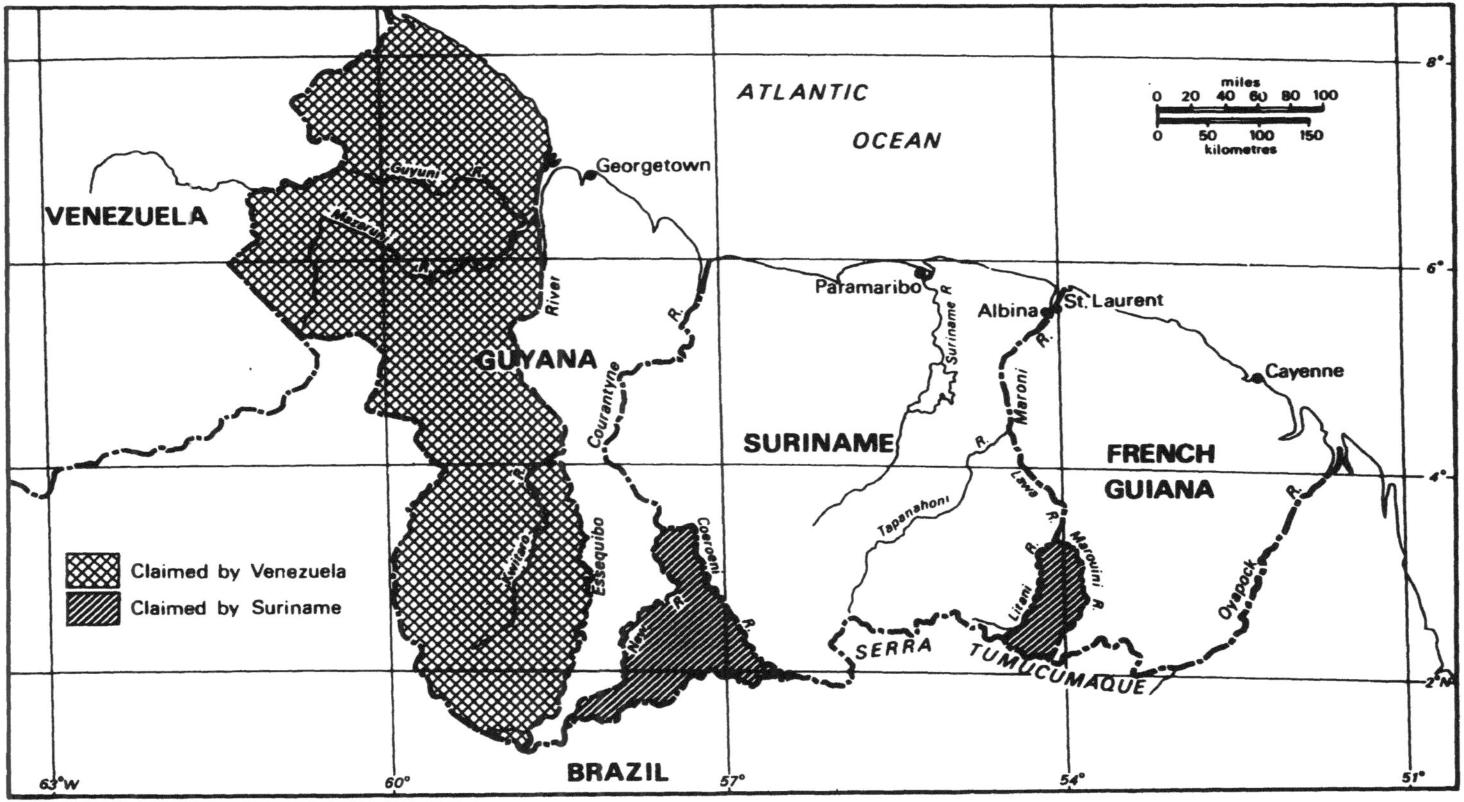

Figure 8.5 Boundary disputes involving Venezuela, Guyana, Suriname, and French Guiana.

Kutari rivers as the boundary between their respective possessions. This offer was made on 4 August 1930, but before the final version could be signed in 1939 war broke out and the matter remained unsettled. After the war the Netherlands again demanded that the boundary should follow the New River and Guyana and Suriname have now inherited this Anglo-Dutch dispute. This will not be an easy dispute to solve because the issue turns on what was understood to be the Courantyne River in 1800, if it is accepted that the act signed by two Dutch governors in the temporary service of Britain has any standing in international law. If that act is accepted, perhaps Suriname could request the return of 'the Post established on the west bank'. If this act is deemed to have no validity, the two sides will have to base their case on the boundary which was understood to exist between the two territories throughout their history.

There is a similar problem to the one just described between Suriname and French Guiana. Since 1688 the boundary between these two territories has been the Maroni River (Marowijne in Dutch). However, when France and the Netherlands first reached this decision they were only concerned with the fertile plains about the broad estuary of this river. Vessels of very shallow draught can sail 43 nautical miles up this river, but vessels with a draught of 15 feet can only reach Albina in Suriname or St Laurent in French Guiana, which are 15 nautical miles from the sea.

As settlement proceeded southwards, a doubt arose about the continuation of the Maroni River. At a point 165 kilometres from the sea the Maroni is formed by the union of two tributaries. The Tapanahoni River drains country south-west of the confluence and the Lawa River drains land to the south-east and south (Fig. 8.5). Discussions between the two governments and between their representatives in the colonies failed to resolve this dispute (Ireland 1938, pp. 243–4). In 1861 the two governors appointed a mixed commission to consider this matter, but the commission found that the name Maroni could only be applied to the river below the confluence of the Tapanahoni and Lawa rivers.

In 1890 France and the Netherlands decided to refer the issue to arbitration by Czar Alexander III of Russia. The result of the arbitration was that the Lawa River was nominated as the proper continuation of the Maroni River. The Netherlands' case was judged to be stronger because Dutch military posts had been maintained on the Lawa since 1700 and because the French government had agreed on occasions that the Dutch controlled the indigenous population in the disputed zone.

Within ten years it was discovered that there was uncertainty about the location of the upper course of the Lawa River. This river has two principal tributaries called the Litani River, which flows from the south-west, and the Marouini, which flows from the south-east. France and the Netherlands were unable to agree on which of these tributaries should form the international boundary before Suriname became independent in

1975. Day (1982, p. 377) reports that Suriname and France reached an agreement by which France would give Suriname a large development grant in return for Suriname's recognition of French sovereignty over most of the disputed area. However, this agreement was not ratified and this area of 4600 square kilometres remains in dispute.

Resource development disputes

Although the Rio Pilcomayo was selected as the boundary between Argentina and Paraguay in 1878, it was only in 1945 that a mixed commission was able to fix the course of the boundary through the poorly drained swampy section in its middle reaches (The Geographer 1979).

In 1974 the two countries agreed to study the Rio Pilcomayo in order to discover how its annual flood could be used most effectively. Argentina then constructed some levees to reduce the risk of flood damage on the Argentinian side. In 1980 Paraguay alleged that the Argentinian works were shifting the course of the Rio Pilcomayo to Argentina's advantage and that this was part of a deliberate plan (Day 1982, p. 339). Argentina has denied this charge, but the matter is still unresolved.

In Chapter 4 the disagreement between Bolivia and Chile over the Lauca River was described. One-third of the Lauca River flows through Chile and the remaining two-thirds flows through Bolivia, before ending in the Lago de Coipasa. In 1962 Chile completed works which enable part of the discharge of the Lauca River to be diverted to the valley of the Azapa. Bolivia has objected that these diversions will adversely affect the climate of the valley of the Lauca River in Bolivia and will increase the salinization of the Lago de Coipasa. Chile tried without success to link this dispute to the question of a corridor for Bolivia to the sea, but the attempt failed and both disputes remain unsolved.

Conclusions

International boundaries in the Americas carry the stamp 'Made in Europe' to a much slighter extent than Asia, Africa, Antarctica, and the Middle East. Britain as a European power had a major influence on the boundaries in North America, but even here the independent colonists of the United States had an equal responsibility for the outcome. Spain in Central America and Spain and Portugal in South America laid the general framework of limits by the beginning of the 19th century, but it was the independent colonists of countries such as Brazil and Paraguay which played the decisive rôle in transforming that general framework into the more or less precise structure which exists today.

The construction of boundaries was more difficult in Central and South

America than in North America. There are three main reasons for this situation. First, there were many more states in Central and South America and therefore more opportunities for disagreements to arise. Secondly, the division of the former Spanish Empire in Central and South America was bedevilled by disputes over the areas controlled by various Spanish civil, legal, and ecclesiastical authorities which provided the buildings blocks for independent states. Such disputes still underlie the problems between Ecuador and Peru and Colombia and Nicaragua. Thirdly, the physical geography of South America was imperfectly known for much of the 19th century, and this ignorance was particularly evident in respect of some rivers' courses on the fringes of the basin of the Amazon River.

Unfortunately, the process of boundary evolution is incomplete in Central and South America and a number of boundary disputes on land remain. In North America the problems which arose over land boundaries have been solved and good progress is now being made in solving those boundary problems which have occurred offshore.

References

Barston. R. P. & P. W. Birnie 1983. The Falkland Islands/Islas Malvinas conflict: a question of zones. *Marine Policy* **7**, 14–24.

Bloomfield, D. L. M. & G. F. Fitzgerald 1958. *Boundary water problems of Canada and the United States*. Toronto: Caswell.

Braveboy-Wagner, J. A. 1984. *The Venezuela–Guyana boundary dispute: Britain's colonial legacy to Latin America*. Boulder, Colo.: Westview.

Burns, E. B. 1980. *A history of Brazil*. New York: Columbia University Press.

Chapman, C. E. 1938. *Colonial Hispanic America: a history*. New York: Macmillan.

Christie, E. W. H. 1951. *The Antarctic problem*. London: Allen & Unwin.

Cunningham, C. H. 1919. *The Audiencia in the Spanish colonies as illustrated by the Audiencia of Manila 1583–1800*. Berkeley: University of California Press.

Day, A. J. (ed.) 1982. *Border and territorial disputes: a Keesing's reference publication*. Detroit: Gale Research.

Delper, H. (ed.) 1974. *Encyclopedia of Latin America*. New York: McGraw Hill.

Dennis, W. J. 1931. *Tacna and Arica*. New Haven, Conn.: Yale University Press.

Fifer, J. V. 1966. Bolivia's boundary with Brazil: a century of evolution. *Geographical Journal* **132**. 36–72.

Fisher, L. E. 1926. *Viceregal administration in Spanish colonies*. Berkeley: University of California Press.

Fisher, L. E. 1929. *The intendant system in the Spanish colonies*. Berkeley: University of California Press.

Goebel, J. 1927. *The struggle for the Falkland Islands: a study in legal and diplomatic history*. New Haven, Conn.: Yale University Press.

Harrisse, J. 1897. *The diplomatic history of America*. London: Stevens.
Herring, H. 1965. *A history of Latin America*. New York: Alfred Knopf.

Ireland, G. 1938. *The possessions and conflicts of South America*. Cambridge, Mass.: Harvard University Press.

Jones, S. B. 1932. The forty-ninth parallel in the Great Plains. *Journal of Geography* **31**, 357–67.
Jones, S. B. 1937. The Cordilleran section of the Canada–United States borderland. *Geographical Journal* **89**, 439–50.

Lambert, A. F. 1965. Maintaining the Canada – United States boundary. *Canadian Cartographer* **2**, 67–71.

Menon, P. K. 1979. The Anglo-Guatamelan territorial dispute over the Colony of Belize (British Honduras). *Journal of Latin American Studies* **11**, 343–71.

Nicholson, N. L. 1954. *The boundaries of Canada, its Provinces and Territories*. Memoir 2, Department of Mines and Technical Surveys, Geographical Branch, Ottawa.

Parry, C. 1969a. *Consolidated treaty series*, Vol. 38. New York: Oceana.
Parry, C. 1969b. *Consolidated treaty series*, Vol. 46. New York: Oceana.
Paullin, C. O. 1923. The early choice of the 49th parallel as a boundary line. *Canadian Historical Review* **4**, 127–31.
Paullin, C. O. 1932. *Atlas of the historical geography of the United States*. New York: Carnegie Institution and the American Geographical Society.
Perl, R. 1983. *The Falkland Islands dispute in international law and politics: a document sourcebook*. London: Oceana.
Piper, D. C. 1967. *The international law of the Great Lakes: a study of the Canadian – United States cooperation*. Durham, NC: Duke University Press.
Polar Record 1948. Antarctic claims – recent diplomatic exchanges between Great Britain, Argentina and Chile, **5**, 228–44.
Prescott, J. R. V. 1985. *The maritime political boundaries of the world*. London: Methuen.

Revised laws of Guiana 1949. Vol. 3. Georgetown: Government Printer.

Scott, J. B. 1922. The Swiss decision in the boundary dispute between Colombia and Venezuela. *American Journal of International Law* **16**, 428–31.

The Geographer 1961. *Dominican Republic – Haiti boundary*. International Boundary Study No. 5, Department of State, Washington, DC.
The Geographer 1978. *Bolivia–Paraguay' boundary*. International Boundary Study No. 165, Department of State, Washington, DC.
The Geographer 1979. *Argentina–Peru boundary*. International Boundary Study No. 166, Department of State, Washington, DC.
The Geographer 1980. *Peru-Ecuador boundary*. International Boundary Study No. 172, Department of State, Washington, DC.
Thomas, A. B. 1956. *Latin America: a history*. New York: Macmillan.

Veliz, C. 1968. *Latin America and the Caribbean: a handbook*. New York: Praeger.

Wilgus, A. C. 1943. *Latin America in maps*. New York: Barnes & Noble.
Wright, L. A. 1941. A study of the conflict between the Republics of Peru and Ecuador. *Geographical Journal* **98**, 253–72.

Zook, D. 1964. *Zarumilla-Maranon: the Ecuador–Peru dispute*. New York: Bookman Associates.

9 *Asia*

This chapter is concerned with that part of mainland Asia which lies east of meridian 60° east. That line lies close to the boundaries separating Iran from Afghanistan and Pakistan. The evolution of political boundaries in this extensive region finds a unity in the sets of relationships between China and the colonial powers of Russia, Britain, France, and Japan.

China formed the hub about which these other states operated. Russia occupied the sector from Vladivistok to Afghanistan; the British theatre stretched from Afghanistan to Thailand and included the Malayan peninsula; France dominated Indochina and Japan held sway in the Korean peninsula. Britain, France, Germany, and Portugal secured small, important footholds on China's coast, but the duration of even Hong Kong was but a moment in the perspective of China's history.

Asia's political boundaries evolved in three phases. In the period before 1914 the colonial powers carved out their Asian empires and set the limits of those Asian states such as Afghanistan, Bhutan, China, Nepal, and Thailand, which were not directly controlled. Apart from the Sino-Russian treaties of 1689 and 1727, these Asian states invariably negotiated from a position of weakness. Indeed, with the exception of China, it seems likely that the other states were not totally annexed because of Britain's preoccupation with the need to avoid common boundaries with other colonial powers and China. That British policy of self-denial did not stop Russia from acquiring large areas of northern Afghanistan or prevent France from compelling Thailand to disgorge extensive tracts in the Mekong valley. Britain's ineffective attempt to create the buffer state of Tibet from a province of China was successfully duplicated by Russia, which detached Outer Mongolia in 1910 and subsequently prevented any resumption of that area by China.

The second phase lasted until the end of World War II, and it was characterized by the preservation of the *status quo*. There were minor boundary alterations involving Afghanistan and Thailand, but Japan's major efforts to redraw boundaries in Indochina and Manchuria failed entirely. During this period the internal boundaries of the British and French empires were maintained. These unilateral domestic limits provided the lines of cleavage along which those empires split in the third phase. The replacement of those empires by a number of independent Asian states is the chief characteristic of this latest period. However, this third phase is marked by three other developments. A number of independent states have negotiated definitive boundary agreements with each other. China has been in the forefront of this activity and has reached settlements with Mongolia, Afghanistan, Pakistan, Nepal, and Burma. In each case the new agreements have either confirmed existing international or traditional limits or altered them only slightly. A number of boundary disputes have emerged and fighting has occurred between China and the Soviet Union, India and Pakistan, India and China, and Thailand and Laos and Cambodia. Conflicts within the region yielded three cease-fire lines in Kashmir, Korea, and Vietnam which became international boundaries. That in Vietnam disappeared in 1975 when the country was unified.

The political boundaries of Asia did not evolve in an orderly fashion through the stages of allocation, delimitation, demarcation, and administration identified in Chapter 3. For example, the boundary along the northern watershed of the Amur River which allocated territory between Russia and China in 1689 was never demarcated. Sections of the boundary between Afghanistan and Iran were precisely delimited by arbitrators and were demarcated very quickly once their decisions were accepted. The delimitations of the period before 1914 were not always free from ambiguities, and the boundary monuments erected at that time were sometimes not maintained carefully. This has meant that in the latest period some of the independent states of Asia have faced three major types of boundary problems. First, it has been necessary to agree on a meaning of imprecise descriptions inherited from colonial administrations. Such a problem led Cambodia and Thailand to the International Court of Justice in 1961. Secondly, where new states have been formed by elevating internal boundaries to the level of international limits there has often been a problem of deciding where these internal limits lay. That difficulty once bedevilled relations between India and Pakistan in the Great Rann of Kutch. Finally, states have had to try to negotiate lines to close the gaps east and west of Nepal which Britain was unable to close through agreement with China.

The Russian sector from Vladivostok to Afghanistan

With one exception, the international boundaries in this region were either established before 1917 or were foreshadowed by arrangements reached before that date. The solitary exception is provided by the sector of boundary between the Soviet Union and Mongolia from Tengis Gol to Huyten Orgil. That line was fixed in 1945 when the Soviet Union acquired the headwaters of the Yenisey and Khemchik rivers, which today form the Tuvinskaya Autonomous Soviet Socialist Republic.

The Sino-Russian line was defined in three different sectors, and in each case the direct cause of negotiations was the contact or threat of contact between Russians advancing eastwards and southeastwards and groups which China considered to be subjects of the emperor. This contact occurred first in the Amur valley and then proceeded southwestwards; the process can be likened to the blades of shears closing while one of the blades is stationary (Fig. 9.1).

The treaty of Nerchinsk, in 1689, defined the boundary from the Argun River to the Pacific Ocean. The Russians were mainly concerned with the commercial provisions of this treaty, which they hoped would afford opportunities to replenish depleted exchequers. The Chinese were glad to have a settlement which prevented Russian interference in border communities with a propensity for rebellion. Because the Chinese attached paramount importance to excluding Russia entirely from the Amur River, the line was fixed along the northern watershed of that river. The main territorial concession won by Russia was the declaration that territory between the River Uda and the watershed of the Amur River would remain neutral pending final allocation. There is no reason to believe that either side realized at that time how this simple provision was going to be used to China'a extreme detriment nearly 200 years later.

From the middle of the 18th century, Russians such as Muller in 1741, Myetlet in 1753, Yakoff in 1756, and Shemelin in 1816 had been urging the Czar to acquire the right of transit through the Amur valley either by negotiation or conquest. By the 1850s Russians sailed regularly along the Amur River and they built settlements, established stockpiles, and stationed detachments of troops. By 1855 the growth of Russia's influence in this region and the sapping of Chinese strength as that country was

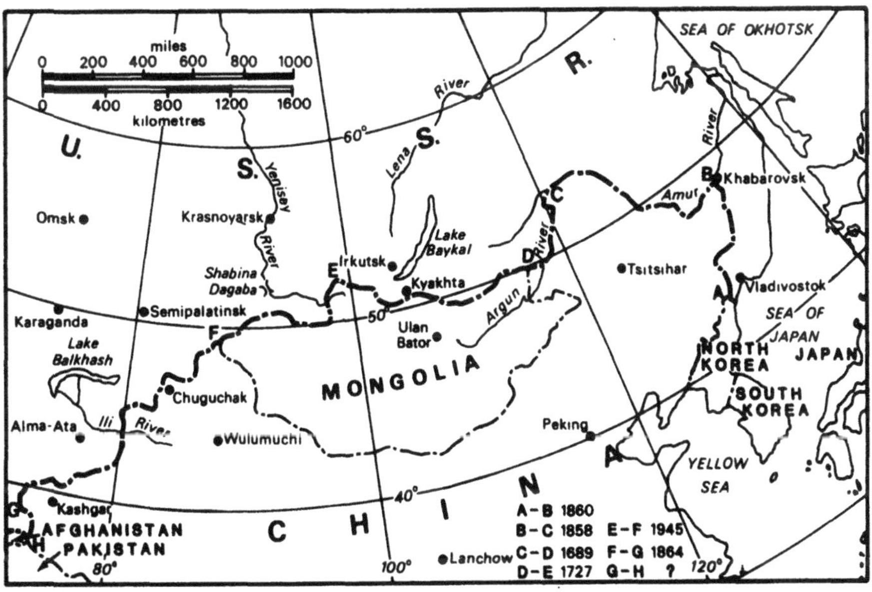

Figure 9.1 The Russian sector.

beset by the Taiping rebellion and external pressure from Britain and France produced a situation when the issue could be forced. Governor-General Muraviev of Eastern Siberia presented China with a memorandum which urged the cession to Russia of all territories on the north bank of the Amur River. The document began by asserting quite falsely that the whole area had been designated neutral in 1689, and subsequent arguments became more spurious. But there was nothing counterfeit about Russia's military superiority, and in 1858 China ceded the area north of the Amur River.

Quite remarkably, China allowed a fresh neutral zone to be created. It extended from the River Ussuri to the sea. The Russian version of the Treaty of Aigun in 1858 makes it apparent that this description applied to the south bank of the Amur River from Khabarovsk to the sea, and that was China's understanding. Once again Russia had a different view, which prevailed. For Russia the neutral area constituted the entire Maritime Provinces bounded by the Ussuri and Amur rivers and the Pacific Ocean! That large area was transferred to Russia by the Treaty of Peking on 2 November 1860.

The central boundary section from the headwaters of the River Argun to Shabina Dagaba was settled in 1727. Once again the states had different motives and China was prepared to give Russia commercial advantages, providing Russia avoided contact with nomadic tribes in the Chinese borderland. The failure of the Russian authorities to return some Chinese deserters according to the terms of the 1689 agreement caused China to suspend trade relations and expel the Russian agent in Peking. This prompted speedy negotiations by the Russians. The commercial provisions of the new treaty were agreed after discussions lasting from November 1726 to April 1727. The delegates then moved to Kyakhta near the border and by August 1727 had defined the boundary for 1690 miles in general terms (Prescott 1975, p. 18). Joint survey teams were immediately sent to define the boundary according to the new agreement to the east and west of Kyakhta (Chen 1949). The survey work was completed in time for final definitions to be exchanged and signed in October 1727. This was a remarkable achievement at that time in central Asia, especially in the area west of Kyakhta where terrain was very rugged and where there were no Chinese frontier posts.

After the treaty had been signed one of the Russian delegates boasted in the following terms:

> . . . much land was delimited which had never before been in Russian possession, namely: from the Khan Tengeri river a distance of approximately eight days horseback ride in length and in width three days. to the Abakana River, and these places had never been under the domination of the Russian Empire. (Mancall 1971, p. 301)

So long as there was an unclaimed territory southwestwards of the terminus of the boundary at Shabina Dabaga it would eventually be necessary to resume the boundary negotiations.

This need was tackled in the treaty signed in 1860. Although it focused mainly on the area between the Ussuri River and the coast, it also defined the continuation of the interior boundary in general terms. Provision was made for a new commission to settle this boundary, and this was done at Chuguchak in 1864. Two factors encouraged China to reach an agreement quickly on this question. First, another Muslim revolt had started at K'u-ch'e in June and within one month Ma-na-ssu, So-ch'e, Ying-chi-sha, and Su-fu were in the hands of the rebels. None of the previous 12 rebellions in this area in the previous century had spread so quickly. Secondly, Russian forces had launched a major assault against Kokand in May, and by September they had captured Dzhambul, Turkestan, and Chimkent. If China failed to reach agreement with Russia, there was a risk that an independent Muslim state would be established with which the Russians would be prepared to negotiate.

The commissioners, having been instructed by the 1860 treaty to draw a boundary related to the direction of mountains, the flow of large rivers, and the line of Chinese pickets, proceeded to define a line which zigzagged westwards along ranges or rivers and southwards across plains. The treaty was signed on 7 October 1864, but its provisions relating to demarcation could not be implemented because of the continued success of the Muslim rebellion. When T'a-ch'eng fell to the rebels on 11 April 1866 Chinese influence was restricted in Singkiang to small areas around Pa-li-k'un and Ha-mi in the east.

Soon afterwards the effects of the rebellion began to be felt in the Russian sphere (Hsu 1965). Trade through the Ili valley was being disrupted; Russian property in T'a-ch'eng and I-ning was either damaged or confiscated; and the influx of thousands of refugees into the Russian border began to create serious administrative problems. To avoid the continuation of these difficulties Russia occupied the Ili valley and promised China that the territory would be returned when China had restored its authority in Singkiang.

That situation occurred in 1878 and China sent an agent to St Petersburg to secure the return of the Ili valley. Quite naturally, Russia did not waste this opportunity to extract some reward for looking after China'a interests for the previous seven years. The Chinese ambassador agreed to transfer two areas to Russia and one of them was in the Tekes valley south of I-ning. China refused to ratify this treaty of Livadia and a new agent was sent to St Petersburg. The treaty named after that city was signed in February 1881. Territorial concessions were still made, but they were not in the Tekes valley, and this was considered by China to possess great strategic significance. The total area ceded to Russia was about 3600 square miles (Prescott *et al.* 1977, p. 10).

This new boundary was delimited by May 1884 when five protocols containing the descriptions were exchanged. The southern terminus of the boundary was located at Pereval Uch-bel', although this pass was not visited by the commission. It was noted that the Chinese and Russian boundaries diverged at this point to the south and south-west respectively. Although that may have been the situation in 1884, today the boundary continues for another 192 miles to the tri-junction with Afghanistan. This sector does not appear to have been defined in any treaty which has been published, although neither country seems to be in any doubt about the location of the boundary through this wild region.

Once again Russia gained most in territorial terms by the treaties of 1864 and 1881. However, some of the areas conceded by China had been held tenuously in the past and sometimes had slipped beyond control. At least these treaties imposed a limit on further encroachment by Russia, and the boundary has served China well ever since.

The boundary that separates China and Mongolia stretches for 2920 miles and it was first defined in December 1962. The publication of the description of its demarcation on 30 June 1964 revealed a line which has been defined with greater precision than any other Asian boundary. A text of 68 000 words describes the location of 678 cement and rock monuments, which are also shown in an atlas of 105 maps at a scale of 1 : 100 000 and six maps at a scale of 1 : 10 000.

The final definition of this boundary completed a process that started in November 1911. In that year Outer Mongolia declared itself independent at a time when disorder was spreading throughout China. China's policies designed to keep Chinese and Mongolians separated had been too successful in preserving and encouraging the distinctiveness of these border peoples.

Russia quickly concluded an agreement with the new state of Mongolia which noted that 'the old relations between Mongolia and China thus came to an end' (McMurray 1921, p. 992). Then, in 1915, Russia persuaded China to sign an agreement which recorded that China exercised suzerainty over Outer Mongolia but that this region was autonomous.

China enjoyed a final brief period of dominance during the confusion in this region at the time of the Russian revolution, but by 1924 Outer Mongolia was clearly independent of China as the Mongolian People's Republic. Mongolia first entered into boundary negotiations with Japan in the period 1935 to 1942, but the decisive arrangement was reached on 14 August 1945 when China agreed with the Soviet Union to recognize the independence of Outer Mongolia providing a majority of the population approved of that course in a plebiscite. The poll was conducted on 20

October 1945 and Mongolia's independence was recognized by China on 5 January 1946.

China insists, with much justification, that the treaties with Russia were signed under duress and that the treaties are unequal in their effects on the two countries. However, there is no reason to believe that the Chinese authorities entertain any hopes of regaining the Maritime Provinces, the areas north of the Amur River or various sectors in central Asia. The only serious positional dispute occurs at the confluence of the Amur and Ussuri rivers and it has been described in Chapter 4.

The British sector from Afghanistan to Thailand

Britain was very successful in gaining a large area occupied by a numerous indigenous population and endowed with a variety of agricultural and mineral resources. Occupation of Ceylon, Singapore, and the northern shore of the Strait of Malacca also provided important strategic advantages. But Britain was unsuccessful in defining clear, peaceful boundaries around this extensive colonial region. In the west, Britain's indecision about its chief aims allowed Russia to annex large areas of northern Afghanistan and advance much closer to British India than the strategists in London thought desirable. Britain's borderland with Afghanistan was the scene for endless raids by fierce tribesmen into the lowlands west of the Indus River. In the north, a satisfactory boundary with Nepal was flanked by two areas in the Himalayas where Britain was unable to persuade China to agree to any boundary. In the east, Britain was able to persuade China to settle a line marking the limit of Burma, but it failed entirely to prevent France from advancing round the north of Thailand and establishing posts in the headwaters of the Mekong valley.

The boundary between Russia and Afghanistan was defined in Anglo-Russian treaties between 1872 and 1895. In October 1872 the two countries agreed to use the Amu Darya as the boundary from Oz Zorkul in the east to Kwaja Salar (Fig. 9.2). The line between Kwaja Salar and the tri-junction of Russia, Afghanistan, and Persia was settled in a series of agreements between 1884 and 1888. The boundary east of Oz Zorkul to the tri-junction with China was fixed in 1895 (Prescott 1975. pp. 99–144).

The Amu Darya was an obvious feature in the landscape and on maps of the region, so its selection as a boundary was not surprising. In fact it did not correspond exactly to the political division of this area. The kingdom of Bukhara controlled Darwaz, which lay south of the river, and Afghanistan exercised authority in Roshan and Shignan, which extended north of the Amu Darya. When the Emir of Afghanistan was reluctant to cede Roshan and Shignan in 1893, Sir Mortimer Durand was sent to Kabul to persuade him. The Emir agreed to honour the 1872 line and Durand took the opportunity to make two other territorial arrangements on behalf of the British government. First, the Emir was persuaded, with great difficulty, to accept the Vakhan valley so that Russian territory would not share a common boundary with British India. Secondly, Durand negotiated the boundary, which now bears his name, between Afghanistan and British India.

It was in the region west of Kwaja Salar, between the Amu Darya and the Persian border, that Britain suffered its worst diplomatic defeat in the negotiations with Russia. The definition of the 1872 line became less precise as it proceeded westwards from Oz Zorkul to the Persian border. Indeed, the final section from a point 50 miles west of Kwaja Salar was not delimited because 'it is well known and need not be here defined' (United Kingdom, Foreign Office 1872–3, pp. 743–4).

The incompetence of the British officials can be shown by three facts. First, appeals by the Emir for assistance in negotiating a boundary with Russia were politely rejected on the ground that the boundary was fixed in 1872 and it would be unwise to reopen the question (Singhal 1963, p. 107). Secondly, when Russia volunteered to negotiate a line from Kwaja Salar westwards to Sarakhs in 1883 Britain rejected the offer by saying that 'the proposal did not in any way meet the requirements of the case' (United Kingdom, Foreign Office 1883–4, p. 947). This was the best offer Britain received during the negotiations and the final terminus of the boundary was at Zulfikar Pass which lies 60 miles south of Sarakhs on the Hari Rud. Thirdly, British officials ignored warnings given by the British ambassadors in Persia and Russia in 1870 and 1872 respectively that Russian policy in this region required the establishment of a safe commercial route, through inhabited areas, from the Caspian Sea to the Amu Darya. Such a route had to pass through areas south of the line between Sarakhs and Kwaja Salar. The British officials ignored these excellent reports believing Russian protestations that they had no designs on Mary. It was only after 14 February 1884 when British officials received news that the Turkoman tribes around Mary had submitted to Russia that they were prepared to undertake serious negotiations.

Even then, Russia continued to whittle away at Afghanistan's territory in the Morghab valley until Britain identified three salient points which could not be ceded. They were Zulfikar Pass, Mari Chaq on the Morghab River, and Kwaja Salar. A line connecting these three points was agreed in September 1885 and the boundary was demarcated with some difficulty by January 1888.

The Durand Line, which extends for 1510 miles from the snowfields of the Hindu Kush to the baked stony deserts of Baluchistan, did not coincide with any clear physical feature and did not separate distinct political organizations. In the foothills of the Afghan plateau there was

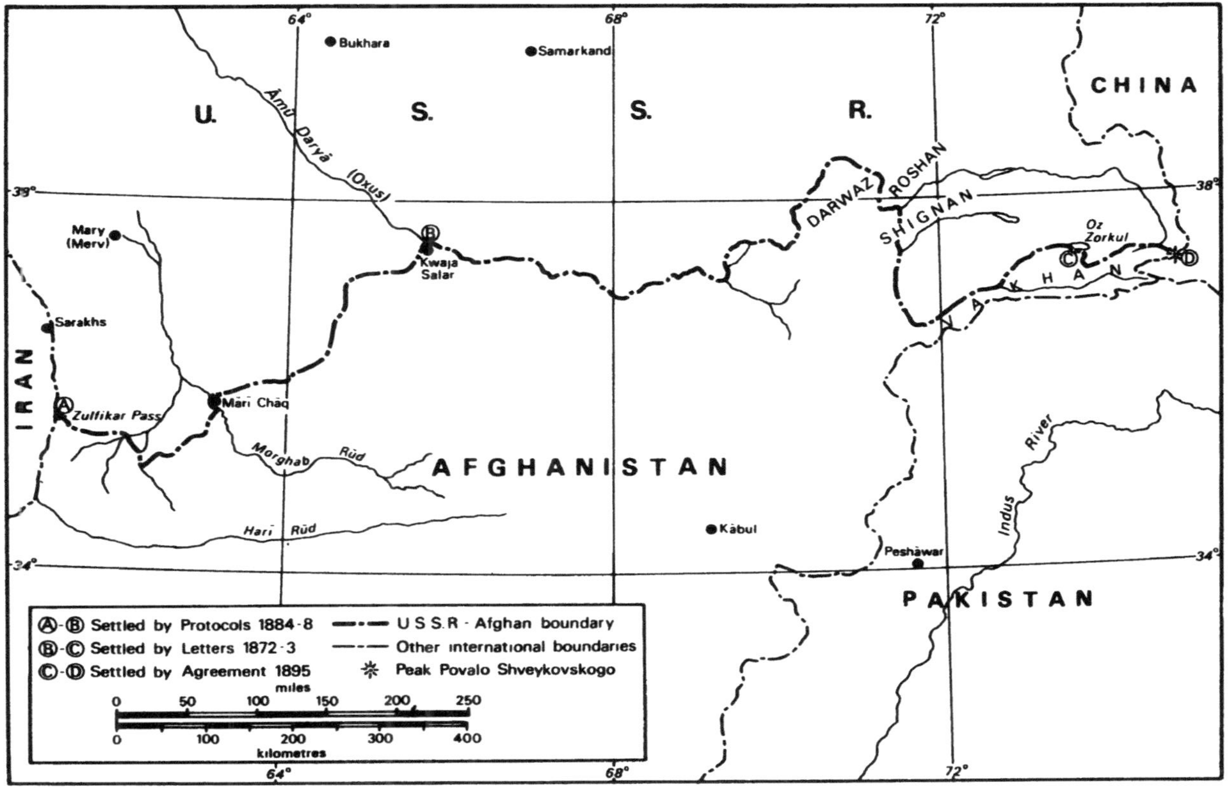

Figure 9.2 The Russo-Afghan boundary.

nothing to suggest one watershed would make a better boundary than the next and the main rivers flowed transverse to the direction the boundary had to follow. The successive waves of conquerors, frontiersmen, and nomads through this borderland had produced a mosaic of small communities which were often fiercely independent. There was thus a basic conflict between the order of the British Empire seeking a fixed, defensible line behind which life could be peaceful and productive and the undisciplined nature of tribesmen living in barren, almost waterless wastes who secured many of the necessities of life by raiding adjacent, fertile lowlands (Davies 1932, p. 179).

The boundary which Durand had agreed with the Emir in 1893 was demarcated with extreme difficulty by May 1896, with the exception of a section on either side of the Khyber Pass (Prescott 1975, pp. 182–5). This final section was redefined after the Third Afghan War in 1921.

The definition of Nepal's boundary with India was one of the few successes British authorities enjoyed. After Nepal breached the terms of peace treaties signed in 1792 and 1801 the Nepalese army was defeated by British forces and a British ultimatum was accepted and enforced. The eastern and western limits were easily fixed along the Mechi and Kali rivers respectively. The longer southern limit was first fixed at the northern edge of the Terai, a malarial forested tract about 20 miles wide between the marshy lowlands of the Ganges plain and the foothills of the Himalayas. In December 1816 and November 1860 Britain returned the Terai to Nepal and by 1875 the boundary had been demarcated.

Despite Indian claims to the contrary, Britain failed to persuade China to agree to any boundaries in northern Assam east of Nepal and in the Aksai Chin to the west of that country. In northern Assam, India relies on the McMahon Line, which was the product of British attempts, after 1904, to limit Chinese expansion and define the precise area of British responsibility in the Himalayas.

There are two versions of the McMahon Line and neither can be considered to be binding on China. First the line is marked on a map in two sheets, at a scale of 1 : 500 000; the map accompanied a secret agreement between Britain and Tibetan officials. Such an agreement cannot be taken seriously since China considered Tibet to be part of China on 25 March 1914 when the agreement was signed. Secondly, the McMahon Line was shown on a map which accompanied a Convention signed by British, Tibetan, and Chinese representatives. The scale of this map was 1 : 3 800 000, which means that the thickness of the line portraying the boundaries represents a zone 4 miles wide! The Chinese authorities repudiated this Convention when copies of the map reached Peking. It has always been the Chinese view that the Convention signed on 27 April 1914 was concerned with improving relations between authorities in Lhasa and Peking. This was achieved by confirming Chinese suzerainty over Tibet and recognizing the autonomy of Outer Tibet. The only purpose of the map, according to the Chinese, was to show the limit between Inner and Outer Tibet.

In Aksai Chin both countries agree that there is a traditional boundary, but they disagree on its location. China quite properly rejects claims that a boundary was drawn by Britain and China in 1899. Lamb (1966, 1973) has shown that Britain offered a boundary at that time but China not only did not accept the offer, it refused to engage in negotiations about the boundary.

Britain did have some success in persuading China to draw a boundary in northern Burma in the 1890s. In 1894 the two countries agreed on a line which trended north-west–south-east from the headwaters of the Irrawaddy River to the Mekong River. Where the terrain was deeply dissected in the north-west the boundary was closely defined by rivers and watersheds. Nearer the Mekong River the pattern of low plateaus offered no obvious boundaries and a line was defined in principle to separate various political groups. When China made concessions to France in northern Indochina in 1896, Britain demanded and received about 1500 square miles north of the 1894 line.

Britain was able to agree on a boundary with Thailand without very much difficulty. When Britain acquired Tenasserim in 1826 it was discovered that the boundary between this eastern province of Burma and Thailand was well known. This line was confirmed in an agreement in February 1868. As British forces penetrated into the headwaters of the Salween River it became necessary to make a decision on whether to halt progress at the Salween River or to annex states which straddled this river. The local authorities in Burma had no doubts about the correctness of crossing the Salween River and that course was adopted by Britain (Mangrai 1965, pp. 220–2). An extension of the 1868 line was surveyed in 1892–3 and confirmed in maps on 17 October 1894.

By that time the British authorities must have realized that there was little chance of avoiding a common boundary between Britain's Burma and France's Laos. Instead of replying positively in 1889 to France's proposal that a corridor of Thai territory should extend northwards to the Chinese border, Britain sent a map showing its impression of Thailand's boundaries. This prevarication was due to the fact that France had proposed the Salween and Nam Oo rivers as the limits of the Thai corridor and Britain had already decided to cross the Salween River. Sensibly, the French did not continue this correspondence, and Britain suddenly discovered in March 1893 that France now considered that Thailand did not own any territory on the east bank of the Mekong River. There was a flurry of diplomatic activity to try to revive the concept of a neutral zone between the two colonies, but it proved fruitless and the countries agreed on 15 January 1896 to a common boundary along the thalweg of the Mekong River.

The closing years of the first period of boundary construction in the British sector was marked by negotiations which produced boundaries between British India and Persia and between Malaya and Thailand. The boundary with Persia had been drawn from the coast at Gwatar Bay to the Rud-i-Mashkel in 1871 when a telegraph line was being established. This boundary was completed to the Afghan tri-junction in December 1895 (Prescott 1975, pp. 212–29). On 29 November 1899 Thailand and Britain agreed to a common boundary across the Malayan peninsula. In 1907, when it appeared likely that a German company would be involved in building a railway through Thailand's southern provinces, Britain reopened negotiations with Thailand with a view to excluding Germany. These negotiations resulted in a northward extension of British Malaya and in return Britain abandoned the extra-territorial rights in Thailand conferred by an earlier treaty.

In addition to the adjustment of the boundary in the vicinity of the Khyber Pass which has been mentioned, there were only two other boundary adjustments in the period between the world wars. During the period 1931–40 the authorities in Burma and Thailand reached a number of agreements to deal with sudden changes in the courses of the Meh Sai and Pakchan rivers which mark the northern and southern sections of their common boundary. In 1935 Britain and China decided to seek assistance from the League of Nations in settling difficulties in the area of the Wa states in the upper reaches of the Salween valley. The Iseln Commission presented its report to both sides in 1937, but agreement was delayed until June 1941 (Toller 1949).

Since World War II several new boundary agreements have been reached in this sector. First, China was very successful in reaching agreements with Burma and Nepal in 1960 and with Afghanistan and Pakistan in 1963. In each case negotiations seem to have been entirely amicable, and where there were different views about the proper location of the boundary the final boundary was based on mutual concessions.

In 1947 and 1948 the emergence of the independent states of India, Pakistan, and Burma elevated former limits of British administrative areas to the level of international boundaries. The boundary between West Pakistan and India created by Lord Radcliffe in 1947 caused a number of problems. First, fighting in Kashmir produced an extension to the line when a cease-fire was arranged. That cease-fire line was changed a number of times until it became fixed in 1972. In the Great Rann of Kutch the disagreement was resolved with the aid of an independent tribunal, which issued its award in 1968 after two years' work (Lagergren 1968). The boundary between India and East Pakistan, which was defined by reference to provincial limits as they existed in 1947, required supplementary agreements in 1952, 1954, 1958, and 1959. In 1964 and 1967 respectively, Pakistan and India reached boundary agreements with Burma confirming lines which had evolved over a long period during the colonial period. Finally, Iran and Pakistan signed a treaty on 6 February 1958 to resolve an ambiguity contained in the 1896 agreement between Britain and Persia. The revised boundary was demarcated in 1958 and the final protocol was signed on 8 December 1959.

The chief dispute remaining in this sector involves India and China in the areas west and east of Nepal. Since the war in 1961 both sides have consolidated their positions and titles on their side of the cease-fire lines. There has been evidence in recent years that the two sides are moving towards better relations, which will allow discussions about the location of the lines. The best chance of successful negotiations would be found by taking both Aksai Chin and northern Assam as two parts of a single package so that the scope for mutual concessions is as large as possible. It is also certain that little progress will be made if India persists with its unqualified insistence on the correctness of the McMahon Line (Murty 1983). It is quite possible that negotiations based on the history and human and physical geography of this region will produce a boundary which in some parts will lie close to the McMahon Line.

The French sector between the Gulfs of Tongking and Thailand

Having considered the short boundary along the Mekong agreed between Britain and France in January 1896, it is now necessary to examine the longer boundaries which France settled with China to the north and Thailand to the west.

China's boundaries with Vietnam and Laos extend for 1060 miles, and they were settled in agreements with France during the decade ending in 1895. This eastern extremity of China's borderland with southern Asia exhibits considerable physical and cultural complexity. Ranges composed of granite and trending south-east overlook plateaus and valleys formed by limestones and sandstones. The rejuvenated rivers, with their rectangular pattern of tributaries, have cut deep gorges in their headwaters. When France entered the scene this region was effectively a frontier between China and Tonkin where both governments exercised fitful control and where brigands flourished. They raided the settled populations in both countries for buffalo and women; these commodities were then exchanged for opium and weapons, which provided the basis for future depredations.

After Garnier and Lagrée had demonstrated the unsuitability of the Mekong River for navigation, France switched its attention northwards to Tonkin and the Red River, since it was considered important to be close to China to take advantage of the commercial opportunities it afforded (Rambaud 1903, pp. 332–3). Commercial agreements secured by French merchants in Hanoi in 1873 proved unenforceable, despite a naval campaign which resulted in the death of Garnier, the French commander. The assault against Tonkin was renewed in 1882 and this time it succeeded. Negotiations began with China and a preliminary convention of peace was signed on 11 May 1884. The most important clauses dealt with the withdrawal of Chinese forces from Tonkin, the guaranteeing of Chinese borders by France, and the freedom of French merchants to trade across the boundary. There was some delay in implementing these arrangements and a fresh treaty was signed on 9 June 1885. This agreement provided for the appointment of commissioners charged with the identification and demarcation of the boundary. The commissioners completed their work in time for a Convention to be concluded on 26 June 1887. The commissioners had managed to agree on the division of islands in the Gulf of Tongking and settle the boundary astride the Claire River and between the Red and Black rivers.

When France gained control of the east bank of the Mekong River in 1893 it became necessary for the Sino-French boundary to be extended westwards from the Black River to the Mekong River. This was achieved in a treaty dated 20 June 1895, and the opportunity was taken to amend certain sections of the boundary agreed in 1887 which had proved unsatisfactory.

The treaties France negotiated with China can be distinguished from those boundary agreements reached with other European powers because they were not regarded by the Chinese as being unequal treaties. China negotiated these boundaries from a position of equivalent strength and the lack of any Chinese challenge to these lines can be taken as proof of their acceptability.

Turning now to the eastern boundary of Thailand, which extends for 1600 miles from the Gulf of Thailand to the upper reaches of the Mekong River, it is necessary to begin the history in 1862. In that year France secured a foothold at the mouth of the Mekong when Annam ceded the provinces of Bein Hao, Gia Dinh, and My Tho. At the same time Annam renounced any claims over Cambodia, which was then a weak state subject to demands and impositions from both Annam and Thailand. Judging that France had inherited Annam's rights in Cambodia, French officials signed a secret treaty with the Cambodian ruler which provided for French protection of that state in return for France being given exclusive influence over Cambodia's foreign relations. Because Emperor Napoleon III delayed ratification of this secret treaty, the Cambodian ruler hastened to sign a treaty with Thailand which stated that Cambodia was tributary to that state. This treaty was ratified on 4 January 1864. When France eventually ratified the secret treaty in April 1864, Cambodia was in the position of having two conflicting treaties with its powerful neighbours.

Predictably, France and Thailand solved the problem. On 15 July the two countries signed a treaty which recognized France's protection of Cambodia and which confirmed that the Cambodian provinces of Batdambang and Siemreab were part of Thailand. The treaty made provisions for officers to demarcate the boundary between Thailand and Cambodia under French protection, but this was only done near the Boeng Tonle Sab. A reliable map prepared by McCarthy (1888) showed the alignment of the Franco-Thai boundary. It .consisted of three main segments. The first segment followed the alignment of the coast of Indochina facing the South China Sea and lay about 35 miles from that coast. In the vicinity of parallel 13° south, the boundary then trended eastwards to the Boeng Tonle Sab. McCarthy shows both French and Thai versions of this boundary segment and the area in dispute was about 4800 square miles. South of the Boeng Tonle Sab the third boundary segment trended south-west to terminate at Laem Samit. In July 1870 a treaty giving nationals of both countries equal fishing rights in the Boeng Tonle Sab identified the rivers which marked the boundaries of Batdambang and Siemreab.

There are three physical features that stand out on McCarthy's map; they are the Mekong River, the Thui Khao Phanom Dongrak (Chaine des Dangrek), and the Chuor Phnum Kravanh (Cardamones montagnes). There is no evidence that French officials set themselves the target of establishing the boundaries of Indochina along these features, but if that was the case it would not have been surprising for the concept of *les limites naturelles* (natural boundaries) flourished in France to a greater extent than in any other European country (Pounds 1954).

Just as France claimed inherited rights from Annam to extend its influence into Cambodia at the expense of Thailand, so the submission of Tonkin in 1882 enabled France to start exerting rights which Tonkin had claimed over tribes in what is now Laos. The fragmented political structure of Laos allowed France to annex individual chieftaincies on a piecemeal basis as opportunities arose. Britain watched this advance of French influence after a warning from the British Ambassador in Bangkok that it was probable that France would eventually claim up to the eastern watershed of the Mekong River (United Kingdom, Foreign Office 1895–6, p. 192). Britain was not unduly worried because at the same time, in 1889, the French Ambassador in London indicated that there was no wish to annex Louangphrabang.

Four years later the authorities in Paris told the British Ambassador quite a different story:

> . . . the left [east] bank of the Mekong was the western limit of the sphere of French influence, and that this opinion was based on the incontestable rights of Annam, which had been exercised for several centuries. (United Kingdom, Foreign Office 1895–6, p. 210)

France acted swiftly to put this policy into effect. A quarrel was forced upon Thailand in February 1893 when allegations were made of Thai aggression against Annam. Stoeng Treng on the Mekong was occupied by French forces, and Thai resistance to this move led to a French ultimatum on 13 April 1893. French pressure and demands were increased until Thailand capitulated and signed a peace treaty on 3 October 1893 (Luguenin 1948, Hall 1968, Tate 1971).

France gained all Thai territory east of the Mekong River and all the islands in that river (Fig. 9.3). Thai naval vessels were prohibited from sailing on the Mekong River and the Boeng Tonle Sab and Thailand agreed that no military posts would be maintained in Batdambang, Siemreab, and a zone 25 kilometres wide along the west bank of the Mekong River. France was also guaranteed any depots of coal or wood which it might require on the west bank of the river, and it was authorized to maintain control of Chantaburi until the terms of the treaty had been satisfied.

Only seven weeks after this great success French officials were drawing attention to the need to preserve the territorial integrity of Louangphrabang, which could only be achieved by Thailand ceding to France those parts of the kingdom which lay *west* of the Mekong River. This fresh target was achieved in a convention signed on 13 February 1904. The new boundary west of the Mekong River followed its watershed with the Mae Nam Nan north of the Nam Huang. In addition, new areas were gained west of the Mekong River, south of its confluence with the Mae Nam Mun. For the first time France managed to secure a boundary along the crest of the Thui Khao Phanom Dongrak. This fault scarp marks the southern edge of the Khorat Plateau, which consists of nearly horizontal beds of Triassic sandstone. Fisher (1967, p. 418) has noted that this scarp provided a definite barrier to communication in the last century. A protocol dated 29 June 1904 produced further French gains. The area on the west bank of the Mekong River opposite Louangphrabang was slightly increased, but there was a major gain south of Boeng Tonle Sab. The coastal terminus of the line on the Gulf of Thailand was moved north from Laem Samit to Laem Ling.

Before the demarcation of this boundary was complete the final boundary agreement was signed. On 23 March 1907 a treaty was signed which provided for the exchange of territory and the relinquishment by France of jurisdiction of French Asians in Thailand. Thailand gained a small area in the north at the head of the Nam Huang valley, a larger area around Trat in the south, and the islands of Ko Chang and Ko Kut in the Gulf of Thailand. France gained all Batdambang and Siemreab. The demarcation of the new sections of boundary was completed by June 1908. The joint commission decided at a number of points to modify the boundary in the interests of convenient administration for both sides.

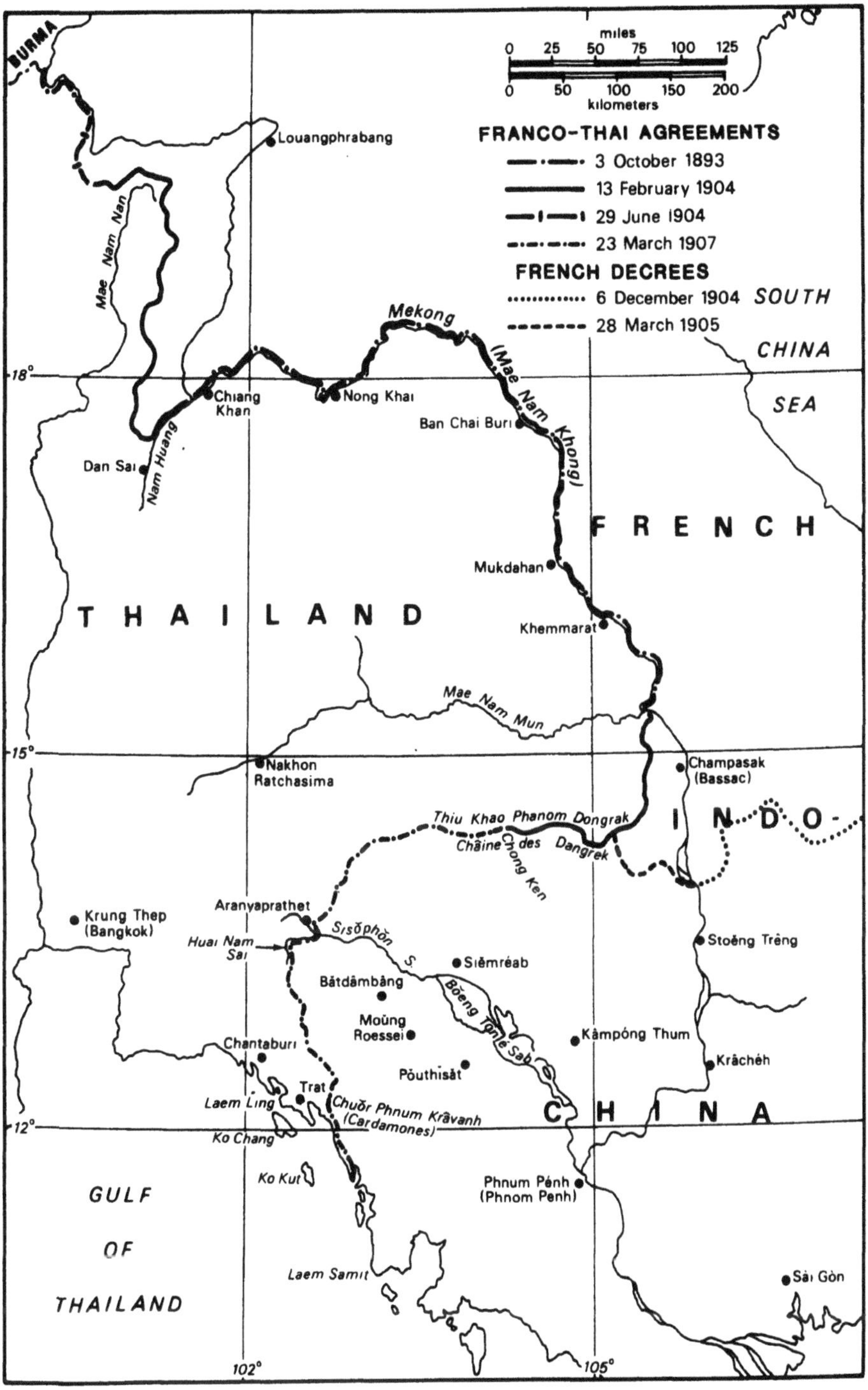

Figure 9.3 The Franco-Thai boundary.

These mutual concessions were authorized by the 1907 treaty. These variations from the boundary described in the treaty mean that the maps produced by the various commissions are the only authoritative definition of the line. This fact was made painfully obvious to the Thai authorities when they lost the case before the International Court of Justice in 1962 which dealt with the temple of Preah Vihear (Prescott 1985, pp. 13–15).

Subsequent agreements between France and Thailand simply confirmed the position of this boundary, and efforts by Japan to redraw these boundaries in favour of Thailand during World War II failed entirely.

The boundaries which separate Laos, Vietnam, and Cambodia were settled during the period 1873 to 1929. They were mainly the result of work by French officers and could reasonably be considered to be the internal boundaries of French Indochina. Cambodian authorities were involved in the early years in negotiating the boundary with Vietnam, but an analysis of the negotiations and agreements suggests that France was always negotiating from a position of strength. However, these boundaries appear to have survived without serious challenge for many years. It seems likely that in view of the hegemony which Vietnam has achieved in Indochina that the boundaries will not be altered unless Vietnam seeks territorial gains to reduce the strategic risk posed by the Cambodian salient called the Parrot's Beak west of Ho Chi Minh City.

The Japanese sector in the Korean peninsula

The boundaries associated with the Korean peninsula were mainly fashioned through wars. The Yalu River was selected as the boundary between Korea and an area of China ceded to Japan after a short war between China and Japan, which the latter country won. This boundary was defined in April 1895 after Japan had successfully challenged China for the dominant influence in Korean affairs. The other major powers were displeased with Japan's acquisition of Chinese territory on the Liaotung peninsula and a fresh agreement in November 1895 returned the territory to China and confirmed the Yalu river as the northern boundary of Korea. In 1909 Japan and China extended this line eastwards along the Tumen River, thus linking up with the boundary along the Tumen established by China and Russia in 1860 when Russia obtained the cession of the trans-Ussuri provinces. The best accounts of Sino-Japanese relations at this time are provided by Conroy (1960), Hulbert (1962), and Kim & Kim (1967).

The division between North and South Korea was produced after arrangements associated with the end of World War II in the Pacific and with the end of the Korean War in 1953. After the Soviet Union entered the war against Japan in August 1945 it was agreed that the Soviet Union and the United States should accept the surrender of Japanese forces north and south of parallel 38° north respectively. This line was hastily selected at the end of the war and was not expected to become a permanent division. However, McCune (1949) noted that this parallel had played earlier rôles in Korean history. In 1896 Japan had proposed to Russia that the peninsula should be divided by 38° north, and in 1904 Russian forces were instructed to engage any Japanese units which ventured north of the same parallel.

After the Korean War a demilitarized zone 4 kilometres wide was constructed across the Korean peninsula in the vicinity of parallel 38° north. By the same agreement South Korea secured the islands called Paengnyong-do situated off the west coast south of 38° south (Prescott 1975, pp. 509–18).

Conclusions

The main lines of political cleavage in Asia were established by European powers in two stages. First, before World War I they carved out Asian empires by negotiating boundaries with China, with each other, and with those Asian states such as Afghanistan, Nepal, and Thailand which were not annexed. Secondly, during the period of control Britain and France established or confirmed existing internal divisions within their empires, which later were raised to the status of international boundaries as countries such as India and Laos became independent.

The colonial powers left an inheritance of boundary problems to independent Asia. In some cases, as in the Himalayas, Britain had failed to negotiate any firm boundaries with China. In other cases the internal boundaries were not defined with precision and were ill suited to become international limits.

Although disagreements over boundaries have produced serious friction between countries such as India and China and minor disagreements between countries such as Thailand and Laos, it appears that the worst difficulties might be past. This satisfactory state of affairs has been encouraged by many factors. They include China's reasonable attitude to boundary negotiations with weaker states such as Mongolia, Afghanistan, Nepal, and Burma; the willingness of countries such as India and Pakistan and Cambodia and Thailand to accept the results of arbitration; and the readiness of countries to negotiate over functional problems such as that which arose over the Farraka Barrage between India and Bangladesh.

References

Chen, A. F. C. 1949. Chinese frontier diplomacy: Kiakhta boundary treaties and agreements. *Yenching Journal of Social Studies* **4**, 51–205.

Conroy, H. 1960. *The Japanese seizure of Korea 1868–1910*. Philadelphia: University of Pennsylvania Press.

Davies, C. C. 1932. *The problem of the northwest frontier 1890–1908*. Cambridge: Cambridge University Press.

Fisher, C. A. 1967. *Southeast Asia*. London: Methuen.

Hall, D. G. E. 1968. *The history of southeast Asia*, 3rd edn. New York: Macmillan.

Hsu, I. C. Y. 1965. *The Ili crisis: a study of Sino-Russian diplomacy 1871–81*. Oxford: Clarendon Press.

Hulbert H. B. 1962. *History of Korea*. London: Routledge & Kegan Paul.

Kim, C. I. E. & H. Kim 1967. *Korea and the politics of imperialism*. Berkeley: University of California Press.

Lagergren, G. 1968. Judicial decisions: the Indo-Pakistan Western Boundary Case Tribunal. *Indian Journal of International Law* **8**, 247–65.

Lamb, A. 1966. *The McMahon Line*. London: Routledge & Kegan Paul.

Lamb, A. 1973. *The Sino-Indian border in Ladakh*. Canberra: Australian National University Press.

Luguenin, P. 1948. Auguste Pavie diplomate, la question franço-siamoise des États Laotien 1884–96 (Auguste Pavie diplomat: the Franco-Thai question over the Laotian states 1884–96). *Revue histoire des Colonies* **35**, 200–30.

McCarthy, J. 1888. Siam. *Proceedings of the Royal Geographical Society* **10**, 188.

McCune, S. 1949. The thirty-eighth parallel in Korea. *World Politics* **1**, 223–32.

McMurray, J. V. A. 1921. *Treaties and agreements with and concerning China 1894–1919*, Vol. 2. New York: Oxford University Press.

Mancall, M. 1971. *Russia and China: their diplomatic relations to 1728*. Cambridge, Mass.: Harvard University Press.

Mangrai, S. S. 1965. *The Shan States and the British annexation*. Data Paper No. 57, Department of Asian Studies, Cornell University, Ithaca, New York.

Murty, T. S. 1983. *Paths of Peace: studies on the Sino-Indian border dispute*. New Delhi: ABC Publishing House.

Pounds, N. J. G. 1954. France and *les limites naturelles* from the seventeenth to twentieth centuries. *Annals, Association of American Geographers* **44**, 51–62.

Prescott, J. R. V. 1975. *Map of mainland Asia by treaty*. Melbourne: Melbourne University Press.

Prescott, J. R. V. 1985. *A study of the delineation of the Thai–Cambodian boundary*. Canberra: Office of National Assessments.

Prescott, J. R. V., H. J. Collier & D. F. Prescott 1977. *Frontiers of Asia and southeast Asia*. Melbourne: Melbourne University Press.

Rambaud, A. 1903. *Jules Ferry*. Paris: Plou-Nourrit.

Singhal, D. P. 1963. *India and Afghanistan*. Brisbane: Queensland University Press.

Tate, J. J. M. 1971. *The making of Southeast Asia*, Vol. 1. Kuala Lumpur: Oxford University Press.

Toller, W. S. 1949. The Burma–Yunnan boundary commission. *Eastern World*, Parts 1 and 2.

United Kingdom, Foreign Office 1872–3, 1883–4, 1895–6, *British and Foreign State Papers*, 63, 75, 88. London: HMSO.

10 *Africa*

Africa is the most partitioned continent in the world. There are 46 countries which have received international recognition – nearly three times the number in Asia. These countries, together with the former territory of Spanish Sahara, which was annexed by Mauritania and Morocco in 1976 are bounded by 103 individual segments of international boundaries.

Boundaries in Africa were drawn by colonial powers to a greater extent than on any other continent. There is no African equivalent to the dominant rôle played by the indigenous state of China in Asian boundary-making. Although the boundaries of South America owe much to the division between Portuguese and Spanish areas and the subdivision of the Spanish domain, these colonial boundaries were never marked as precisely as the limits of African colonies. Thus, the independent states of South America in the 19th century put a clearer individual stamp on international boundaries than has been achieved by African states in the post-colonial period.

More colonial powers were involved in drawing international boundaries in Africa than on every other continent. Spain, Portugal, France, Britain, Belgium, Germany, Italy, and Turkey all played a decisive rôle in fixing three or more of Africa's international boundaries.

The establishment of the British, French, and German empires in Africa was greatly assisted by large commercial firms. The British South Africa Company in southern Africa, the Royal Niger Company in Nigeria, the Deutsche Kolonialgesellschaft für Süd West Afrika in Namibia, the Nord West Kamerun Gesellschaft in Cameroun, and the numerous French companies in the area of Congo, Gabon, and the Central African Republic made important contributions to the consolidation of territorial claims (Coquery-Vidrovitch 1969, p. 188).

The proliferation of boundaries in Africa and the involvement of so many European governments has promoted much detailed research in the evolution of boundaries and their associated problems by a multitude of scholars. They have collectively published hundreds of books and articles on these subjects, which would probably take a lifetime to read and understand. The long, detailed studies of famous travellers such as Barth (1857), Monteil (1895), and Staudinger (1889) were succeeded by surveyors such as Detzner (1913), Nugent (1914), and Olivier (1905), whose reflections were in demand by the editors of the major geographical journals. Then came the administrators, who had to organize life for many people with differing cultural characteristics in the territories which had been created. Lugard (1923), Johnston (1897), and others provided useful accounts of the practical problems some African boundaries presented. Throughout the colonial and post-colonial period there have been commentators such as White (1890), who provided a contemporary account of the partition of Africa; Rudin (1938), who described in detail German activities in Kamerun; Widstrand (1969), who edited a useful collection of essays on Africa's boundary problems; and Sortia (1978), who considered the geographical developments influenced by the establishment of the boundary between Zaire and Zambia. Scholars interested in African boundaries owe a considerable debt to two writers who compiled collections of relevant treaties. Hertslet (1909) provided an invaluable tool in the first place and, more recently, Brownlie (1979) has published a comprehensive compendium of material relating to treaties and maps associated with the evolution of Africa's boundaries and their problems.

The evolution of Africa's boundaries

There is a general view amongst critics of European colonization of Africa that international boundaries were drawn in a few years at the end of the 19th century by career diplomats whose greed for national wealth was matched only by their lack of knowledge about the nature of African geography, politics, and society. This view is summarized in the phrase 'the scramble for Africa'.

Touval (1972, pp. 3–17) and Gann and Duignan (1969, pp. 100–31) have shown the falsity of this view in elegant essays. Gann and Duignan explore the motives for the annexation of colonies in Africa and reach the following conclusion:

> The history of European imperialism can therefore be written only in a polycentric fashion. Whether we deal with white expansion in Africa, or with any other great instance of cultural diffusion on a continental scale, no unitary theory will ever untangle for us the richness and variety of the historical skein. (Gann & Duignan 1969, p. 128)

This sound judgement is reached after looking at Africa through the eyes of the decision-makers and the political commentators of the time. They saw Africa as an El Dorado of unknown resources; a place for guaranteed profitable investment of surplus wealth; a limitless market for manufactured goods; a continent where middle-class public servants and soldiers could find employment which would advance their status; a people that desperately needed deliverance from the evils of slavery and spiritual oppression; an area with only a few, key strategic locations in terms of communications and the balance of naval power. Gann and Duignan note that Salisbury, Rosebery, and Bismarck made rational decisions about whether a territory should be claimed or not. They were aware that onerous responsibilities accompanied the acquisition of colonies and that the careless construction of boundaries sowed a harvest of problems which would be reaped later.

Through an examination of treaties between European powers and African chiefs, Touval (1972, pp. 4–11) has shown how some African leaders were able to influence the location of boundaries, and the care that most powers exercised in trying to ensure that the boundaries did not tear the fabric of existing political and cultural institutions. It is now well known that lofty aim was defeated on more than one occasion, but this was sometimes due to conflicting territorial claims by African rulers. In other areas, as Anene (1970, p. 290) has shown, the ethnic and political fragmentation of some frontiers between larger states formed a human mosaic with a complex pattern which defied rational separation by a single line.

In some cases the boundaries were delimited in terms of existing political organizations. For example, the Anglo-French declaration of 1890 specified that the boundary from Say on the River Niger to Barruwa on Lake Chad should be drawn so that the Royal Niger Company secured all that properly belongs to the Kingdom of Sokoto (Prescott 1971, pp. 66–7). The agreement between Britain and France on 8 September 1919 defined the southern part of the boundary between Chad and the Sudan as the watershed between the Nile and Congo River basins as far north as parallel 11° north. Beyond this point the boundary was drawn to separate in principle the countries of Dar Kouti, Dar Sila, Wadai, and Dar Tama from the country of the Taaisha, Dar Masalit, Dar Gimr and other tribes subject to Darfur (Brownlie 1979, p. 626).

Many boundary agreements dealing with Africa contained clauses which allowed the demarcation teams to vary the delimited line if local circumstances made such deviations appropriate. Sometimes a maximum distance from the defined line was specified; on other occasions it was necessary for the gains and losses of territory for each side to match. For example, the following latitude was given to surveyors in the Anglo-German agreement of 1910 dealing with the boundary between Uganda and what is now Rwanda:

> In marking out the boundary along the three straight lines above mentioned between the point (b) and the point (f), the commissioners appointed for the purpose have authority to deviate from the straight lines, so as to make the frontier coincide with natural features where this is possible. The deviations shall not, however, exceed 5 kilom. on either side of the straight lines, and neither the total area of British territory nor the total area of German territory shall be altered thereby. (Brownlie 1979, p. 992)

When it was known that there was some risk that small social units might be divided by a boundary, provision was made to allow the migration of people across the line. Such an article is found in the exchange of notes by Britain and France in respect of the boundary between Guinea and Sierra Leone in July 1911:

> Within six months from the date on which the boundary is finally agreed upon, natives living in territory which has been transferred may, if they so desire, cross over to live on the other side, and may take with them their portable property and harvested crops. (Brownlie 1979, p. 344)

Many African boundaries were drawn along the course of rivers, and it was usual to grant equal fishing rights to residents of both banks. The Anglo-Italian agreement of 15 July 1924, which fixed the boundary between Kenya and Italian Somaliland, made arrangements to alleviate the incidence of drought in Somaliland:

> If, however, the enquiries of the Commission referred to in Article 12 show that in the neighbourhood of the sector of the new frontier running from El Beru to the Jubuland–Tanaland boundary there exists a shortage of pasture for the tribes situated on the Italian side of the frontier, and if these enquiries also show that during the rainy season there is on the British side of that sector and in the region bounded on the east by the new frontier and on the west by the line Goochi–Ribba–El-Tulli–Lakola–Toor Guda–Ramaguda more pasturage available than is required for the tribes in British territory, then the Commission will have the power to decide that for a certain period, not less than five years, Somalis or other natives of the transferred territory may during the rainy season cross the boundary at such a distance and in such numbers as the Commission may prescribe. (Brownlie 1979, pp. 892–3)

There is thus plenty of evidence to show that boundaries were drawn to preserve existing indigenous political and social units when they could be identified; that surveyors were instructed to make the delimited boundary fit new geographical and cultural facts which became available during the process of demarcation; and that rules were developed in many cases to ensure that people living near the border were not unduly hindered in the conduct of traditional subsistence, commercial and social activities.

Boundary evolution before 1914

World Wars I and II conveniently divide the evolution of African boundaries into three periods. In the period before 1914 the majority of African boundaries were delimited and demarcated according to bilateral agreements involving the various colonial powers. Although there was a spate of agreements in the 1880s and 1890s, they were often based on coastal footholds of long duration. The first British fort at the mouth of the Gambia River was built in 1664; Freetown was established in 1791; France's concession at Grand Lahu on the Ivory Coast dated from 1787; a Portuguese fort was built at Luanda in 1575, 70 years after a similar fort was built at Sofala on the east coast of the continent. Shortly after the beginning of the 19th century, Britain had inherited the substantial Dutch possessions in the extreme south of the continent.

From these coastal bases expeditions travelled inland, and treaties were often signed with African chiefs and rulers. Sometimes these treaties dealt only with exclusive trading rights, sometimes they concerned the establishment of trading depots, and sometimes they involved protection of the chief's territory and the payment of subsidies. During the last 30 years of the 19th century, treaties were treated rather like options. Once a treaty had been signed, that territory was reserved for possible annexation in the future. It was always considered to be an advantage to possess a treaty which bore an earlier date than that held by a competitor (Fig. 10.1). Some of these treaties were signed by rulers whose eminence and authority was widely accepted by the European authorities. The Bey of Tunis, the Sarduana of Sokoto, King Lewanika of Barotseland, and Lobengula, leader of the Amandabele, were such eminent rulers. However, there were also minor chiefs who could be bribed to sign treaties and who could then be represented as independent authorities. The archives in Paris, London, and Lisbon bulge with correspondence between governments and between governments and their agents dealing with the extent of particular kingdoms or chieftaincies.

Eventually, when both sides were satisfied that enough information had been accumulated or that no other useful information was likely to be available in the near future, negotiations would begin. Where the area was well known the boundary was usually delimited with some degree of precision. In other cases where knowledge was deficient arbitrary lines were allocated and the principles which would govern the eventual delimitation of the boundary were set out. Sometimes the principle would refer to tribal limits, sometimes to the division of major drainage basins. The Anglo-German boundary of 1890 between South-West Africa and Bechuanaland followed meridian 21° east and then parallel 18° south. However, it was stipulated that Germany had to secure a corridor at least 20 miles wide to the Zambezi River. This finger of territory, called the Caprivi Zipfel after Bismarck's successor, still survives, although Germany never used it as a line of communication as originally intended.

These bilateral boundaries were often demarcated, except when they followed rivers, although the demarcation was often less than comprehensive and frequently pillars and cut lines were not maintained and were soon reclaimed by the African landscape. This was perhaps not inappropriate

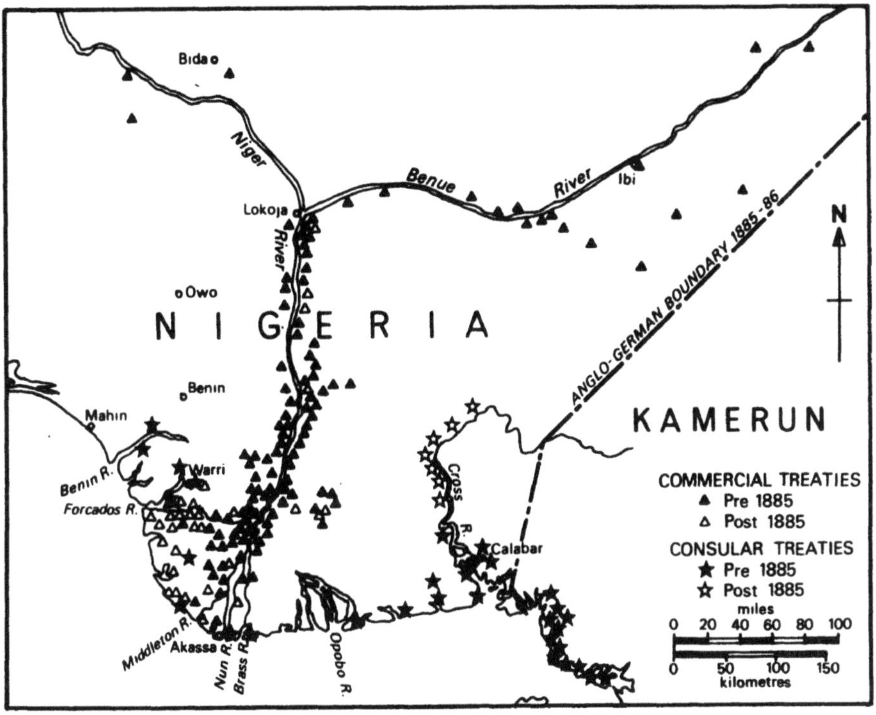

Figure 10.1 British treaties in the Niger Delta.

since the colonial administrations lacked the resources to supervise regulations along the entire boundary and usually concentrated their efforts at the principal crossing points. This *laissez-faire* attitude to long sections of boundary meant that the normal activities of border populations were not seriously disrupted even when the boundary agreements contained no specific dispensation to simplify movement across the line.

Although Africa is better endowed with mountains than Australia, there is no African equivalent to the Himalayas, Pyrenees, or Andes and for this reason ranges and escarpments were rarely used to delimit boundaries. Examples are found mainly in southern Africa. The boundary between the Orange Free State and Basutoland was delimited in 1869 by authorities in the Cape of Good Hope and the Orange Free State. Part of that line followed the Drakensberg. In the same year the South African Republic and Portugal selected the boundary which today separates South Africa and Mozambique along the Lebombo Mountains (The Geographer 1973a, p. 2). Figure 10.2 shows a sketch map of the Manica Plateau which persuaded the British authorities this feature should be used to define the boundary with Portugal in 1891. In fact the edge of the plateau could not be found and the matter was taken to arbitration.

Rivers mark the course of many African boundaries. For example, the boundary between Sierra Leone and Guinea was drawn by Britain and France in the period from 1882 to 1898. It extends for 405 miles and about half this distance is marked by the Great Scarcies, Kita, Wulafu, Meli, and Makona rivers; these river segments are connected by straight lines (The Geographer 1973b, p. 1). In 1892 France and Liberia defined their common boundary in West Africa. By September 1907 it had been realized that this definition was impossible to apply to the terrain of the borderland and so a new boundary was proclaimed. It follows the rivers Nuon, Nimoi, Dain, Boan, and Cavally for all but 2 miles of its length of 445 miles (The Geographer 1973a, p. 1). Although in this particular case the boundary followed a bank of the rivers, in most cases the boundary in the river was not defined or defined as the *thalweg*.

Watersheds of major and subsidiary basins were also used to define boundaries. The boundary of Sudan with Zaire and the Central African Republic coincides with the watershed separating drainage to the Nile and Congo rivers. These boundaries were fixed by an Anglo-Belgian agreement of 12 May 1894 and an Anglo-French agreement of 21 March 1899.

Lakes were evidently prominent features on maps of Africa. The lakes in the Rift Valley of East Africa from Lake Albert to Lake Nyassa provide the sites for boundaries, and other lakes which were involved in boundary definition include Lakes Mweru, Bangweulu, Victoria, Chad, and Rudolph.

In some cases bilateral treaties dealt with countries in various regions of the continent. For example, the Anglo-French treaty of 14 June 1898 dealt with boundaries of the Gold Coast, Lagos, and Northern Nigeria. On 1 July 1890 Britain and Germany signed an agreement which provides the basis for boundaries between Kenya and Tanzania, Malawi and Tanzania, and Namibia and Botswana. Britain and Portugal signed an agreement on 11 June 1891 which underpins the present boundaries of Angola and Zambia and of Mozambique with Malawi and Zimbabwe.

When it proved impossible for two countries to reach a compromise over their differences, arbitration was sometimes sought. In 1897 Mr Paul Vigliani, formerly Chief President of the Court of Cassation of Florence, made an award between Portugal and Britain in the border between Mozambique and what was then Rhodesia. Eight years later King Emmanuel III of Italy performed a similar function by defining the western limit of the Kingdom of Barotseland. This limit had been set at the boundary between what are now Angola and Zambia by the Anglo-Portuguese agreement of 11 June 1891. In 1909 Britain and Germany referred the intractable problem of the boundaries of Walvis Bay to King Alphonso XIII of Spain. He appointed Professor Joaquin Prida as arbitrator and on 23 May 1911 this jurist announced an award which gave Britain all the disputed territory.

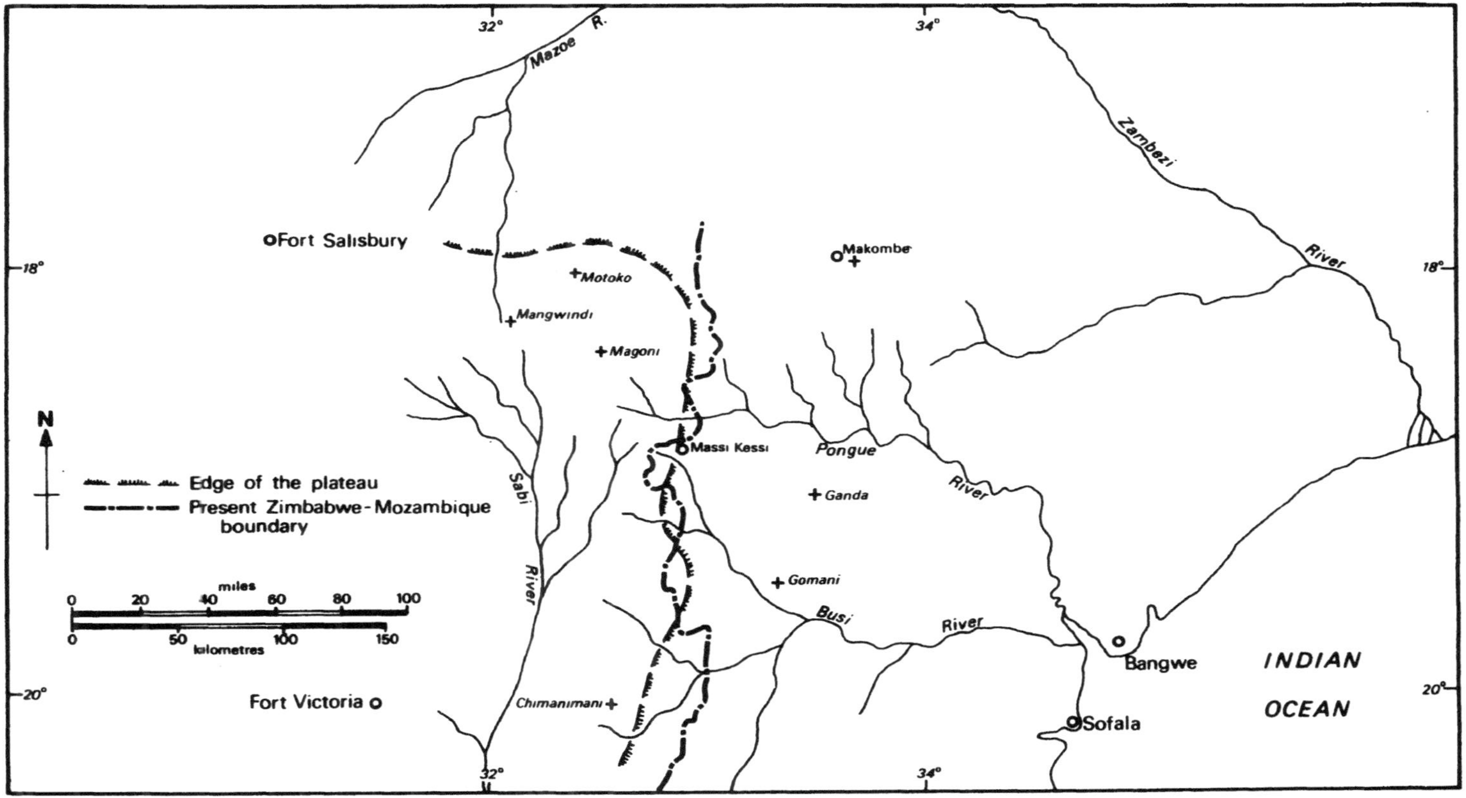

Figure 10.2 Sketch map of the Manica Plateau.

In addition to bilateral boundaries, Britain and France created some unilateral limits to carve their larger possessions into smaller sections for convenient administration. On 18 October 1904 a French decree defined the internal limits of French West Africa. This long decree defined boundaries which were the forerunners of modern boundaries between Senegal and Mali, Guinea and Mali, Guinea and Ivory Coast, Ivory Coast and Mali, Ivory Coast and Burkina Faso, Benin and Burkina Faso, Benin and Niger, and Senegal and Mauritania. Changes in these limits were made from time to time to suit the new needs of the French colonial administration, and uncertainty about the limits which applied at independence has created some difficulties in West Africa. Britain's unilateral boundaries were drawn between territories which are today called Uganda, Kenya, and the Sudan and between Malawi, Zambia, Zimbabwe, and Botswana. The boundary between Kenya and Uganda was significantly altered in the interwar period.

Boundary evolution between World Wars I and II

The first major development in the interwar period was the division of some former German colonies. British and French forces conquered Togoland and Cameroun fairly easily, and British, Belgian, and Portuguese troops eventually conquered Tanganyika after a difficult campaign. South-West Africa was conquered by South African forces alone and was handed over intact as a mandate to South Africa. Although Togoland was divided fairly evenly between Britain and France, the latter country received the lion's share of Cameroun. In 1908 France and Germany had fixed the boundary between Cameroun and French Equatorial Africa. On 4 November 1911 the Agadir crisis over French claims to Morocco was resolved by a Franco-German agreement. Under its terms the French position was recognized in Morocco and Germany received a section of French Equatorial Africa which was added to southern Cameroun. The most important part of France's concession consisted of two corridors down the Sangha River to the Congo River and along the Lobaye River to the Ubangi River. After the defeat of Germany in Cameroun these French concessions were restored to France and the 1908 boundary was reconstructed.

Britain secured most of Tanganyika, but Rwanda and Burundi were excised as a Belgian mandate in 1923, and Portugal obtained the Kionga Triangle. This small area lay south of the Rovuma River, which formed one of its sides. The other sides were formed by the coast and the Minegani River and the parallel which intersected this river 5 miles above its mouth. This southern limit was defined in a description of the territory of the Sultan of Zanzibar by British, French, and German authorities on 9 June 1886. Portugal controlled this triangle from 1887 to 1894 when

Germany seized it. The territory was returned to Portugal by the Allied Supreme Council on 25 September 1919 (McEwen 1971, pp. 207–10). Belgium received Rwanda and Burundi as partial reparations for damage suffered in Europe.

According to Article 13 of the Treaty of London dated 26 April 1915, France and Britain had agreed that if they increased their possessions in Africa at the expense of Germany then Italy might claim some equitable compensation. Italy's reward consisted of two rather barren zones ceded by Britain. By an agreement dated 15 July 1924 Britain and Italy agreed to shift the boundary between Kenya and Somaliland westwards away from the River Juba. The boundary had been fixed along the thalweg of this feature in March 1891. The new boundary consisted of a series of straight lines which were soon defined by 29 pillars (King 1928). In 1934 agreements involving Britain, France, and Italy resulted in the Sarra Triangle being transferred to Italy's territory of Libya. This was a rather awkward extension of the Sudan left over from Anglo-French negotiations in 1899 (Brownlie 1979, pp. 133–5).

During the interwar period a number of bilateral boundaries were adjusted. For example, on 22 July 1927 Belgium and Portugal exchanged small areas of territory near boundary pillar 25 on the line between the Belgian Congo and Angola. This was done to give mutual proof of good neighbourliness and to encourage the development of their territories. In January and February 1925 British and French surveyors found that pillar 76 on the boundary between Senegal and The Gambia was about 2 kilometres east of its correct position. An adjustment was made to the boundary, which in this sector was supposed to be an arc of a circle. On 26 April 1934 British, French, and Ethiopian delegates finally settled the tri-point at which the boundaries of what were then British and French Somaliland joined Ethiopia (Brownlie 1979, pp. 770–3). Twelve years earlier, in December 1922, Britain and Italy found it necessary to mark more closely the boundary between Eritrea and the Sudan. The original demarcation had relied on a few pillars erected on prominent hills. Difficulties had arisen in the intervening areas, especially where there was rich grazing or cultivable land.

Some unilateral boundaries were also adjusted during this period. For example, in 1926 and 1927 parts of the *cercles* of Say and Dori were transferred from Upper Volta to Niger. On 28 December 1936 the boundaries of French Equatorial Africa were modified for the third time since their origin in December 1903. After 1919 British authorities in Kenya administered part of Rudolf Province of Uganda on a *de facto* basis. That practical transfer was legally confirmed by an Order in Council dated 1 February 1926.

Boundary evolution since 1945

The ebb and flow of fighting in north Africa and Somaliland did not produce any new boundaries after World War II. Since 1945 there have been a number of boundary changes in the continent. First, a number of territories and some of their boundaries have disappeared. The small Spanish coastal territory of Ifni was ceded to Morocco on 30 June 1969. This territory had been made available for Spanish fishermen sailing between their home ports and the Islas Canarias. The territory had only been officially occupied in 1934, which was 73 years after the right had been acquired. In exchange for this cession Spain gained fishing rights in part of Morocco's territorial waters (Africa Research 1969, pp. 1303–4). In 1976 Spain handed the territory of Spanish Sahara to Morocco and Mauritania; they then partitioned it. The partition line passed from the head of the Bahia de Rio de Oro along the parallel to Bir Enzaran and then followed a southeasterly course to reach the original eastern boundary of Spanish Sahara near the intersection of 13° west and 23° north. On 5 August 1979 Mauritania abandoned its ruinous war against the liberation movement Polisario, which is fighting for the freedom of the Saharan Arab Democratic Republic, and renounced its claims to the southern part of Spanish Sahara. Morocco then annexed the area which Mauritania had renounced and continued the battle against Polisario.

Eritrea became an Italian colony in 1889 and remained in that condition until occupied by Britain during World War II. Italy formally renounced its claims to this territory by Article 23 of the 1947 Peace Treaty, and after five years of administration by the United Nations Eritrea was handed to Ethiopia on 11 September 1952. The international boundary between Ethiopia and Eritrea was retained for a time as a federal boundary between the two territories, but on 14 November 1962 the federation was abolished and was replaced by a unitary state. At the same time the rebellion began in Eritrea and has continued with varying degrees of success ever since.

British Somaliland and Italian Somaliland united to form the Somali Republic on 1 July 1960, and the former international boundary which separated them was reduced to the status of an administrative line.

In March 1957 British Togoland was united with the Gold Coast as the independent state of Ghana. This meant that the former Anglo-German boundary which had separated these two areas ceased to exist. The arrangements in Cameroun were more complex. During the period of British administration the British Cameroons were divided into the Northern and Southern British Cameroons by an extension of the boundary between Northern and Southern Nigeria along parallel 7°10′ north (Prescott 1971, p. 60). Shortly after this initial decision this geometric boundary was replaced by a more southerly line which followed a section of the Donga River. In February 1961 a plebiscite was held in the two portions of the British Cameroons. The Northern Cameroons decided to remain with Nigeria and the Southern Cameroons elected to join Cameroun, the former French trusteeship which had become independent on 1 January 1960. This means that Nigeria's western boundary consists of a southern section which was part of the Anglo-German boundaries of 1884 and 1886 and a northern section consisting of part of the Anglo-French inter-Cameroons boundary of 1920; these two sections are linked by the former internal boundary between the Northern and Southern Cameroons of Britain.

There have been a number of recent arrangements by African countries to demarcate their common boundaries. In most cases where changes in location have occurred they have been very small. For example, in 1972 Ethiopia and the Sudan agreed to move their common boundary from the base of hills called Halawa, Umdoga, El Mutan, and Jerok to their crests. These features are located between Matam and Bumbodi. The need for closer demarcation of this line was provided by rebels against both governments crossing the boundary to find sanctuary. The cause of a fresh demarcation of the boundary between Ghana and the Ivory Coast was smuggling:

> It is common knowledge that vital border pillars – 31–38 – along the cocoa-growing areas of Dormaa district are frequently removed, making the daily patrolling of our border guards extremely difficult. In such a situation, professional smugglers are emboldened to sneak through to the other side of the border with our cocoa and coffee, with the resultant loss to the country in foreign exchange. A lot of aliens are cultivating farms along the border and, in the absence of these pillars, they can easily trespass on Ghana land, as was the case recently. (The *Ghanaian Times* of 4 February 1970 quoted in Africa Research 1970, p. 1665)

This critical stretch measuring 50 miles was carefully demarcated in August 1970.

On 4 September 1973 a treaty came into force concerning the boundary between Botswana and South Africa. It is an interesting fact that this treaty made use of a suggestion made by a British surveyor 39 years before. Colonial Crossthwait suggested that international boundaries should be photographed from the air and that the photographs should be attached to the protocol (Peake 1934, p. 279). The South African – Botswana Treaty used air photographs to show the boundary in the vicinity of the confluence between the Shashe and Limpopo rivers and the intersection of the boundary with the Marico River (Republic of South Africa 1973).

Other agreements concerning demarcation of existing boundaries have been reached by Upper Volta with Ghana and Niger, by Ghana and Togo, by Chad and Cameroun, by Algeria with Morocco, Mali, Niger, and Tunisia, Ethiopia and Kenya, Malawi and Tanzania, and Burundi and Tanzania.

A major change in the alignment of a unilateral French boundary occurred in February 1963 when Mali and Mauritania signed the Kayes Treaty. This agreement shifted the long north–south segment between parallels 25° and 15°30′ north as much as 40 miles westwards in Mali's favour (Brownlie 1979, p. 406).

The final development in the postwar period concerns the creation of four independent territories in South Africa. Under the policy of separate development Ciskei, Transkei, Bophutatswana, and Venda have been granted independence. This independence has not been recognized by any country other than South Africa. Even though these boundaries have not secured international recognition, they have been carefully surveyed and marked.

African boundary problems in the 1980s

When Algeria and Mali signed a boundary agreement on 8 May 1983 the Algiers newspaper *El Moudjahid* marked the occasion with a comment which included the following sentence:

> It is clear that the artificial frontier problems handed down from colonialism can constitute one of the major obstacles to the development and natural evolution of balanced relations between neighbours. (Africa Research 1983, p. 6825)

President Chadly of Algeria had graphically described Africa's boundary problem as 'delayed action bombs left by colonialism' in a speech three months earlier.

At its first ordinary session in Cairo in July 1964 the Organization of African Unity recognized and tried to avoid the difficulty to which President Chadly referred. In a declaration on 21 July 1964 members of the Organization pledged themselves to respect the borders existing on their achievement of national independence (Brownlie 1979, p. 11). Considering that this pledge was based on the misleading belief that the boundaries of all African states 'constitute a tangible reality', it is not surprising that there has been a variety of boundary problems on the continent. However, it is encouraging there have not been more boundary problems to bedevil relations between African states.

An earlier article on Africa's problems (Prescott 1979) likened Africa's boundaries to the old wine skins into which the new wine of African nationalism had been poured. It concluded with the view that if countries followed the lead given to that time by Libya, Uganda, and Somalia then the wine skins would split and much of the wine would be wasted. The fact that most of the skins are still intact owes most to two causes. First, the boundaries were better drawn in the colonial period than many commentators admitted. Secondly, African states behaved responsibly when boundary problems arose and there might have been an opportunity to pursue narrow sectional interests.

The general location of African boundary problems in the 1980s are shown in three categories in Figure 10.3. Territorial disputes occur where the proper ownership of land is in question rather than the location of the boundary. Positional disputes arise when there is disagreement about the correct interpretation of delimitation treaties or when the previous demarcation of the boundary has been allowed to lapse. Functional disputes emerge when a state applies regulations at its boundary in a way which adversely affects the normal activities and interests of a neighbour.

Territorial disputes

There are six current territorial disputes. It was noted earlier that Morocco occupied the whole of Spanish Sahara by 1979 and resisted efforts by the local population to create an independent country called the Saharan Arab Democratic Republic. In an effort to thwart raids by forces based in Algeria, Moroccan authorities have built a series of walls, which extend for about 880 miles from Zag in the north via Guelta-Zammur on the Mauritanian boundary to Imilili on the Atlantic coast. This revetment of sand and stone stands 9 feet high and is equipped with electronic sensors (Africa Research 1985, p. 7753).

Since 1975 Libya has occupied a strip of northern Chad about 50 miles wide. It appears that Libya is relying on a Franco-Italian treaty signed on 7 January 1935 by Pierre Laval and Benito Mussolini but never ratified (Brownlie 1979, pp. 129–30). This agreement shifted the boundary southwards in Italy's favour, presumably as part of the equitable compensation which Italy had been offered in Africa by the Treaty of London.

A territorial dispute has been shown between Egypt and the Sudan, but it has not been pressed with any vigour by either side. The potential problem arose in the following way. An Anglo-Egyptian agreement dated 19 January 1899 fixed the boundary between Egypt and the Sudan along parallel 22° north. On 4 November 1902 an *arrêté* created a special régime whereby Egyptian authorities ruled some tribes south of 22° south and the Sudanese authorities controlled some tribes in a larger area north of that parallel. The terminus of the administrative boundary lies about 115 miles north of parallel 22° north. It is to the advantage of the Sudan to retain the administrative boundary as the international line, though Egypt gains more territory if its insistence on parallel 22° north is successful (Reyner 1963).

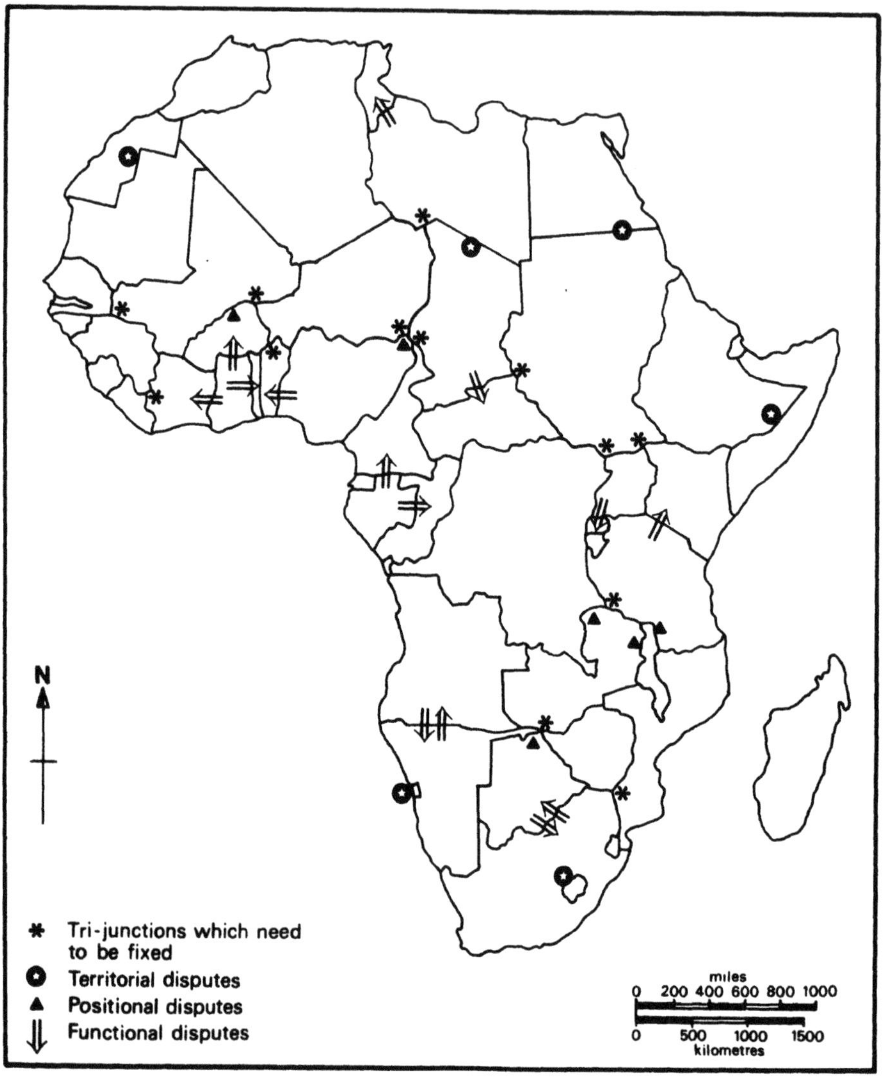

Figure 10.3 Africa's boundary problems in the 1980s.

Somalia's claim to the Haud and Ogaden areas of Ethiopia has caused much conflict, with many casualties. The diplomatic and legal history of this borderland has been set out in a number of excellent studies (Mariam 1964, Africa Research 1966, pp. 530–2, The Geographer 1975, Brownlie 1979, pp. 826–51). The northern boundary with former British Somaliland was clearly demarcated in 1932–5 (Clifford 1936). Only about 80 miles of boundary between Dolo and Iet were demarcated in respect of Italian Somaliland in 1910. It would be impossible to determine the line north of the Scebeli River unless one of the maps by which the boundary was defined in

1897 can be found. Emperor Menelik drew the boundary on two maps on 24 June 1897 and one copy was sent to the Italian government. The Italian government accepted this proposal, but both these maps have disappeared. Since the 1908 agreement defines the boundary as 'the line accepted by the Italian government in 1897', it is not helpful.

Although Ethiopia rests its case on the various treaties with Britain and Italy, Somalia argues that the colonial powers ceded territory to Ethiopia without authority and breached previous protectorate agreements.

Lesotho claims a strip of territory up to 30 miles wide along its western limit with South Africa. This zone, which is called 'the conquered territory', extends from the Orange River northwards to the northernmost part of Lesotho. The present ownership of this territory was settled in three wars between the Basuto and the Orange Free State in 1858, 1865, and 1867, and the present boundary was defined in the second Treaty of Aliwal North, dated 12 February 1869, after Britain had extended protection to Basutoland. The best account of these events from South Africa's viewpoint is provided by Eloff (1979) in a carefully researched study. Lesotho's argument is that these lands were annexed by the Orange Free State by military action. South Africa argues that the wars were necessary because Basuto were raiding into areas settled by white farmers.

The South-West Africa People's Organisation on behalf of Namibia has claimed the South African territory of Walvis Bay. Walvis Bay is the only good port on the coast of Namibia. Luderitz is small and its rocky seabed cannot be dredged. Walvis Bay has 1400 metres of wharves, serviced by electric cranes and storage sheds with a capacity of 35 000 cubic metres. This area was claimed for Britain by Captain Dyer of the *Industry* on 12 March 1878. Letters patent were issued on 14 December 1878 and the area was incorporated into the Colony of the Cape of Good Hope on 25 July 1884. Earlier Britain had annexed 12 guano islands off the coast of Namibia in 1861 and 1866. They were included in the Colony of the Cape of Good Hope on 16 July 1866.

The United Nations' Security Council by Resolution 432 (1978) has declared that the territorial integrity of Namibia must be assured through the reintegration of Walvis Bay into Namibia. This is a case that although South Africa has a clear legal title to Walvis Bay, many members of the international community have decided that this right should be set aside in favour of Namibia when it becomes independent (Joint Committee on Foreign Affairs and Defence 1982, pp. 73–82).

Positional disputes

There are a number of positional disputes in Africa and it is certain that other similar difficulties will arise in the future. It is known that 13 tri-junctions have not been exactly surveyed; they are shown in Figure 10.3.

In terms of area, the largest positional dispute concerns Mali and Burkina Faso. Uncertainty about the location of the former French administrative boundary has created a disputed zone about 10 miles wide and 100 miles long, which is occupied by about 60 000 people. Conflict in the period 1974–76 has been replaced by negotiations which, in 1983, led the two countries to approach the International Court of Justice while still maintaining parallel negotiations.

In April 1983, fighting broke out between Nigeria and Chad on the islands called Hadide and Kinasara in Lake Chad (Africa Research 1983, p. 6827). The straight line boundary between the two countries across the Lake is defined by the Anglo-French agreements of 29 May 1906 and 9 January 1931 and the Franco-German treaty of 18 April 1908. Because the location and extent of islands vary in Lake Chad it is possible that disagreements could arise over their ownership when that is decided by a fixed straight line.

Since 1980 Zaire has occupied Kaputa in northern Zambia. There is some uncertainty about the exact location of Cape Akalunga, which is nominated as the eastern terminus on Lake Tanganyika in the Anglo-Belgian agreement of 12 May 1894, and Kaputa appears to be very close to the location of Akalunga as shown on a map of Lake Tanganyika by Hore (1882).

Brownlie (1979, pp. 1098–107) provides an interesting analysis of the question whether there is a common boundary between Zambia and Botswana in the Zambezi River. The matter has a practical strategic significance so long as South Africa commands the Caprivi Strip. It seems likely that there is a common boundary between these two states because, according to a map prepared by representatives of South Africa and Northern Rhodesia at Katima Mulilo on 8 August 1931, the tri-junction of Zambia, Namibia, and Botswana is located in the middle of the Zambezi River west of the terminus of the boundary between Botswana and Zimbabwe. If in the future South Africa relinquishes control of Namibia, this will become only an academic question.

Functional boundary disputes

The functional boundary disputes which afflict Africa do not usually last long. An exception was the closure of Tanzania's boundary with Kenya from 1977 to 1983. The problems are usually associated with the expulsion of refugees or aliens or punitive raids by one country against rebels seeking sanctuary in a neighbouring state. In October 1982 Uganda expelled 25 000 Rwandan refugees to their country of origin; in November 1984 and June 1985 forces from Chad raided into the northern area of the Central African Empire.

Since 1981 there has been an encouraging tendency for neighbouring states to create joint committees to solve minor boundary problems. This has been done by Mali and Niger, Benin and Nigeria, Kenya and Somalia, Kenya and Uganda, Zambia and Zaire, Tanzania and Kenya, and Lesotho and South Africa.

Conclusions

The concept that the colonial powers were only motivated by national greed and drew boundaries without due regard for the structure of African social, cultural, and political structures is wrong. In most cases the boundaries were drawn after obtaining the best information available, often from tribal leaders. Frequently the boundaries were demarcated and adjustments were made during that process to create better functional boundaries. Often the boundary treaties included provisions to reduce the hardship faced by people living near the lines. This view does not deny that the colonial boundaries created problems of varying severity for some individuals and tribes. Instead, it asserts that some efforts were made to avoid such problems or to solve them when they became evident.

Territorial, positional, and functional boundary disputes have occurred, but fortunately not on the scale that marred the post-colonial period in South America or Asia. Resource boundary disputes have not been prominent, although they might become common as the seas and seabed are claimed and explored. Already arbitration has settled maritime disputes between Libya and Tunisia and Guinea and Guinea-Bissau.

When President Chadly asserted that the colonial powers had left delayed-action bombs in the form of boundary disputes, he should have added that only the African states can detonate them. There is no doubt that a belligerent African state could manufacture a *causus belli* from any of its borderlands; fortunately, none has done so, except in the Horn of Africa.

Boundary problems also offer an opportunity for international co-operation, and it is evident that in the first half of the 1980s there is an increasing tendency for African states to settle boundary problems through peaceful negotiation and compromise.

References

Africa Research 1966, 1969, 1970, 1983, 1985. *Africa Research Bulletin*, Political social and cultural series, Vols 3, 6, 7, 20, 22. Exeter: Africa Research.

Anene, J. C. O. 1970. *The international boundaries of Nigeria 1885–1960: the framework of an emergent African nation*. London: Longman.

Barth, H. 1857. *Travels and discoveries in north and central Africa* (5 volumes). London: Longman.

Brownlie, I. 1979. *African boundaries: a legal and diplomatic encyclopaedia*. London: Hurst.

Clifford, E. H. M. 1936. The British Somaliland–Ethiopian boundary. *Geographical Journal* **87**, 289–307.

Coquery-Vidrovitch, C. 1969. French colonization in Africa. In *Colonialism in Africa 1870–1960*, L. H. Gann & P. Duignan (eds), Vol. 1, 165–98. Cambridge: Cambridge University Press.

Detzner, H. von 1913. Die Nigerische Grenze von Kamerun zwischen Yola und dem Cross-fluss (The Nigerian boundary with Kamerun between Yola and the Cross River). *Mitteilungen aus den deutschen Schutzgebeiten* **26**, 317–38.

Eloff, C. C. 1979. *The so-called conquered territory disputed border area between the Orange Free State (Republic of South Africa) and Lesotho (Basutoland)*. Pretoria: Human Sciences Research Council.

Gann, L. H. & P. Duignan 1969. Imperialism and the scramble for Africa. In *Colonialism in Africa 1870–1960*, L. H. Gann & P. Duignan (eds), Vol. 1, 100–31. Cambridge: Cambridge University Press.

Hertslet, E. 1909. *Map of Africa by treaty* (3 volumes). London: HMSO.

Hore, E. C. 1882. Lake Tanganyika. *Proceedings of the Royal Geographical Society*. New Series **IV**, 1–28.

Johnston, H. A. 1897. *British central Africa*. London: Methuen.

Joint Committee of Foreign Affairs and Defence 1982. *Namibia*. Canberra: Australian Government Publishing Service.

King, L. N. 1928. The work of the Jubuland commission. *Geographical Journal* **72**, 420–35.

Lugard, Sir F. D. 1923. *The dual mandate in British tropical Africa*, 2nd edn. London: Blackwood.

Mariam, M. W. 1964. The background of the Ethiopian – Somaliland boundary dispute. *Journal of Modern African Studies* **2**, 189–219.

McEwen, A. C. 1971. *International boundaries of East Africa*. Oxford: Clarendon Press.

Monteil, P. L. 1895. *Saint Louis à Tripoli par le lac Tchad* (Saint Louis to Tripoli by Lake Chad). Paris: Alcan.

Nugent, W. V. 1914. The geographical results of the Nigeria-Kamerun boundary demarcation commission. *Geographical Journal* **43**, 630–51.

Olivier, E. 1905. La délimitation de la frontière Niger-Tschad (The delimitation of the boundary from the Niger to Lake Chad). *Revue de Géographie* **29**, 52–6.

Peake, E. R. L. 1934. Northern Rhodesia – Belgian Congo boundary. *Geographical Journal* **83**, 263–80.

Prescott, J. R. V. 1971. *The evolution of Nigeria's international and regional boundaries 1861–1971*. Vancouver: Aldine.

Prescott, J. R. V. 1979. Africa's boundary problems. *Optima* **28**, 2–21.

Republic of South Africa 1973. Boundary treaty between the Republic of South Africa and the Republic of Botswana. *Treaty Series*, 2/1973. Pretoria: Government Printer.

Reyner, A. S. 1963. Sudan – United Arab Republic (Egypt) boundary: a factual background. *Middle East Journal* **17**, 312–16.

Rudin, H. R. 1938. *Germans in the Cameroons 1884–1914: a case study in modern imperialism*. New Haven: Yale University Press.

Sortia, J. R. 1978. La frontière zairo–zambienne et ses conséquences géographiques (The Zaire–Zambia boundary and its geographical consequences). *Geo-Eco-Trop* **2**, 391–400.

Staudinger, R. 1889. *Im Herzen der Haussaländer* (In the heart of Hausaland). Oldenburg: Schulzesche Hof-Buchhandlung.

The Geographer 1973a. *Ivory Coast – Liberia*. International Boundary Study 132, Department of State, Washington, DC.

The Geographer 1973b. *Guinea – Sierra Leone*. International Boundary Study 136, Department of State, Washington, DC.

The Geographer 1975. *Ethiopia–Somalia*. International Boundary Study 153, Department of State, Washington, DC.

Touval, S. 1972. *The boundary politics of independent Africa*. Cambridge, Mass.: Harvard University Press.

White, H. B. 1890. The partition of Africa. *Scottish Geographical Magazine* **6**, 561–74.

Widstrand, C. G. (ed.) 1969. *African boundary problems*. Uppsala: Scandinavian Institute of African Studies.

11 *The Middle East*

There has been much debate amongst academics about the correctness of identifying as a separate subcontinental region the zone where Europe, Asia, and Africa meet (Fisher 1961, p. 1). Further, even those who agree such a region exists are divided on what it should be called. For the purposes of this study it is sensible to distinguish this area, which consists of the Arabian peninsula, Turkey, and Iran. The boundaries that evolved in this area owed much to local factors and conditions which were either absent in Europe, Asia, and Africa or present in only a weak form. The collapse of the Ottoman Empire in western Arabia and the Levant, the growth of Arab nationalism, and the involvement of Britain and France to protect perceived strategic interests such as communications and supplies of oil provide the main reasons why this region is considered separately. There are three arguments for associating Iran with this region. It is an important country in affairs connected with the Persian Gulf; from time to time it has displayed territorial ambitions in the Arabian peninsula; and the boundary dispute over the Shatt el Arab was the cause of the war with Iraq which started in 1980. Although this region could have been called South-West Asia (Brice 1966), the term Middle East has been retained because of its well known political connotations.

Although disagreeing on nomenclature, Fisher (1961) and Brice (1966), who have written useful regional descriptions of the Middle East, concur with the view that there are two prime zones within this region, which is transitional between its flanking continents. To the north there is a zone of fold mountains; to the south there is the ancient crystalline block of Arabia. The division between the two corresponds roughly with the southern boundaries of Turkey and Iran. The fold mountains were created in the period stretching from the Cretaceous to the Pliocene from the thick sediments which underlay the ancient sea called Tethys. Arabia was the anvil against which the hammer of central Asia fashioned these fold mountains. Some of the insular fragments lying off the northern block were enclosed by the folds and survive as basins such as those associated with Tuz Golo.

The Arabian peninsula has been tilted towards the north and north-east so that the highest sections are found in Yemen. The northern parts of this stable block have been overlain with considerable thicknesses of more recent sandstone and limestone. Thick sediments have filled in the depression which previously existed along the line of the present Tigris and Euphrates and they have recreated a fairly flat surface with low gradients.

In the vicinity of the Persian Gulf and its structural continuation along the lowlands of the Tigris and Euphrates rivers, geological conditions favoured the formation of oil and gas fields. The thick porous strata which had been gently deformed and tilted created traps within which hydrocarbons could accumulate from the rich deposits of remains of creatures which once lived in the warm Tethys Sea.

Since most of this region experiences a Mediterranean climate with summer drought and winter rains, the climatic division does not accord exactly with the structural zoning. However, it is generally the case that winter rains in the mountainous region support the growth of winter crops whereas irrigation is almost invariably necessary in the lowlands.

A map showing the distribution of languages (Fisher 1961, p. 103) reinforces the division into northern and southern zones. The predominance of Arabic throughout the southern region contrasts with the distinct subregions in the north where Turkish, Kurdish, Baluchi, and Persian are spoken.

Historically, the northern and southern regions can be distinguished on political grounds. In the north there were only two states, and both at different times were the centre of major empires. In addition, although the imperial powers of Western Europe had an interest in these areas and infrequently played a direct rôle there, it was in the southern region of Mesopotamia, the Levant, and Arabia where Britain and France played a long and direct part in political affairs. There are now 12 states in the areas south of the fold of mountains, and in historic times there were often a multitude of small political organizations, which often had a feudal relationship to each other and coalesced into varying groups for short and long periods.

Boundary evolution in the northern region

The chief characteristic that distinguishes the boundaries of the northern region from those of the southern zone is that they were determined by negotiations, often following hostilities, between the indigenous states of Persia, Russia, and Turkey without significant involvement of outside powers. The northern boundaries were also established in their main outlines before the beginning of the present century, whereas the boundaries of the southern zone were settled in the 20th century.

In 1800 the ruler of Georgia transferred his allegiance from Persia to Russia and precipitated a conflict which was to cost Persia further territory. Operating from fortified positions in the Aras valley at Erevan and Nakhichevan, Persian armies tried to recover the lands of Georgia in the lowlands of the Kura River. After the battle of Aslanduz, west of the confluence of the Kura and Aras rivers, Persia sued for peace, which was granted by the Treaty of Gulistan on 12 October 1813. This treaty, which was ratified on 15 September 1814, ceded all the territory north of the Aras River, except for the areas around Erevan and Nakhichevan (United Kingdom, Foreign Office 1817).

A dispute over the interpretation of this treaty led to fresh fighting in 1825 and at first the Persian forces won a number of victories. However, after Russia had captured Erevan and Tabriz it was evident that Persia would have to sign another treaty ceding even more land west of the Caspian Sea. The Treaty of Torkaman dated 10 February 1828 transferred Erevan and Nakhichevan to Russia, thereby removing all Persian authority from north of the Aras River. At the mouth of the Kura River on the western shore of the Caspian Sea, Persia ceded the area of Lenkoran. Russia also acquired a small bridgehead south of the Aras in the vicinity of Nakhichevan. This bridgehead was retained until 1893 when it was exchanged for the oasis of Firyuza east of the Caspian Sea (United Kingdom, Foreign Office 1828).

In the first half of the 19th century Russia had been slowly advancing into Kazakhstan east of the Caspian Sea to protect its route eastwards to Siberia from raids by fierce tribesmen. Once the initial line of forward posts had been established at Irgiz, Turgay, and Atbasar, the Russians found that there were strong reasons for a further advance, as the British Ambassador in Saint Petersburg noted:

> His Majesty [the Czar] further remarked that there was no intention of extending the Russian dominions; but it was well-known that, in the east, it is impossible always to stop when and where one wishes. (United Kingdom, Foreign Office 1872–3, pp. 680–1)

The advance southwards into this area north of Persia was complemented after the war in the Crimea by an eastwards drive from the eastern shore of the Caspian Sea. By 1869 bases had been established at Krasnovodsk and Chikishlyer. This latter fort is at the mouth of the Atrek River and on 13 December 1869 this river was defined as the boundary between Russia and Persia for 30 miles.

It is not known whether the Persian authorities understood Russia's aims, but they had been predicted by the British Ambassadors to Russia and Persia:

> The principal object of Russia it may be presumed, in any military operations she may now undertake on the frontier, or eventually against Khiva, is to secure a safe commercial route to Central Asia from the Caspian and her Transcaucasian provinces. (United Kingdom, Foreign Office 1872–3, p. 731)

The Ambassador to Persia was almost exactly right in his guess about future developments:

> . . . in order to open a road to the Oxus [Amu-Darya] from the Caspian, the Russians would have to construct forts and station troops within the Turkoman country through which it will pass, and this being done the Turkoman tribes will sooner or later be brought under the protection and authority of Russia. The desert across which the Russians now propose to establish a line of communication with central Asia is ill adapted for the purpose, the supply of water being insufficient for caravans traversing the plains, and the heat in summer being excessive. It is possible that before long they will find these difficulties insurmountable, and they may then seek a more practicable route, which will be found by starting from Hassan Kooli [Gasan-Kuli] at the embouchure of the Attrek, in the bay of Asterabad, near Ashoorada, the Russian naval station in the southeast of the Caspian; following the course of that river eastwards, and then skirting along the hills to the north of Bojnoord [Bojnurd] and Kochan [Quchan] in the direction of Merv [Mary], which is not more than 4 marches from the Oxus [Amu-Darya], and within 10 easy stages of Herat. By that line the road would pass for nearly the whole distance from the Caspian through an inhabited tract of country, where an abundant supply of water exists together with rich pasturage, and a salubrious climate at all seasons. (United Kingdom, Foreign Office 1872–3, pp. 687–8).

The Russians followed this script faithfully and by a treaty dated 9 December 1881 the boundary was extended eastwards to Babadurmaz along the northern flanks of the Kopet Dag, which lies north of Bojnurd and Quchan. Twelve years later, on 27 May 1893, the boundary was continued along the northern slopes of the Kopet Dag as far as the Tedzhen River at Serakhs and then southward along that river to the tri-junction with Afghanistan at the Zulfikar Pass. This boundary gave to Russia the fertile lower slopes of the range and included oases such as Yaradzha and Archman, which are nourished by water draining from the Kopet Dag.

The boundaries established in the 19th century by Russia and Persia have survived, apart from some very small changes set out in agreements reached in February 1921 and May 1955.

When Russia defeated Turkey in 1828 the Treaty of Adrianople, signed on 14 September 1829, defined the boundary between the two countries across the isthmus separating the Black and Caspian seas. The line was fixed north of the Turkish provinces called Batumi, Ardahan, and Kars. With the aid of European allies, Turkey managed to escape from the Crimean War without conceding territory, but the war with Russia in 1877 was fought without assistance and lost. The price exacted in Armenia was the provinces of Batumi, Ardahan, and Kars, and this adjustment of the boundary was confirmed by the Treaty of Berlin, signed on 12 July 1878.

After the Russian revolution and the withdrawal of that country from World War I, Turkish forces were able to drive northwards and regain Batumi, Kars, and Ardahan. However, this recapture was short lived because the collapse of other fronts made it impossible for Turkey to prevent the establishment of an independent state of Armenia. Because the allies were not prepared to defend the new state, it did not survive for long. In 1920, attacks from the east and west by the Soviet Union and Turkey were successful, and by the Treaty of Aleksandropol on 2 December 1920 the two countries fixed the boundary which still exists today. Turkey regained the provinces of Ardahan and Kars and the southern part of Batumi. In recompense for the northern part of Batumi, which remained with the Soviet Union, Turkey was given that part of Yerevan lying south of the Aras River. The new line was confirmed in subsequent treaties signed on 16 March 1921 and 13 October 1921.

Iran's western limit with Turkey and Iraq was negotiated as a single line in the period from 1639 to 1914. After World War I, when Iraq emerged as a separate country, small modifications were agreed to Iran's limits with both countries. After one of their periodic wars the rulers of the Turkish and Perisian empires agreed in Zohab on 17 May 1639 on the general alignment of the boundary from the upper reaches of the Aras River to the Persian Gulf. This line, which descends from the lava plains around Mount Ararat to the thick sedimentary plains of lower Mesopotamia, was drawn to separate recognized Turkish and Persian provinces such as Kars and Yerevan. In the south the Shatt al Arab, which carries the combined discharge of the Euphrates and Tigris rivers, was identified as the boundary.

Although Turkish forces conquered territories east of this boundary in the first half of the 18th century, the 1639 line was confirmed in treaties signed by the two countries in 1746 and 1823. Shortly after this, Britain and Russia became involved in the discussions about the demarcation of the boundary. On 31 May 1847 the second Treaty of Erzurum defined the boundary in general terms and made arrangements for a joint commission assisted by British and Russian surveyors to demarcate the line. Progress was not swift, and the onset of the war in the Crimea effectively halted operations in the 1850s. Finally, in 1869, the various parties were able to produce an agreed map, which was made up covering the entire length of the boundary with a zone 25 miles wide. This map was later cited in the 1913 agreement as the basis for the subsequent demarcation in 1914, but Ryder, a British member of the 1914 joint commission, described it as being 'insufficiently accurate for purposes of frontier demarcation' (Ryder 1925, p. 237).

After the Russo-Turkish war of the 1870s, one of the provisions of the Treaty of Berlin signed by most of the European major powers on 13 July 1878, was that Turkey should cede Qotur to Persia. This is a small region situated west of Khvoy in north-western Iran. This revised boundary was delimited in July 1880, but it was not demarcated. In December 1911 Turkey and Persia agreed that their boundary should be finally demarcated, and on 17 November 1913 a protocol was agreed defining the line as closely as possible and making arrangements for its demarcation by a joint commission, which once again would include Russian and British surveyors. It was provided that if the Turkish and Persian officers could not reach agreement they had to submit the dispute to arbitration by the Russian and British leaders, and they had only two days in which to prepare arguments to justify their particular view. The British commissioner recorded that the effect of this provision was that the Turkish and Persian commissioners would only agree on unimportant matters. On any other issue they would play safe and submit it to arbitration rather than take the risk of unpopularity with their superiors by making the wrong decision (Ryder 1925, p. 239). Apart from a section of about 40 miles in the vicinity of Qotur, the demarcation of the boundary with 227 pillars was completed by October 1914. Several boundary pillars did not last longer than 24 hours in the uplands occupied by Kurds on both sides of the boundary. However, it was the view of one British official that the rebuilding of the pillars every two years in the period from 1919 to 1925 fixed the boundary more clearly than those limits which are demarcated once and then never altered (Ryder 1925, p. 238).

It is now necessary to look at the small alterations which occurred after World War I on Iran's border with Turkey and Iraq. By a treaty dated 23 January 1932 Iran and Turkey agreed to modify their boundary in three areas. Although each of the areas was small, they had been a serious irritant in relations between the two countries. In the extreme north the transfer of former Russian territory to Turkey by the Turkish–Soviet agreement of October 1921 made it necessary for Iran and Turkey to extend their boundary to the Aras River. This amendment was accomplished by using the former Russo-Persian boundary along the Kara Su, which is a tributary of the Aras River. The boundary was finally demarcated in the vicinity of Qotur, and a marked Persian salient created by the protocol of 1880 was eliminated when the settlement of Heratil was returned to Turkey. West of Rezaiyeh the boundary was adjusted so that Iran gained Mazbicho (Crone 1938, p. 59). The boundary in the vicinity of Mazbicho had to be revised again in an agreement of 26 May 1937 when it was discovered that the commission of 1914 had made an error by not relying on the map produced in 1869!

The second Treaty of Erzurum gave to Persia the territories on the east, or left bank, of the Shatt al Arab, together with Khizr Island, which is sometimes called Abadan Island on modern maps; it lies at the mouth of the Shatt al Arab and lies between that waterway and the Bamishir River. It was decided by the demarcation team in 1914 that the boundary between what was then Turkey and Persia lay along 'the high-water line on the left bank' (Ryder 1925, p. 229). The only exception to this rule occurred off the port of Khorramshahr. In 1847 this settlement was called Mohammerah, and Persia was granted the anchorage of this port. This anchorage lay in the Shatt al Arab, and so for a short section opposite Khorramshahr the boundary left the high-water line to follow the thalweg, which is the line of the deepest continuous channel (Sevian 1968, p. 220).

After World War I the Iranian authorities sought an alteration of the boundary from the left bank to the thalweg along the entire course of the Shatt al Arab. As part of this campaign Iran refused to recognize the new territory of Iraq until 1929 and persistently flouted the instructions of the Basra Port Directorate, which legally controlled the Shatt al Arab, apart from the anchorage of Khorramshahr. The matter was settled on 4 July 1937, when Iraq agreed to move the boundary off the port of Abadan from the bank to the thalweg. This new segment of boundary measured 6 kilometres.

In 1969 a new phase began in this dispute. Iran declared that the 1937 agreement was void because Iraq had not implemented certain technical requirements in connection with navigation on the Shatt al Arab and the use of Iranian pilots. On 13 June 1975 the two countries agreed that the boundary would follow the thalweg along the Shatt al Arab. This new arrangement did not last long. On the last day of October 1979 Iraq called for the abrogation of the 1975 treaty and at the same time demanded that Iran should evacuate the islands it had seized from the United Arab Emirates in 1970 and give a measure of autonomy to the Kurds, Baluchis, and Arabs living in Iran. Not surprisingly, this demand was rejected the following day. Iraq abrogated the 1975 treaty on 17 September 1980 and the war with Iran started.

It is difficult to see how this matter will be resolved. Iran clearly regards guarantees of navigations without hindrance contained in treaties from 1847 to 1937 as unsatisfactory. Iraq plainly finds joint ownership unacceptable. This dispute appears to be a classical illustration of the fact that if relations between states are cordial then boundary problems can be settled fairly easily, whereas if states are hostile towards each other in respect of national philosophy then simple boundary problems are intractable.

Boundary evolution in the southern region

In terms of boundary evolution the contrast between the northern and southern regions is sharp. In the north two large states negotiated their boundaries with each other and with Russia before the beginning of World War I, with only slight involvement of other European powers. In the south the boundaries of 12 countries were settled almost entirely after World War I, and in that process France and Britain played a decisive rôle.

The boundaries of the Middle East have been classified in Figure 11.1 according to their origin. First, there are the limits created by bilateral negotiations, even though sometimes one party was in a dominant position. A distinction has been made according to whether the participants in the bilateral negotiations were external or indigenous powers. In the south most of the bilateral boundaries were negotiated between external powers or by one external and one indigenous authority. Only short sections of boundary have been delimited through bilateral negotiations between two indigenous parties and, with the exception of the boundary between Yemen (Sana'a) and Saudi Arabia, they were all settled after World War II.

Some lines were fixed unilaterally and in each of these cases France or Britain was the responsible power. The unilateral boundaries included in this class are those which have been accepted by the separated states such as Oman and the United Arab Emirates or Lebanon and Syria. Unilateral claims which might be enforced by one powerful country such as Saudi Arabia against a weaker neighbour such as Yemen (Aden) are considered to be undefined.

Finally cease-fire lines mark some of the limits of Israel, and they have shown a tendency to change significantly in the period since 1967.

Before World War I the only boundaries which had been determined by bilateral treaties in the southern region separated Turkey's domains from Egypt and the British colony of Aden, which had been established in 1839. In 1892, when Turkey permitted Egyptian authorities to administer Sinai, Egypt was effectively a British protectorate. So immediately the British authorities announced a unilateral boundary which extended in a straight line from a point just east of El Arish to a point near the head of the Gulf of Aqaba. Some years later Turkey proposed two alternative boundaries (which are shown in Figure 11.2) but neither of these satisfied the twin British aims of providing a protective zone east of the Suez Canal and of controlling the western shore of the Gulf of Aqaba. On 14 and 15 May 1906 the British and Turkish authorities exchanged notes which provided for a boundary to be drawn from Rafah to a point on the Gulf of Aqaba at least 3 miles from Aqaba. This boundary was defined in an agreement signed by Turkey and Egypt on 1 October 1906 (Parry 1980, pp. 190–1, 1981, pp. 19–21). The eastward shift of this line was caused by the discovery that this line coincided with the traditional limit of the Ottoman Empire (Naval Intelligence Division 1943, p. 2, The Geographer 1965, p. 3). With the exception of the Gaza Strip, the 1906 line forms the cease-fire line between Egypt and Israel.

On more than one occasion in the 1890s, difficulties at the southern tip of the Arabian peninsula with tribesmen caused the British authorities to

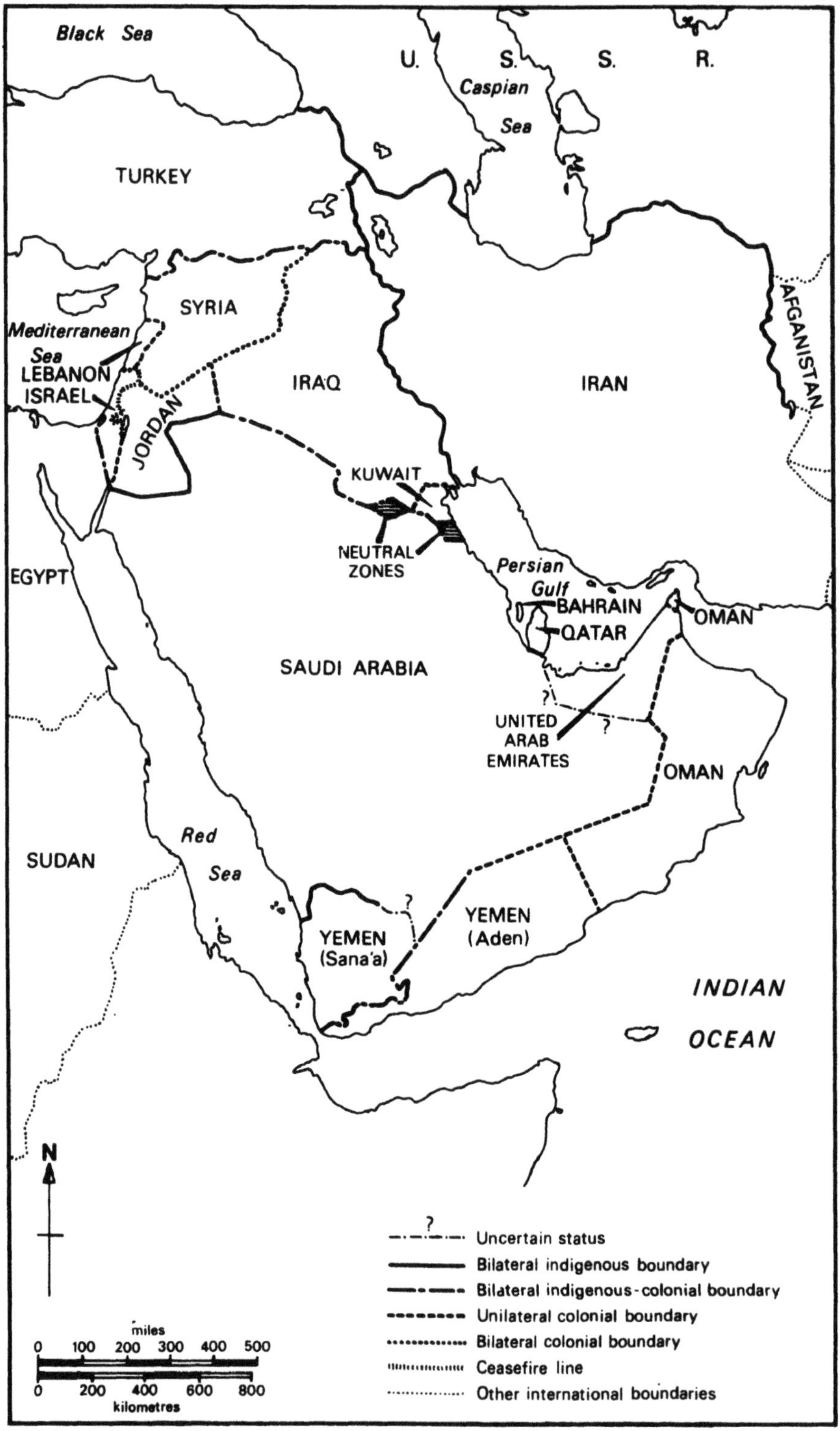

Figure 11.1 The origin of boundaries in the Middle East.

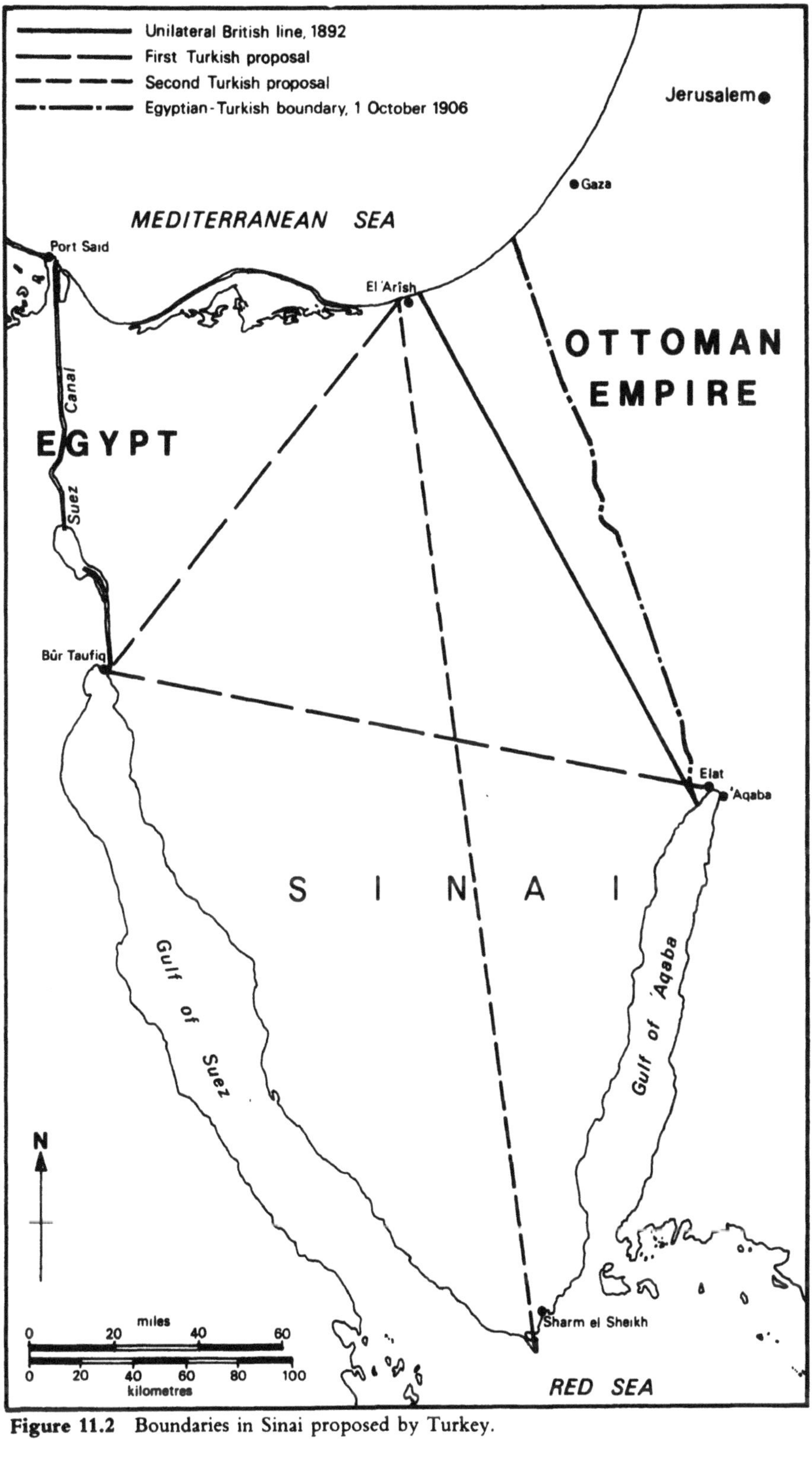

Figure 11.2 Boundaries in Sinai proposed by Turkey.

propose delimitation of a boundary to Turkey. The Turkish rulers demurred because they had not entirely given up hope of extinguishing this British colony. However, in 1900 and 1901 Turkish authorities experienced such problems with tribes on the Dali plateau that they agreed to delimit a clear line (Wenner 1967, p. 44). The joint commission worked hard for two years until May 1902 when, at a point near Harib, they began to enter territory where neither side had any pretensions to authority. From this point they drew a line north-east to the desert called Rub al Khali (Hickinbotham 1958, pp. 55–6).

After World War I the boundaries of the southern region were shaped in four important stages. First, the boundaries of the zones acquired by France and Britain were defined. Secondly, Britain and France at different times drew lines to divide their territories into separate states. The third stage occurred mainly after the British and French occupation was ended and involved indigenous states altering existing boundaries or agreeing on lines through zones where no boundary had ever been delimited. After World War II the fourth stage occurred, when some of Israel's boundaries were created as cease-fire lines in various wars.

During World War I British authorities sought to encourage an Arab revolt in Arabia and they opened negotiations with Sherif Husain, who since 1908 had been the official custodian of the sacred cities of Mecca and Medina (Glubb 1959, p. 57). In a series of letters, Sherif Husain and Sir Henry McMahon, the latter fresh from his boundary negotiations in the Himalayas, reached agreement on an Arab contribution to the war and British promises about support for an independent Arab state after the war (Hurewitz 1975, Vol. 1, pp. 46–56). The areas of the Levant lying west of the districts of Damascus, Homs, Hama, and Aleppo were excluded from this agreement by Britain because these areas were not considered to have an Arab majority and because of France's known interests in Syria and Lebanon.

This correspondence ended in March 1916, and two months later Britain and France concluded the Sykes–Picot Agreement, which set out the territorial rewards France and Britain would receive in the Middle East at the end of the war (Hurewitz 1979, Vol. 1, pp. 60–4). Russia was later joined in this agreement; the proposed arrangements are set out in Figure 11.3. In the southern region there were five zones. In two of them Britain and France could establish direct control. The French area included the Levant and parts of southern Turkey; the British area included lower Mesopotamia and the shores of the Persian Gulf. Between these two areas there was a zone within which Britain and France would recognize and protect one or more Arab states. This intermediate zone was divided into zones where Britain and France would have priority rights of enterprise and the sole right to provide advisers and foreign functionaries to the Arab states. Finally, in the general area of Palestine there was an international zone.

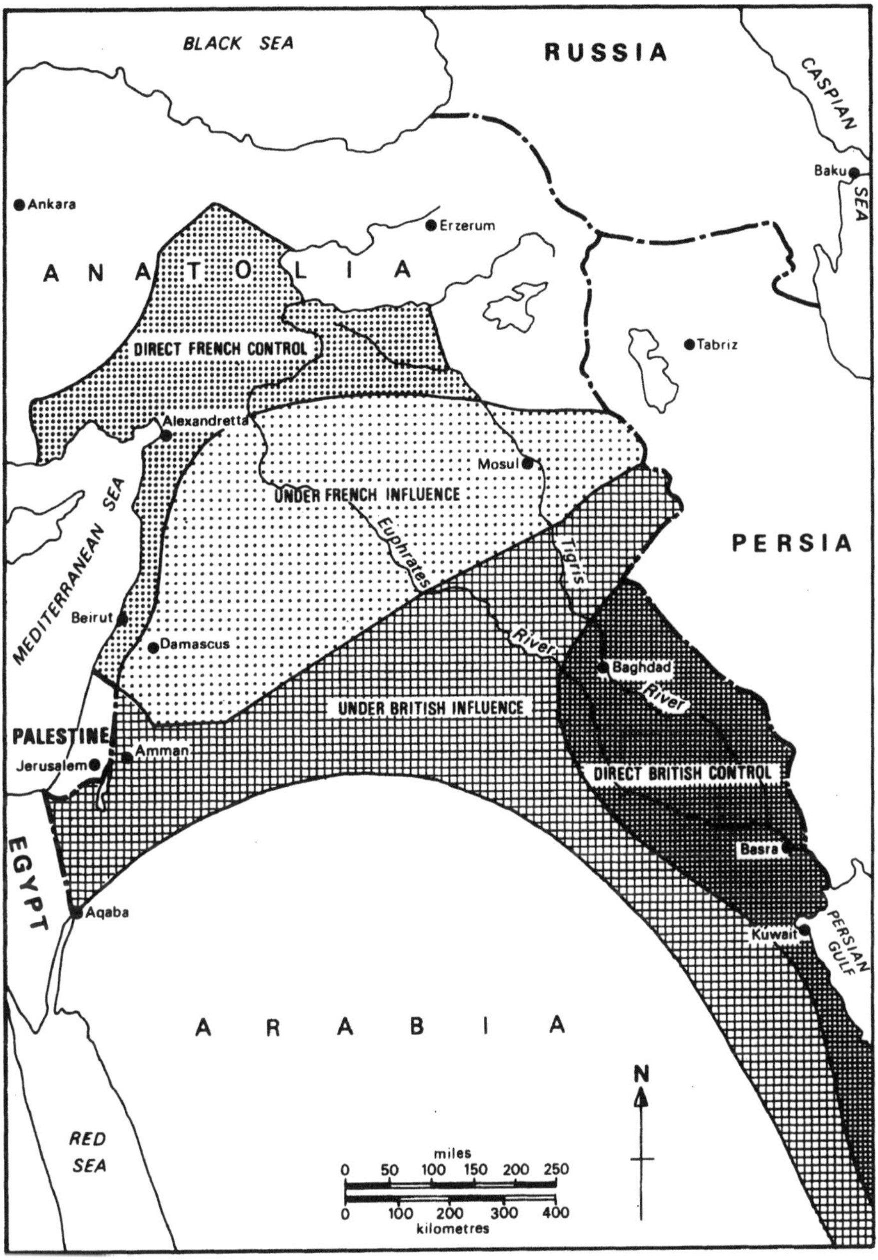

Figure 11.3 The Sykes–Picot Agreement.

Glubb (1959, pp. 67–75) has analysed with the aid of maps the problem of reconciling the British undertakings given to Sherif Husain and the terms of the Sykes–Picot Agreement. In the event, the neat division into areas of direct and indirect zones of British and French influence did not work and the two countries acquired direct control over the entire area. Temperley (1924, pp. 134–92) has provided a detailed account of the confused events which made it necessary for Britain and France to begin negotiations with each other and with Turkey to the north and Saudi Arabia to the south to settle the limits of their territories. The main limits of the Anglo-French zone in the Middle East were settled by 1927; but after that date there were some alterations, though most of them were minor. Only in the north did France face the problem of negotiating a boundary with an indigenous state.

By 1920 Turkish nationalists led by Kemal were demonstrating a military capacity which was threatening further French casualties if an effort was made to hold Cilicia and Sivas. Unwilling to face this risk and the resulting domestic unpopularity at home, French authorities signed the Ankara Agreement on 20 October 1921 (Temperley 1924, pp. 166–7). By this agreement the two countries defined the boundary between Turkey and French Syria from a point on the Iskenderun Korfezi just south of Payas to the Tigris River. The line was defined in general terms and for much of its length followed the Baghdad railway, which remained on Turkish territory as far east as Nusaybin.

The Ankara Agreement also made special provision for the district of Alexandretta which is now called Hatay (Hurewitz 1975, pp. 263–4). This arrangement would allow the Turkish majority to retain and develop their culture, and the Turkish language was given official recognition. When a programme for Syrian independence was put in train in September 1936 Turkey objected that the population of Alexandretta should remain outside Syria. The matter was referred to the League of Nations on 16 December 1936 and a commissioner was sent to investigate the situation (Glubb 1959, p. 115). In fact, Turks formed only 40 per cent of the population of the district, although they formed the largest single group. In 1938 Turkish troops entered Alexandretta to share security responsibilities with French forces. Many Arabs and Armenians fled to Syria and Lebanon and when electoral lists were compiled soon afterwards Turks constituted 63 per cent of the population (Hurewitz 1975, Vol. 1, p. 545). Turks won 22 of the 40 seats in the legislative assembly and they filled all the important administrative positions. The names of the town and the district were changed to Iskenderun and Hatay respectively. At about the same time, a joint Franco-Turkish commission demarcated the limit of the boundary between Syria and Hatay and reported its work on 1 November 1938. The new segment of boundary turned south from the previous boundary near Meidan Ekbes and followed an irregular arc to the

Mediterranean coast just north of Er Ras el Basit. The district of Hatay became part of Turkey on 13 July 1939.

The boundary between Iraq and Turkey was eventually settled by the League of Nations. The Council established a boundary known as the Brussels Line in 1924, which lay close to the northern boundary of the former Turkish district of Mosul. Turkey took the matter to the International Court of Justice in the hope of persuading the Court that it should be granted Mosul. The Court found in favour of Britain and therefore Iraq in November 1925, and the final boundary was settled in these terms by a treaty dated 5 June 1925 (Toynbee 1927, pp. 471–530).

The southern borders of Iraq were attacked in early 1922 by puritanical Moslems of the Akhwan brotherhoods (Naval Intelligence Division 1944, p. 318). These incursions were ended by a treaty signed by British and Saudi representatives at Khorramshahr on 5 May 1922. The final boundary was then delimited by a protocol signed at Al Uqayr on 2 December 1922. Dickson (1956, pp. 274–5), who acted as translator for the British High Commissioner at the meetings at Al Uqayr, has given a remarkable account of how Sir Percy Cox, sick of the endless debate between Ibn Sa'ud and other Arab leaders from Kuwait and Iraq, first took Ibn Saud aside and reprimanded him like 'a naughty schoolboy' and then announced that he would decide the type and direction of the boundary! At the next meeting Cox carefully drew the boundaries which separated Iraq, Kuwait, and Saudi Arabia and created two neutral zones between Kuwait and Saudi Arabia and between Iraq and Saudi Arabia. Those lines survived until the neutral zones were divided between the flanking states after 1965.

Serious raids from Saudi Arabia into Transjordan in 1924 were defeated only through the use of aeroplanes and armoured cars by British forces. Ibn Sa'ud was persuaded to sign an agreement defining a boundary between the two territories from the terminus of the Saudi–Iraq boundary to the intersection of parallel 29°35′ north and meridian 38° east. That agreement was signed on 2 November 1925 (Toynbee 1927, p. 343), and eighteen months later, in an exchange of letters, the boundary was carried westwards to the head of the Gulf of Aqaba (Naval Intelligence Division 1943, p. 400).

The boundary between the British and French territories was defined from the Tigris River to the Mediterranean coast on 23 December 1920 (League of Nations 1924, pp. 355–61). This agreement fixed the principal alignment of the boundary, which was slightly modified in subsequent years. The present boundary between Israel and Lebanon was demarcated in 1922 and its description incorporated in an Anglo-French agreement dated 7 March 1923. The boundary between what was then Palestine and Syria was fixed by the same agreement, but recent wars have resulted in a cease-fire line which does not accord with that line drawn in 1923. The boundary between Jordan and Syria was finally delimited in an Anglo-

French protocol dated 31 October 1931, and it differed only slightly from the 1920 definition. The easternmost section of the boundary is a straight line drawn from a peak called Tell Romah at the southern end of the Jebel ed Druz to the centre of Abu Kemal on the Euphrates River. The eastern part of this line forms the western part of the boundary between Syria and Iraq. This straight line west of the Euphrates was defined in the 1920 Anglo-French agreement. Between the Tigris and Euphrates rivers the boundary drawn in 1920 was altered by a commission appointed by the League of Nations, which issued its report on 10 September 1932. The main change shifted the boundary northwestwards so that Iraq possessed the entire Jabal Sinjar west of Mosul (Naval Intelligence Division 1944, p. 321).

Once in control of the Levant, France lost no time in achieving a goal which General Gouraud noted had first been identified in 1862 (Temperley 1924, pp. 167–8). France issued an *arrêté* on 31 August 1920 creating a Greater Lebanon State. To the original district of Lebanon were added Beirut, Tripoli, and the valley called El Beqa'a. Toynbee (1927, 355–6) drew attention to the considerable Arab population thus included in the new state and hinted at the problems which overtook Lebanon in the 1980s.

On 2 September 1922 Britain drew a boundary separating Palestine from Transjordan (Naval Intelligence Division 1943, p. 2) and ten years later a line was drawn between Iraq and Transjordan which still survives. This straight line links Jabal Unayzah in the south with Jebel Tanf in the north on the straight line linking Tell Romah and Abu Kemal.

The construction of boundaries within the British territories fringing the Arabian peninsula required a change in the territorial philosophy of the local rulers:

> In contrast [to the coastal divisions], the inter-state boundaries did not emerge until the 1930s. Before then the concept of a state with clearly defined boundaries was totally alien to the political notions of the rulers and tribes of the area. Political boundaries were dependent on tribal loyalties to particular sheikhs and consequently were subject to frequent change. Therefore, the frontier between the Trucial States and the Sultanate of Muscat and the inter-state boundaries changed frequently during the nineteenth and twentieth centuries as it was based on the *dirah* of the tribes. *Dirah* in Arabia at this time was a flexibly defined area, changing in size according to the strength of the tribe which wandered within it. In addition, a tribe's loyalty was determined by its own interests and could, and at this time often did, alter . . . (Abdullah 1978, p. 291)

Thus, throughout the peninsula the tribal areas varied in extent and the agglomerations of tribes into alliances also varied. Heard-Bey (1982,

p. 300) has noted that this arrangement suited Britain until the search for oil began in earnest in 1935. This development put a sudden end to the convenience of the existing arrangement because the concessions had to have limits which were precisely defined.

Abdullah (1978, pp. 291–316) has provided the best account of the evolution of the boundaries between Oman and the United Arab Emirates. Mr Julian Walker of the British Foreign Office completed his report in 1960, and in 1963 and 1967 additional fieldwork by Foreign Affairs officers settled the remaining undelimited sections. Abdullah (1978, pp. 300–1) makes it clear by reference to the work of Lorimer that the Sultanate of Oman consisted of two distinct areas before 1903.

Saudi Arabia has negotiated boundaries with at least four neighbours. Wenner (1967, pp. 142–6) has described the events in the relations between Saudi Arabia and Yemen Sana'a which led to war in 1933. This conflict was ended by the Treaty of Taif (Taif was the name of a small settlement south of Al Hudaydah). This agreement defined the boundary from the coast south of Abu Arish to a hill called Thar south-east of Najran. Philby (1962) has provided a useful description of this boundary, together with maps of its eastern terminus and a complete list of the 239 boundary pillars.

After reaching agreement in July 1960 that the neutral zone between their countries should be divided, Kuwait and Saudi Arabia signed an agreement on 7 July 1965 arranging for surveys and demarcation. The demarcation report was signed on 18 December 1969 and the formal division of the neutral zone occurred on that date. The equal division of this area did not extend to the islands off the shore of the former neutral zone, so it is not clear how the offshore boundary will be determined. On 2 June 1975 Iraq and Saudi Arabia agreed to divide their neutral zone by a straight line.

On 7 November 1965 Jordan and Saudi Arabia redefined their common boundary, which Saudi Arabia had negotiated in the 1920s with Britain. Saudi Arabia gained the largest area by the transference of the hills called At Tubayq from Jordan. These remarkable features consist of sandstone mesas capped with black volcanic rock, in a desolate landscape. Jordan gained a small area when the boundary was moved eastwards to coincide with the Wadi Sirhan. However, Jordan's main gain occurred on the eastern shore of the Gulf of Aqaba when the coastal terminus of the boundary was moved about 10 miles south. One article in this agreement provided for any future oil revenues from a defined zone to be shared.

Israel's boundaries with Syria, Jordan, and Egypt have been settled in various wars. Apart from the Gaza Strip, the boundary with Egypt is the Anglo-Turkish line of 1906. It was most recently confirmed in the Camp David Concord of 18 September 1978. This provided detailed arrangements for Israel's withdrawal from Sinai (Africa Research 1979, p. 5285). The armistice agreement between Israel and Jordan signed on 3 April 1949 gave Jordan control of a large salient on the west bank of the River Jordan (Dearden 1958, pp. 201–8). During the short war of June 1967 Israel conquered this salient and re-established the boundary with Jordan along the Jordan River which the British High Commissioner for Palestine and Transjordan had declared to be the boundary between the two provinces on 1 September 1922 (Naval Intelligence Division 1943, p. 2).

Israel's boundary with Syria, inherited from the Anglo-French agreements of the 1920s, was also overrun by Israeli forces in 1967, and an agreement to separate forces was signed on 31 May 1974. This arrangement provided for a buffer zone policed by the United Nations and flanking zones in which the strength of Israeli and Syrian forces was precisely defined (Sachar 1981, p. 800).

It was noted earlier that Saudi Arabia had reached boundary agreements with at least four of its neighbours. The uncertainty in this statement arises from the fact that it has not been possible to find specific agreements with other neighbours, although such agreements are mentioned by reliable authors. For example, Zahlan (1979, p. 85) states that the Saudi–Qatar boundary was resolved by 1965, without giving any details of its alignment. Abdullah (1978, pp. 200, 212) refers to a satisfactory frontier agreement between Saudi Arabia and the United Arab Emirates in 1975 without providing any details. This lack of information is in sharp contrast with Abdullah's full and masterly account of the unsuccessful negotiations between Saudi Arabia and Britain in the 1930s and 1950s, which ended with the clash at Buraimi Oasis. During the second half of the 1930s, as it became necessary to settle clear lines for petroleum concessions Britain and Saudi Arabia proposed a variety of lines to separate their territories in eastern Arabia. Abdullah (1978) and Kelly (1964) have described those negotiations in detail and recorded the various lines on a series of useful maps. Wilkinson (1971, p. 365) has also provided a useful map showing most of the proposed lines. Britain boldly began by trying to secure the 'blue line' defined in an agreement with Turkey on 29 July 1913 (Hurewitz 1975, p. 569), but they finally had to settle for a unilateral declaration of a boundary on 26 October 1955 (Heard-Bey 1982, p. 305). This unilateral boundary followed closely the Riyadh Line, which Britain had offered to Saudi Arabia on 25 November 1955 (Wilkinson 1971, p. 365, Heard-Bey 1982, p. 463).

Although there is some uncertainty whether Saudi Arabia has boundary agreements with Qatar and the United Arab Emirates, there is no apparant reference to boundaries between Saudi Arabia and Oman or Yemen (Aden). Further, it does not appear that the boundary with Yemen (Sana'a) drawn in 1934 has been continued throughout the remainder of their borderland.

Conclusions

In terms of boundary construction the Middle East is transitional between Asia and Africa. The competition between major European powers in this region was much more muted than the conflict between the interests of Russia and Britain in Asia or Britain and France in Africa. Although tribal concepts of territory and sovereignty placed important difficulties in the path of negotiators in the Middle East and Africa, there was no African equivalent to Saudi Arabia. Further, the long history of boundary evolution in Africa was not matched in Arabia, where negotiations and agreements were compressed into a short period after World War I.

Within the Middle East there is a clear division between the processes of boundary evolution in the northern and southern regions. The rôle of external powers was much more important in the southern region than in the north, and in the southern region there are still some boundaries that have not been delimited. However, it appears that it is not the unsettled boundaries in eastern Arabia which present the main problems for the future; they are still to be found in the vicinity of the Shatt al Arab and the eastern and northern boundaries of Israel.

References

Abdullah, M. M. 1978. *The United Arab Emirates*. London: Croom Helm.

Africa Research 1979. *Africa Research Bulletin*, Political, social and cultural series. Vol. 16. Exeter: Africa Research.

Brice, W. C. 1966. *Southwest Asia*. London: University of London Press.

Crone, G. R. 1938. The Turkish–Iranian boundary. *Geographical Journal* **91**, 57–9.

Dearden, A. 1958. *Jordan*. London: Robert Hale.

Dickson, H. R. P. 1956. *Kuwait and her neighbours*. London: Allen & Unwin.

Fisher, W. B. 1961. *The Middle East*. London: Methuen.

Glubb, Sir J. B. 1959. *Britain and the Arabs*. London: Hodder & Stoughton.

Heard-Bey, F. 1982. *From Trucial States to United Arab Emirates*. London: Longman.

Hickinbotham, Sir T. 1958. *Aden*. London: Constable.

Hurewitz, J. C. 1975 & 1979. *The Middle East and North Africa in world politics: a documentary record* (2 volumes). New Haven, Conn.: Yale University Press.

Kelly, J. B. 1964. *Eastern Arabian frontiers*. London: Faber.

League of Nations 1924. *Treaty series*, Vol. 22. Geneva: League of Nations.

Naval Intelligence Division 1943. *Palestine and Transjordan*. London: HMSO.

Naval Intelligence Division 1944. *Iraq and the Persian Gulf.* London: HMSO.

Parry, C. 1980. *The consolidated treaty series*, Vol. 201. New York: Oceana.
Parry, C. 1981. *The consolidated treaty series*, Vol. 203. New York: Oceana.
Philby, H. St J. B. 1962. *Arabian highlands*. Ithaca, NY: Cornell University Press.

Ryder, C. D. H. 1925. The demarcation of the Turco-Persian boundary in 1913–1914. *Geographical Journal* **66**, 227–42.

Sachar, H. M. 1981. *A history of Israel*. New York: Knopf.
Sevian, V. J. 1968. The evolution of the boundary between Iraq and Iran in *Essays in political geography*, C. A. Fisher (ed.), 211–23. London: Methuen.

Temperley, H. W. V. 1924. *A history of the peace conference of Paris*, Vol. VI. London: Oxford University Press.
The Geographer 1965. *Israel–UAR armistice line*. International Boundary Study No. 45. Department of State, Washington, DC.
Toynbee, A. J. 1927. *Survey of international affairs 1925*, Vol. 1. London: Oxford University Press.

United Kingdom, Foreign Office 1817. *British and Foreign State Papers*, Vol. 5. London: HMSO.
United Kingdom, Foreign Office 1828. *British and Foreign State Papers*, Vol. 10, London: HMSO.
United Kingdom, Foreign Office 1872–3. *British and Foreign State Papers*, Vol. 63. London: HMSO.

Wenner, M. W. 1967. *Modern Yemen 1918–1966*. Baltimore, Md.: Johns Hopkins Press.
Wilkinson, J. C. 1971. The Oman question: the background to the political geography of southeast Arabia. *Geographical Journal* **137**, 361–71.

Zahlan, R. S. 1979. *The creation of Qatar*. London: Croom Helm.

12 *Australia, Indonesia, and the islands of the South Pacific Ocean*

Geography and the colonial policies of various maritime powers combined to give the history of international boundaries in this region two distinct qualities. First, there are only four international boundaries on land throughout this extensive zone. Secondly, the procedures which resulted in the present territorial division of this region were appropriate to the ocean which Magellan called Pacific.

There are five international boundaries on land and four of them were associated with the former colony of the Netherlands East Indies, which originated before the colonization of Australia, and the islands of the South Pacific Ocean. The fifth boundary, which encloses Brunei, was determined unilaterally by Britain. The lack of any serious challenge to Britain's annexation of Australia and New Zealand, which were the main prizes, meant that colonial competition was for comparatively small islands, which were mainly arranged in compact or distinct groups. The fact that only the Solomon and Samoa Islands were divided between two powers is an indication that the colonial authorities generally accepted that the groups should remain intact when they passed under colonial control.

There were a number of reasons for this view. First, unlike the situation in Africa and Asia, the region had been thoroughly explored and the limited economic potential of the islands had been understood. The exploitation of sandalwood, whales, guano, *bêche de mer* and pearl shell, and the cultivation of sugar and cotton promised personal rather than national profits. It seemed almost certain that any challenge to secure a share of an island group where another country had a prior claim would not be worth the effort. Secondly, in some of the groups, such as Hawaii, Tahiti, Tonga, and Fiji, there was a single recognizable ruler. Thirdly, the colonial activities of Britain, France, and Germany in Africa and of Britain and France in Asia discouraged them from seeking territories which would involve additional quarrels and difficulties. So, when the policy of minimum intervention (Ward 1948, pp. 41–9) was ended by the naval powers they adopted the wide seas and narrow straits in place of the mountain ranges and rivers of Africa and Asia to separate their claims.

With the exceptions of Britain's annexations of Australia in 1788 and New Zealand in 1840 and France's annexation of Tahiti in 1842 and New

Caledonia in 1853, this policy of minimum intervention persisted until the 1870s. It was maintained for a number of reasons, despite requests by the rulers of some kingdoms for protection. The reasons included the fact that the main European commercial activities flourished without official supervision, the opposition of entrenched missionary groups, and the burden of colonial administration in other parts of the world which were perceived to be more important than the South Pacific Ocean.

The policy of minimum intervention was abandoned for three main reasons. First, in some countries, such as Fiji and Hawaii, the domination of government by European settlers made annexation desirable to defend the rights of indigenous groups. Secondly, the expansion of trade between the burgeoning cities of Australia, New Zealand, and California, the improving prospect of commerce with Japan and China, and the rising level of economic activity amongst the islands meant that the Pacific Ocean was becoming even more important as an international highway. This development, together with the perceived strategic concerns of Australia and New Zealand, encouraged the maritime powers to focus on the advantages of coaling stations and naval bases rather than the disadvantages of administrative costs. Thirdly, Germany, which was a late starter in the race for overseas territories, began to capitalize on the economic activities of various German firms in the Pacific Ocean. Germany, which had no colonies in Asia and only some of the less desirable properties in Africa, made clear its determination to secure a major share in this region.

In addition to four international boundaries on land, the colonial authorities agreed on three international maritime limits to separate adjacent claims to islands, and two unilateral maritime limits were proclaimed for the same reason. Although only one of these maritime limits has survived, they will all be examined briefly because of the increasing tendency throughout the world for such lines to be used in an unjustified manner as the basis for modern maritime claims.

The boundary on New Guinea

The boundary which now separates Indonesia and Papua New Guinea lies very close to the limit of Dutch annexations in the first half of the 19th century. Van der Veur (1966) has published a good account of the early history of this line. In 1828 Dutch authorities decided to take action to substantiate vague claims to New Guinea based on the jurisdiction of the Sultan of the Moluccas. An expedition was sent to the western part of New Guinea and a small settlement was established at Teluk Matyan Tutul after the coast had been explored from the most northerly cape called Tandjung Jamursba to the Digul River. On 24 August 1828 a unilateral claim was made to New Guinea from a point on the south coast intersected by meridian 141° east to Tandjung Jamursba (van der Veur 1966, p. 10). The selected meridian lay about 150 miles east of the limit of Dutch exploration at that time. It may have been chosen so that most of the Digul catchment was also annexed, and if that was the reason then 141° east was the correct meridian.

Prompted by British requests for a definite description of the territory occupied by tribes subject to Dutch authority, a new, secret description was prepared. The southern limit remained the same, but on the north coast the terminus was moved eastwards from Tandjung Jamursba to Tandjung Djar, which is the eastern limit of Teluk Jos Sudarso; its meridian is 140°47′ east. When this boundary was made public in 1865 and reaffirmed in 1875 it was described as a straight line from the south coast at 141° east to the north coast at Tandjung Djar; this means that it lay close to meridian 141° east (Fig. 12.1).

Britain and Germany annexed the remaining south-east and north-east parts of New Guinea respectively in 1884, and a year later they agreed on a boundary separating their claims. It followed a series of straight lines from Mitre Point in the east to the intersection of parallel 5° south and meridian 141° east in the west.

Anglo-Dutch negotiations started in 1893 because Britain wanted a clear boundary so that authorities could deal with raids by headhunters from Dutch areas. It was decided that the mouth of the Bensbach River should mark the southern terminus and that was fixed at 141°1′47.9″ east. In return for the British concession the Dutch authorities agreed that when this meridian reached the Fly River the boundary would then follow the river until its second intersection with meridian 141° east; this meridian would then be followed north to the tri-junction with German territory.

Efforts to settle the boundary by Dutch and German representatives were overtaken by World War I, when Australia replaced Germany in the northern part of New Guinea. Since Australia had also succeeded Britain in Papua in 1905, the entire line was now a matter for Australia and the Netherlands.

Cook *et al.* (1968) have provided an account of developments since 1919. Joint survey teams placed a marker near meridian 141° east on the north coast in 1933 and it was agreed that the boundary would follow the meridian to the Fly River. This boundary was not demarcated before World War II, and after the war it was decided that the boundary would be a great circle through the marker on the north coast and the most northerly intersection of the Fly River and meridian 141° east. South of the Fly River, it was agreed that the boundary would follow the meridian 141°1′7″ east, which was the new determination of the mouth of the Bensbach River.

After 1963, when Indonesia replaced the Netherlands in Irian Jaya, the

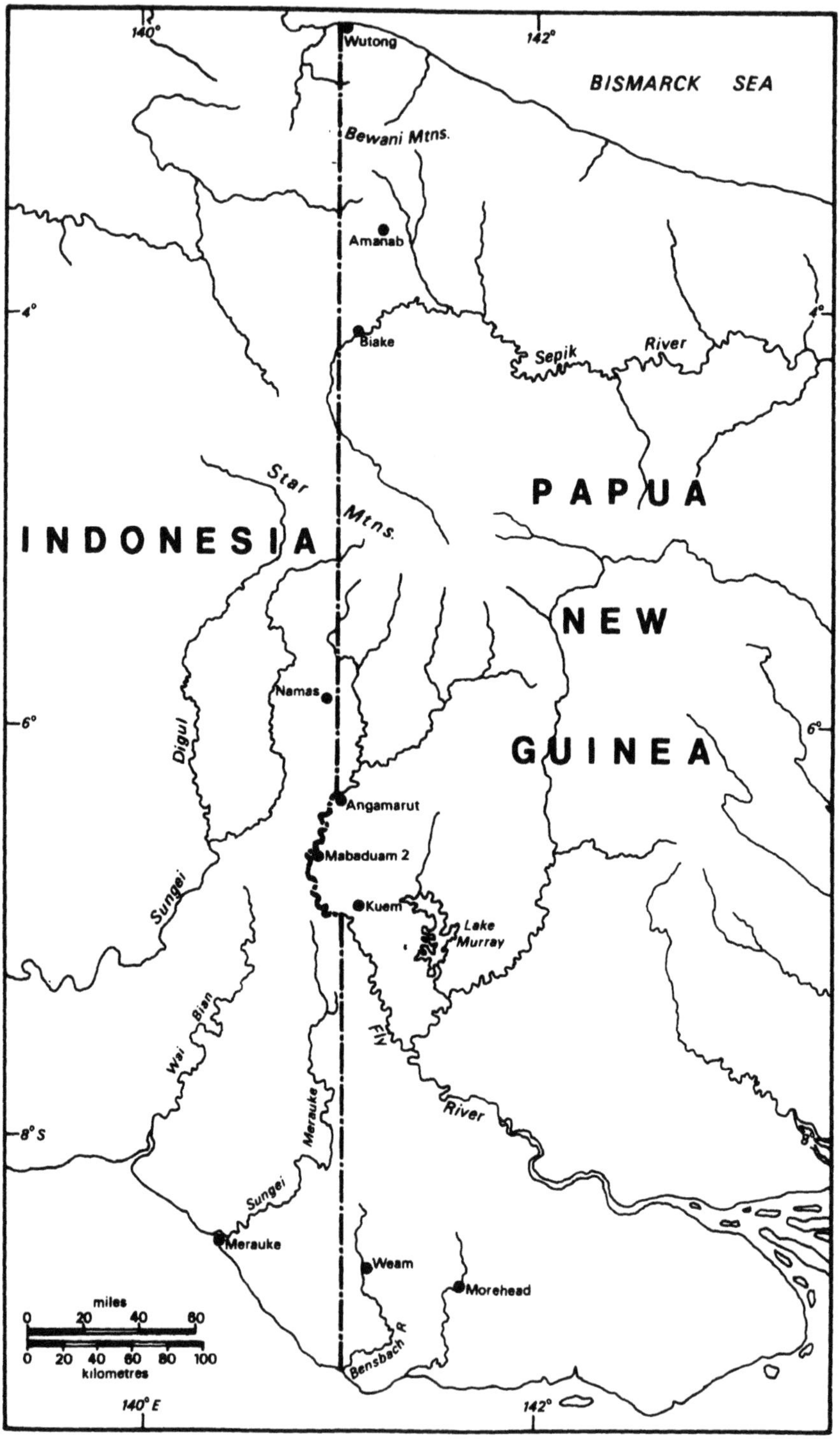

Figure 12.1 The Indonesia – Papua New Guinea boundary.

work had to be done all over again. Within a year the two countries had agreed that north of the Fly River the boundary would be meridian 141° east whereas south of the river it would follow the meridian passing through the mouth of the Bensbach River. The markers were erected in the period 1966 to 1968. Ten markers were placed north of the Fly River and four to the south. Their locations were listed in an agreement dated 12 February 1973, when it was revealed that the third determination of the mouth of the Bensbach River placed it at 141°1′10″ east. It was sensibly arranged that Australia would seek the approval of the House of Assembly of Papua New Guinea before ratifying this agreement; this meant that when Papua New Guinea became independent there was no need for further surveys.

The boundaries on Timor

Portuguese Timor ceased to exist in 1975 when Indonesian forces occupied it. Because some countries and organizations have refused to recognize this incorporation it seems proper to describe the limits which used to define Portuguese Timor.

Portuguese and Dutch authorities signed a number of treaties with tribes on the island in the first half of the 19th century, and it was decided on 20 April 1859 to end existing uncertainties about the territory over which each had jurisdiction. According to the first article of this agreement, Portugal obtained the territories of Cova, Balibo, Lamakitu, Tahakay, and Suai, and the Netherlands received Djenilo, Naitimu, Fialarang, Mandeo, and Lakecune. Although the island was effectively divided into an eastern Portuguese region and a western Dutch sector, each country possessed small areas of territory in the other half of the island. The Dutch retained Maucatar in the Portuguese region and Portugal retained Ocussi, Ambeno, and Noe Muti in the Dutch sector. This agreement also resolved disputes over the ownership of islands lying north of Timor. Portugal obtained Pulo Kambing and the Netherlands secured Flores, Adenara, and Solor.

In 1893 the two countries agreed to define the boundary exactly and to eliminate the enclaves which each held in the other's territory. During the period 1898 to 1899 a joint commission surveyed the boundary across the island and a strip on either side, and this work provided the basis for each side to make specific proposals. Attempts to complete a survey of the limits of Ocussi were thwarted by hostile tribesmen. Negotiations in 1902 tried to reconcile the various conflicting proposals and they lasted two years. Portugal tried to eliminate the Dutch salient of Fialarang, but the Dutch stood firm and threatened to raise the question of whether Ocussi was an enclave which should be eliminated. Because it has a coastline

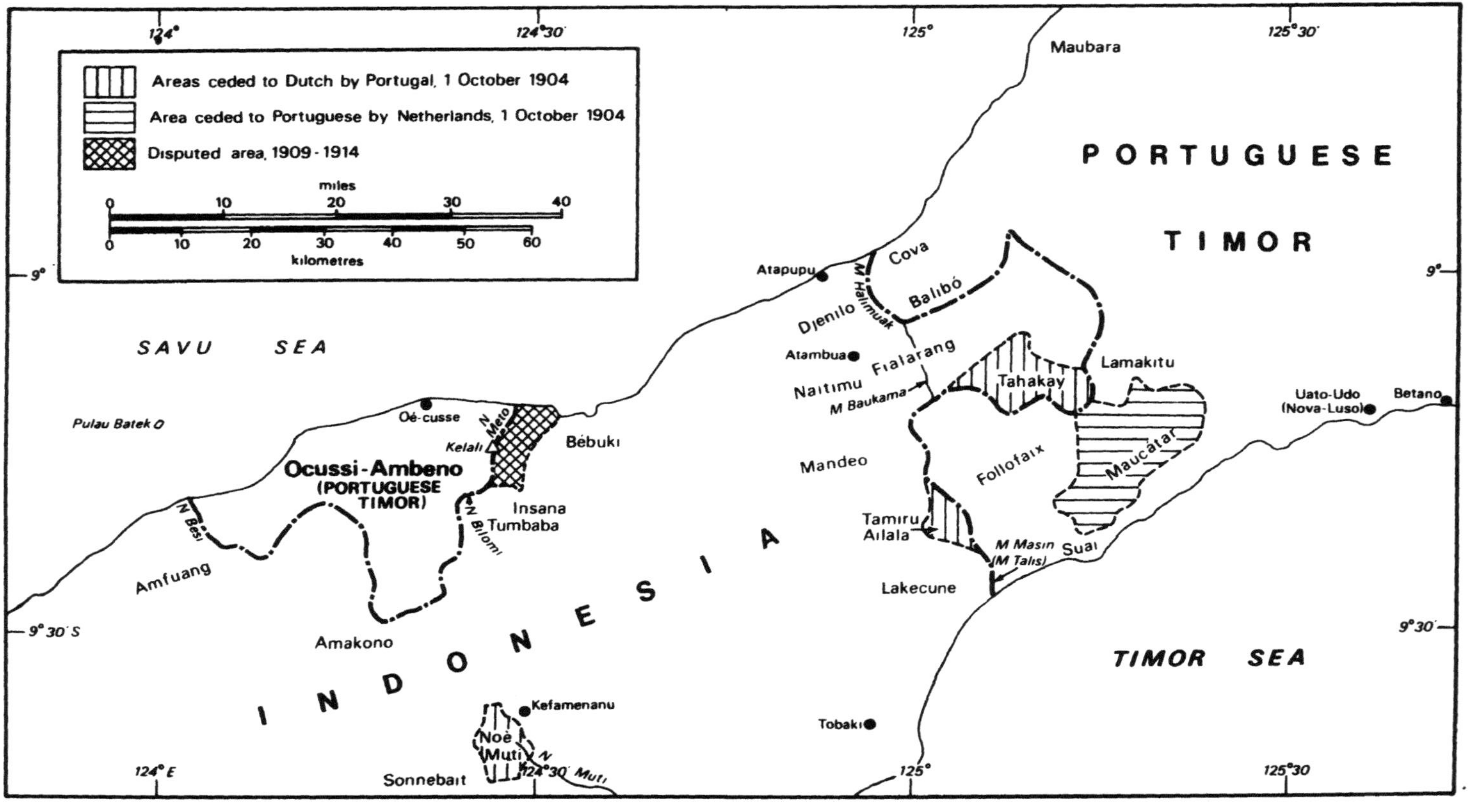

Figure 12.2 The boundaries of Portuguese Timor.

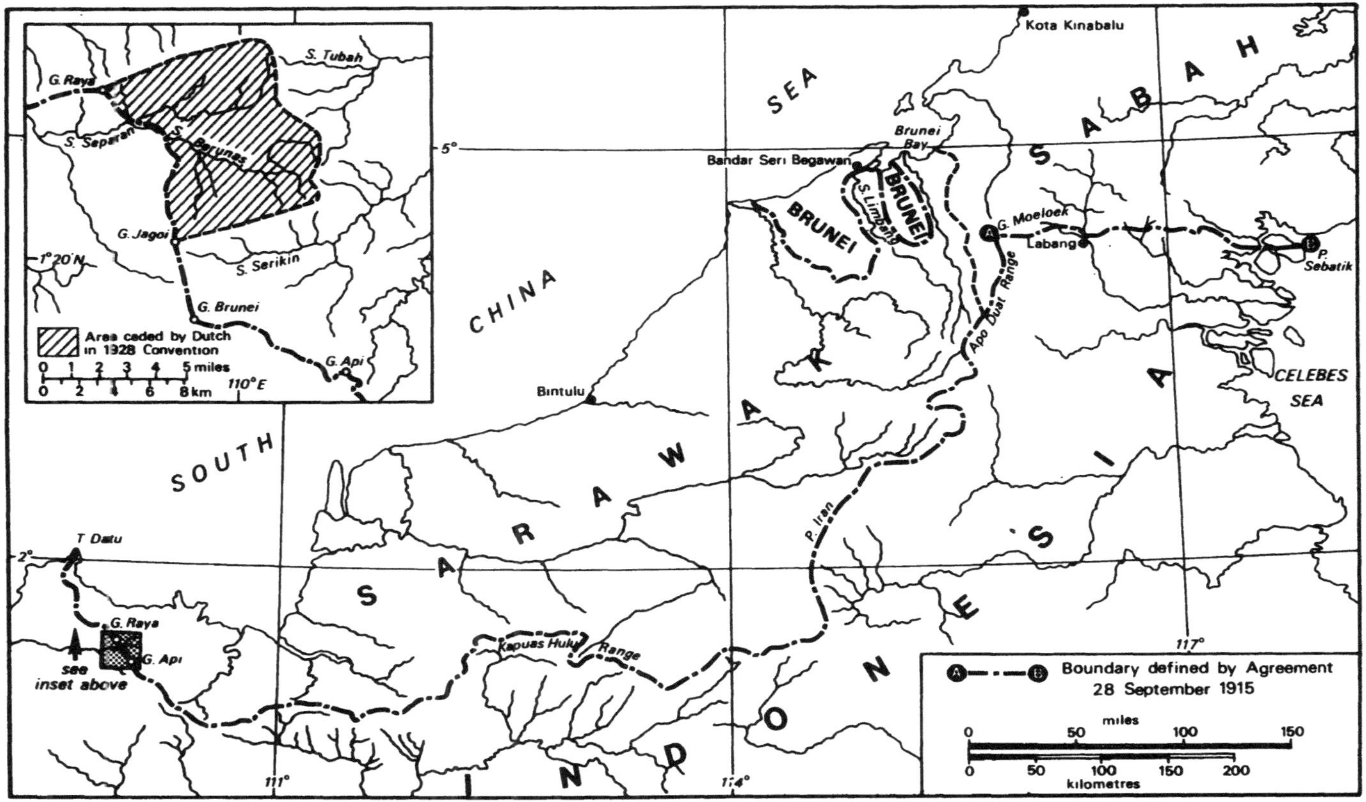

Figure 12.3 The Indonesia–Malaysia boundary.

Ocussi could not technically be described as an enclave, but the Portuguese decided to avoid any threat to Ocussi and Fialarang remained Dutch. The Dutch enclave of Maucatar was exchanged for the Portuguese enclave of Noe Muti, and as a makeweight in this exchange the Portuguese also ceded Tahakay and Tamiru Ailala, which had previously been on the Portuguese side of the boundary across the island (Fig. 12.2).

When a joint commission began to demarcate these boundaries in 1909 a positional dispute developed along the eastern limit of Ocussi. Although the disputed area only measured 22 square miles, neither side was prepared to concede it and the matter was sent to arbitration. On 25 June 1914 the Permanent Court of Arbitration found in favour of the Netherlands, and after the final sector of boundary was demarcated this borderland became a backwater until the conflict between Japanese and Australian forces in 1942. After 1945 the boundaries became effective again for 30 years, until the collapse of Portuguese authority and Indonesia's invasion effectively eliminated them. According to an official handbook of Indonesia published in 1984 (Department of Information 1984), the main boundary across the island has been maintained as the western limit of the new province of East Timor. Ocussi appears to have been incorporated into the province called West Nusa Tenggara, with its capital at Mataram on Lombok Island.

The boundaries on Borneo

Dutch settlements were established on the south coast of Borneo at the beginning of the 17th century, whereas the British intervention on the north coast occurred in the 1840s when James Brooke arrived in Sarawak. In return for assistance to the Sultan of Brunei, Brooke was granted some land. The Sultan had nominal authority over the whole of north Borneo, but he had great difficulty in collecting taxes beyond the immediate area of his capital. The decline in his authority, which had been evident since the middle of the 17th century, accelerated after 1881 when the British North Borneo Company was given a royal charter and began to establish its authority throughout Sabah. Territory subject to Brunei was whittled away from both sides as authorities in Sarawak and Sabah competed with each other and advanced towards a common boundary as they secured the cession of each river valley. Sarawak enjoyed more success than Sabah, and its purchase of the Terusan valley in 1884 and the Limbang valley in 1890 ensured that Brunei was a divided coastal state surrounded by Sarawak. The boundary of western Brunei follows watersheds for most of its length and some of them along the west flank of the Limbang River follow pronounced ridges rising to 4000 feet. The boundary of eastern Brunei follows the Pendaruan River east of the Limbang valley and a watershed to the west of the Terusan valley. The best accounts of Britain's occupation of northern Borneo are provided by Tarling (1971) and Wright (1970).

Britain's authority over the whole of northern Borneo was confirmed when protectorates were established over Sarawak, Brunei, and Sabah in 1888. Three years later, the first definition of the Anglo-Dutch boundary was agreed in a treaty signed on 20 June 1891 (Fig. 12.3). On the east coast the Dutch were literally able to steal a march and push beyond the Sebuku River, which was the southern limit of the original cession secured by Overbeck and Dent from the Sultan of Brunei in 1878. The Dutch success was made possible because the British representatives were delayed in advancing southwards from Sandakan by the presence of pirates in Darvel Bay.

The boundary commenced on the east coast at parallel 4°10′ north, which divided Sebatik Island. The boundary then followed a northwesterly course to a point with the co-ordinates 4°20′ north and 117° east, so that the Simengaris River was left in Dutch territory. The boundary then proceeded westwards to join the main watershed separating rivers flowing northwards to the South China Sea from those flowing east and south to the Celebes Sea and the Java Sea respectively. Provision was made for those short sections where the boundary was marked by straight lines for the demarcation commissions to shift the line as much as 5 miles to avoid cutting off bends in rivers or their headwaters.

The boundary east of Moeloek Mountain was demarcated in 1912 and 1913, and the final description was contained in an agreement signed on 28 September 1915. Examination of the final line shows that the commissioners took advantage of their authority to vary the boundary close to rivers, and it appears likely that this was done on the basis of mutual concessions.

The last Anglo-Dutch boundary treaty was signed on 26 March 1928 when a short section of the boundary between peaks called Api and Raya was moved south-west, enabling Sarawak to gain about 38 square kilometres in the upper Separan and Berunas valleys. In this zone the watershed is comparatively low, and it seems likely that settlers from Sarawak had penetrated across the watershed into unoccupied Dutch territory.

Boundaries in the sea

Although most of the island groups were sufficiently distinct to make any more formal definition of national territory unnecessary, there were some areas where island groups were very close together or where island groups were divided between two colonial powers. In Torres Strait, which stretches for about 80 nautical miles between Cape York in Australia and

Coreigegemuba Point on the coast of Papua, there are about 120 islands. Although these islands vary in size and are unevenly distributed across the strait, it requires only very short journeys by sea to use the islands as stepping-stones from Australia to Papua New Guinea.

When the colony of Queensland was created in 1859 it included 'all and every the adjacent islands, their members and appurtenances, in the Pacific Ocean' (McLelland 1971, p. 679). In 1872 the Governor was empowered to extend his jurisdiction over all islands within 60 miles of the coast of Queensland. This authority placed all islands south of 10° south in Torres Strait within the limits of Queensland and included the Prince of Wales Channel, which is the only safe passage for large vessels. The first regular steamship service started in 1873, and about this time authorities in Brisbane began to press the British government to annex the coast of Papua. This course was urged for a number of reasons. For example, it was argued that British commerce through Torres Strait would be completely protected and Papua would provide markets for British goods. There were also predictions that gold would be found in eastern New Guinea. The colonial administration wished to prevent France or Germany from establishing bases close to Australia, especially if they were to be used for penal settlements. It was also pointed out that if southern Papua was annexed it would be much easier to regulate pearling, fishing, and the recruitment of native labour. To drive these points firmly home Torres Strait was described as the Bosphorus of the Turkey that was Queensland (Prescott *et al.* 1977, p. 80).

Although Britain resisted all these blandishments, Queensland was authorized to annex all the islands in Torres Strait by an Act dated 24 June 1879. The language of the schedule attached to this Act makes it quite clear that the defined boundary was merely a form of geographical shorthand to avoid the tedious task of naming each island.

Australia's ownership of most of the islands in Torres Strait, including some hard against the coast of Papua, created difficulties when Australia and Papua New Guinea started to negotiate the location of maritime boundaries through the strait. Papua New Guinea demanded, and was given, areas of sea and seabed which were closer to Australian islands than to the territory of Papua New Guinea. Indeed, the Australian government managed to cede three islands, which had been considered as part of Australia since 1879, to Papua New Guinea by an ingenious but unconvincing piece of research (Joint Committee on Foreign Affairs and Defence 1979, Appendix II, Griffin 1981).

On 6 April 1886 Britain and Germany agreed to draw a line through the sea, separating their spheres of influence in the western Pacific Ocean. The boundary commenced at Mitre Rock, which was the terminus of the land boundary agreed in eastern New Guinea by the two countries in April 1885, and proceeded via seven turning points to a position north-east of the

Marshall Islands at 15° north and 173°30′ east. This line divided the Solomon Islands, with Germany securing the northern section, and distinguished the Gilbert and Ellice Islands acquired for Britain from the Marshall Islands, which went to Germany.

On 10 December 1898 the United States and Spain signed a peace treaty. The main effect of this treaty was to transfer the Philippines to the United States. The transfer was made by means of a frame drawn between seven defined points in the sea: 'Spain cedes to the United States the archipelago known as the Philippine Islands and comprehending the islands lying within the following line' (Martens 1905, p. 74). Soon after this event it was discovered that the frame excluded some Philippine islands close to the coast of Borneo. A second treaty, dated 7 November 1900, rectified this oversight (Martens 1905, p. 82), but the new definition of the limit was ambiguous and Britain and the United States resolved the matter by a treaty on 2 January 1930 (League of Nations 1933, p. 298). Once again a line was drawn joining 11 defined points. It was stipulated that if these lines intersected any rocks or islands they belonged to the United States. It appears that this boundary has survived as the maritime boundary between Malaysia and the Philippines. The segments of boundary through this region have been incorporated into the limits of the territorial sea claimed by the Philippines on 17 June 1961 and are shown on an official map of Malaysia's maritime claims. This map, which was published by the Director of National Mapping in 1979 at a scale of 1 : 1 500 000, shows this 1930 line as an international boundary.

American, British, and German interests overlapped in the Samoan Islands. The appointment of a consul by the United States in 1844 was followed by similar British and German appointments in 1847 and 1861 respectively. Jostling for political and economic preeminence continued amongst these three powers until March 1889. At that time, when it appeared to the other powers that Germany was making a determined bid for absolute control, there were seven warships in Apia harbour. Three belonged to America, three to Germany, and one to Britain. A powerful storm struck the anchorage and only the British vessel managed to avoid being grounded. This dramatic event persuaded the three governments to negotiate three months later in Berlin and a condominium was created (Naval Intelligence Division 1943, p. 598).

This arrangement proved cumbersome and inconvenient, and in 1899 it was agreed that the islands would be divided amongst the three countries. The United States would receive Tutuila, and Upolu and Savaii would be divided between Britain and Germany. At that time Britain was beset with colonial problems in Africa and was easily persuaded by Germany to abandon its claims in Samoa for adjustments elsewhere. On 14 November 1899 Britain and Germany reached an agreement in which Britain formally renounced its claims in Samoa. In return, Germany renounced its interests in Tonga in favour of Britain, ceded islands in the Solomons south and east of Bougainville Island, gave up its extraterritorial rights in Zanzibar, and gave Britain the boundary it wanted to divide the neutral zone between the Gold Coast and Togoland. On 2 December 1899 the three powers signed an agreement which effectively divided the Samoan Islands between Germany and the United States by meridian 171° west.

This chapter of events was completed on 7 March 1904 when Britain and Germany signed an agreement modifying the line drawn in 1886 to separate their spheres of influence east of New Guinea (van der Veur 1966, p. 39).

Present territorial problems

The ownership of Matthew and Hunter Islands is disputed by Vanuatu and France. Matthew Island is volcanic in origin and has two cones rising to 460 feet and 580 feet; there is no vegetation on its surface. Hunter Island is also volcanic and rises to 1000 feet in height; it is sparsely covered with clumps of grass and a few trees. These islands are not specifically named in the protocol of 1914 which established the Anglo-French condominium over the New Hebrides, which became Vanuatu in July 1980. In 1965 there was an exchange of letters between the New Hebrides Joint Court and the British and French Resident Commissioners in which it was recorded that Britain was content with the view expressed by France that Matthew and Hunter Islands formed part of New Caledonia.

Vanuatu has not accepted French claims and on 9 March 1983 sent a party to land on Hunter Island, remove the French plaque, and raise the national flag. Vanuatu's assertions are understandable because although the islands have no intrinsic economic value they do permit the owner to claim economic rights over 59 400 square nautical miles of sea.

There are some maritime disputes over the claims that can be made from Minerva Reefs by Tonga and from Pocklington Reef by Papua New Guinea, but all other territorial disputes appear to have been solved. Miangas Island, which falls within the seas claimed by the Philippines and which is part of the Indonesian system of archipelagic straight baselines, was the subject of an adjudication by Max Huber in the Permanent Court of Arbitration in April 1928. The dispute was between the United States in the Philippines and the Netherlands in the Dutch East Indies, and the island was awarded to the Netherlands.

The claims by the Philippines to the northern part of Sabah were renounced at a recent meeting of the Association of Southeast Asian Nations, but the appropriate legislation was not introduced into the parliament in Manila.

Conclusions

The colonial powers operating in this region faced fewer problems in drawing boundaries than they did in the Americas, Asia, Africa, and the Middle East. The main reason for this situation was the fact that Britain secured Australia and New Zealand without any significant challenge. When the policy of minimum intervention was ended by the imperial powers in the 1870s most of the islands were arranged in fairly compact, distinct groups. The limited economic potential of these islands was known by then and only two island groups were divided between two colonial authorities.

None of the present boundaries on land are disputed and the governments of this region have turned their attention seawards to establish claims to areas of the seabed and sea. Even in this maritime sphere, it appears that there are very few potential maritime disputes of any consequence.

References

Cook, D., J. C. McCartney & P. M. Stott 1968. Where is the border? *Australian External Territories* **8** (5), 7–18.

Department of Information 1984. *Indonesia 1984: an official handbook.* Jakarta: Government Printer.

Griffin, J. 1981. Territorial implications in the Torres Strait. In *The Torres Strait Treaty*, P. J. Boyce & M. W. D. White (eds), 92–113. Canberra: Australian National University Press.

Joint Committee on Foreign Affairs and Defence 1979. *The Torres Strait Treaty.* Canberra: Australian Government Printing Service.

League of Nations 1933. *Treaty Series*, Vol. 137. Geneva: Publications Department.

Martens, G. F. de 1905. *Nouveau receuil général de traites et autre actes relatifs aux rapports de droit international. Continuation de grand receuil de G. Fr. de Martens par Felix Stoerk.* (A new general collection of treaties and other acts related to international law. A continuation of the large collection of G. Fr. de Martens by Felix Stoerk), second series 32. Leipzig: Librairie Deiterich.

McLelland, M. H. 1971. Colonial and state boundaries in Australia. *Australian Law Journal* **45**, 671–9.

Naval Intelligence Division 1943. *Pacific Islands*, Vol. II. London: HMSO.

Prescott, J. R. V., H. J. Collier & D. F. Prescott 1977. *Frontiers of Asia and Southeast Asia.* Melbourne: Melbourne University Press.

Tarling, N. 1971. *Britain, the Brookes and Brunei*. Kuala Lumpur: Oxford University Press.

van der Veur, P. 1966. *Search for New Guinea's boundaries*. Canberra: Australian National University Press.

Ward, J. M. 1948. *British policy in the South Pacific (1786–1893): a study of British policy in the South Pacific islands prior to the establishment of government by the great powers*. Sydney: Australasian Publishing.

Wright, L. R. 1970. *The origins of British Borneo*. Hong Kong: Hong Kong University Press.

13 *Antarctica*

The political boundaries on Antarctica may be distinguished from the political boundaries on other continents in five ways. First, all the claimed political boundaries coincide with meridians. Secondly, there are no bilateral boundary agreements dealing with national claims in Antarctica, but Britain, Norway, Australia, France, and New Zealand have ensured that their adjoining claims are conterminous. Thirdly, none of the boundaries in Antarctica has been demarcated. Fourthly, none of the countries which claim territory in Antarctica enforces any restrictions on the movements of people, goods, or ideas at the limit of its territory. The issue of sovereignty was placed in abeyance by Article 4 of the Antarctic Treaty:

> 1 Nothing contained in the present Treaty shall be interpreted as:
> (a) a renunciation by any Contracting Party of previously asserted rights of or claims to territorial sovereignty in Antarctica;
> (b) a renunciation or diminution by any Contracting Party of any basis of claim to territorial sovereignty in Antarctica which it may have whether as a result of its activities or those of its nationals in Antarctica, or otherwise;
> (c) prejudicing the position of any Contracting Party as regards its recognition or non-recognition of any other State's rights of or claim or basis of claim to territorial sovereignty in Antarctica.
>
> 2 No acts or activities taking place while the present Treaty is in force shall constitute a basis for asserting, supporting or denying a claim to territorial sovereignty in Antarctica or create any rights of sovereignty in Antarctica. No new claim, or enlargement of an existing claim to territorial sovereignty in Antarctica shall be asserted while the present Treaty is in force. (Lovering & Prescott 1979, p. 203)

Finally, the political boundaries of Antarctica are distinct because all the proclamations dealing with them were made in the 20th century.

British claims

On 23 June 1843 British authorities issued details of the arrangements for the government of the Falkland Islands and their Dependencies (International Court of Justice 1956, p. 41). This proclamation did not define the extent of the Dependencies, nor did subsequent proclamations, governors' commissions, or laws clarify this question in the period before 1907. However, from 1887 the *Colonial Office year book* referred to South Georgia as part of the Dependencies.

In the last decade of the 19th century there was a marked increase in voyages to Antarctic seas for purposes of scientific research and the capture of whales and seals. This activity encouraged the British authorities to set out the limits of the Dependencies in Letters Patent dated 21 July 1908:

> Whereas the group of islands known as South Georgia, the South Orkneys, the South Shetlands, and the Sandwich Islands, and the territory known as Graham's Land, situated in the South Atlantic Ocean to the south of the 50th parallel of south latitude, and lying between the 20th and 80th degrees of west longitude, are part of our Dominions, and it is expedient that provision should be made for their government as Dependencies of our said Colony of the Falkland Islands . . . (*Polar Record* 1948, p. 241)

According to British sources, South Georgia was discovered, or rediscovered, by Captain James Cook on 17 January 1775. Having named them in honour of the King and claimed them for Britain, Cook sailed eastwards and two weeks later discovered the South Sandwich Islands which were named for the First Lord of the Admiralty and claimed for Britain. The South Shetland Islands were discovered by Captain William Smith on 18 February 1819. He claimed them for Britain in October of that year when he called them New South Britain. Graham Land was discovered by Captain Edward Bransfield in company with Smith on 30 January 1820. Some American scholars, including Hobbs (1939, 1941) and Martin (1938, 1940), assert that Captain Nathaniel Palmer first sighted the mainland of Antarctica, but their evidence is shown to be dubious by Gould (1941) and Hinks (1939, 1940, 1941). According to British sources Captain George Powell discovered the South Orkney Islands on 6 December 1821, and he claimed them the following day when he landed on the island to which he gave the name Coronation.

The British authorities must have been embarrassed to discover that the limits set out in the proclamation of 1908 enclosed the southern tip of South America, although there was no intention to claim that area. To avoid any uncertainty the boundaries of the British claim were amended on 28 March 1917:

> The Dependencies of Our said Colony shall be deemed to include and to have included all islands and territories whatsoever between the 20th degree of west longitude and the 50th degree of west longitude which are situated south of the 50th parallel of south latitude; and all the islands

> and territories whatsoever between the 50th degree of west longitude and the 80th degree of west longitude which are situated south of the 58th parallel of south latitude. (*Polar Record* 1948, p. 242)

This definition of the Dependencies was confirmed in The Falkland Islands (Legislative Council) Order in Council dated 26 November 1948 (Great Britain 1948, p. 59). On 26 February 1962 Britain detached part of the Dependencies to create the colony called British Antarctic Territory:

> . . . the British Antarctic Territory means all islands and territories whatsoever between the 20th degree of west longitude and the 80th degree of west longitude which are situated south of the 60th parallel of south latitude . . . (Great Britain 1962, p. 356)

The Dependencies were defined as the remaining area of the 1917 definition (Fig. 13.1).

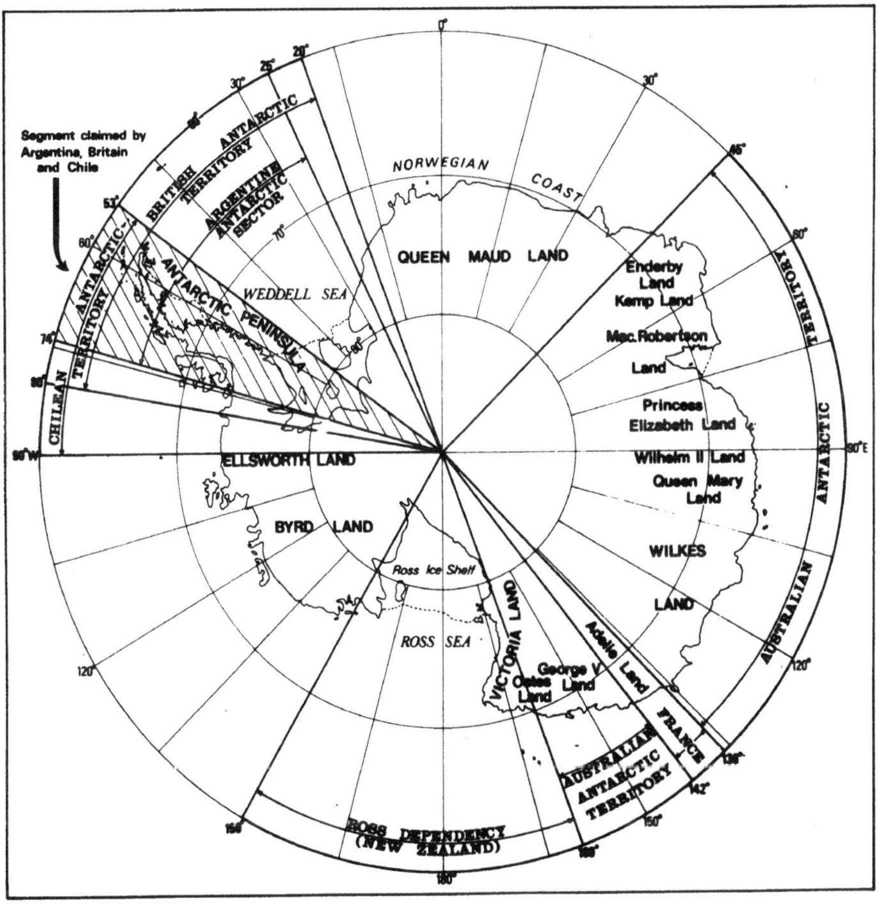

Figure 13.1 National claims in Antarctica.

New Zealand's claim

The boundaries of the Ross Dependency were proclaimed on 30 July 1923 (*New Zealand Gazette* 1923, p. 1). For some time the British authorities had been searching for a way to extend control over the coasts of Antarctica without displaying excessive greed. The Ross Sea was an area of particular concern because Norwegians were seeking rights to catch whales there (O'Connell & Riordan 1971, pp. 312, 314). Eventually it was decided to issue an Order in Council under the British Settlements Act of 1887:

> 1 From and after the publication of this Order in the Government Gazette of the Dominion of New Zealand that part of His Majesty's Dominions in the Antarctic Seas, which comprises all the islands and territories between the 160th degree of east longitude and the 150th degree of west longitude, which are situated south of the 60th degree of south latitude shall be named the Ross Dependency. (*New Zealand Gazette* 1923, p. 1)

The boundaries of this territory have not been altered since that date.

Australian and French claims

The limits of Australia's Antarctic Territory were fixed by an Order in Council on 7 February 1933:

> That part of His Majesty's Dominions in the Antarctic Seas which comprise all the islands and territories other than Adelie Land which are situated south of the 60th degree of south latitude and lying between the 160th degree of east longitude and the 45th degree of east longitude is hereby placed under the authority of the Commonwealth of Australia. (Bush 1982, Vol. 2, p. 143)

This announcement fulfilled a plan which had been set out in the report of the Committee on British Policy in the Antarctic which was made to the Imperial Conference in 1926 (Bush 1982, Vol. 2, pp. 100–4). That report identified six areas of coast which were suitable for Australian control. The six areas were Enderby Land (45° to 52°30′ east), Kemp Land (58°30′ to 60° east), Queen Mary Land (86°30′ to 101° east), an unnamed section (131° to 135°30′ east), King George V Land (142° to 153° east), and Oates Land (157° to 159° east). Adelie Land separated the unnamed section and King George V Land, and there were four other wedges separating the nominated areas of which the largest extended for 30° east of Queen Mary Land. The four wedges which separated the nominated coasts were claimed for Australia by expeditions led by Douglas Mawson in 1930 and

1931. When the limiting meridians were announced the eastern boundary coincided with the western limit of the Ross Dependency.

In 1933 the exact extent of Adelie Land had not been announced by France. Bush (1982, Vol. 2, p. 481 *et seq.*) has provided a useful collection of documents about the definition of Adelie Land. On 20 December 1911 Britain enquired whether France claimed any part of Antarctica and was told firmly that a claim to Terre Adelie was maintained. The French reply of 16 April 1912 explained that Captain Dumont d'Urville's voyage and claims on behalf of France were reported in various newspapers, including the *Sydney Morning Herald* of 13 March 1840.

A year later. on 29 March 1913, Britain advised France that it intended naming part of the coast of Antarctica King George V Land. The co-ordinates of the sector were provided and it was noted that, as Britain understood Adelie Land extended from 136° to 147° east, the new coast would not impinge on French claims. No answer was the stern reply!

By 1933 there were three Anglo-Australian versions of the limits of Adelie Land. The 1913 letter specified meridians 136° and 147° east. The report to the Imperial Conference in 1926 left an area between 135°30′ and 142° east. The gap left by Mawson's claims on 5 January 1931 and 13 February 1931 was bounded by meridians 138° and 142° east.

When France declared the limits of Adelie Land in a letter dated 24 October 1933, the British were dismayed to discover that they had selected 136° and 147° east (Bush 1982, Vol. 2, p. 498). These boundaries were rejected by British authorities and they suggested that 136°30′ and 142° east were the proper limits of Adelie Land because they were shown on a chart based on d'Urville's work published in 1840. In a reply dated 5 October 1936 France trumped the French chart with the British letter of 1913. It was pointed out that the chart had been available for more than 60 years, when Sir Francis Bertie wrote to M. Pinchon, and that it was impossible to understand why the charts were now being interpreted in a different way. However, the French offered a compromise which would leave the western boundary at 136° east while the eastern limit would be settled somewhere between 142° and 147° east.

It took the British authorities a year to reply, which was not surprising at that time in European history. On 13 October 1937 Britain rejected the French claim for a boundary east of 142° east and suggested that the maximum extent of Adelie Land was 136° to 142° east. This letter also pointed out that in 1913 Lord Bertie had relied on the *Sydney Morning Herald*'s account of 13 March 1840 for a definition of Adelie Land, and that this newspaper had printed 147° in mistake for 142°. The correct co-ordinates, 136° and 142° east, were contained in the French report published originally in a Hobart journal on 3 March 1840. The French authorities accepted the British suggestion on 5 March 1938 and on 1 April in the same year a decree was published to that effect:

> Les iles et territoires situes au sud du soixantième parallele de latitude sud et entre le cent-trente-sixième et cent-quarante-deuxième meridiens de longitude est de Greenwich relevent de la souveraineté française. (Bush 1982, Vol. 2, p. 505)

> (The islands and territories situated south of parallel 60° south and between meridians 136° and 142° east of Greenwich are within the jurisdiction of French sovereignty.)

These limits have not been changed since that proclamation.

The British explanation of the change in their opinion between 1913 and 1933 is not soundly based. The account in the *Sydney Morning Herald* appeared on 20 March 1840; it was based on a newspaper published in Hobart and it did cite meridian 147° east. The Hobart newspaper was the *Austral-Asiatic Review: Tasmanian and Australian Advertiser* of 3 March 1840. When this journal was consulted it was found to include a reference to 147° east, and so no misprint was involved.

Norway's claim

The first claim by Norway to Antarctic territory was made on 23 January 1928, when a Royal Decree asserted sovereignty over Bouvet Island. Britain's objection that this island had first been found by a Frenchman in 1739 and rediscovered by Captain Norris, a British sailor, in December 1825 was eventually withdrawn and Norway's claim was reiterated in a law relating to Bouvet Island on 27 February 1930:

> The Bouvet Island is placed under Norwegian sovereignty. (United States Naval War College 1950, p. 238)

A year later, on 1 May 1931, a Royal Proclamation announced that Peter I Island had been claimed by Norway:

> We, Haakon, King of Norway, make known: Peter I Island is placed under Norwegian Sovereignty. (United States Naval War College 1950, p. 239)

Although Norwegian explorers, such as Christensen and Mikkleson, discovered about 2080 nautical miles of coast and mapped from the air about 80 000 square kilometres of the continent in the period 1926 to 1937, no formal claim was made to the sector between the British and Australian claims. In 1938 and 1939 German authorities sent the *Schwabenland*, commanded by Alfred Ritscher, to stake a claim to part of the continent in the vicinity of the Greenwich meridian. In a short time planes were used to photograph 350 000 square kilometres of territory during flights totalling 12 000 kilometres. The German flag was dropped from planes every 25 kilometres to support the intended claim. When the Norwegian authorities learned of this activity they quickly organized a formal claim. After receiving a recommendation from the Ministry of Foreign Affairs on 14 January 1939, the King issued a proclamation on the same day:

> That part of the mainland coast in the Antarctic extending from the limits of the Falkland Islands Dependencies in the west (the boundary of Coats Land) to the limits of the Australian Antarctic Dependency in the east (45° E Long.) with the land lying within this coast and the environing sea, shall be brought under Norwegian sovereignty. (United States Naval War College 1950, p. 239)

Norway is the only claimant state in Antarctica which has not claimed an entire sector. Norway has eschewed such a claim in the Antarctic because sector claims by its neighbours in the Arctic Ocean would place Norway at a disadvantage.

A decade earlier it had seemed possible that Norway would claim a larger area than was defined in the 1939 proclamation. In 1929, when Commander Byrd was making the first of his three expeditions to Antarctica, the Norwegian authorities informed the United States of those parts of the continent which were excluded from possible American claims because of Norway's prior discovery:

> . . . the territory immediately circumjacent to the South Pole, which as will be known, was taken possession of in the name of the King of Norway by Captain Roald Amundsen in December 1911, under the name of Haakon VII's Plateau, nor to comprise the territories on both sides of Captain Amundsen's route to the South Pole south of Edward VII's Land and including Queen Maud's Range. (United States Department of State 1929, p. 717)

The Norwegian note to the American representatives observed that though it was not intended to claim sovereignty at that time Norwegian authorities were convinced that all requirements to justify such a claim had been satisfied.

Claims of Chile and Argentina

In 1906 Chile launched a preliminary investigation into the proper extent of the country's claim to Antarctica. In 1939 Professor Julio Escudero, who later played an important rôle in developing the philosophy of the Antarctic Treaty, completed the Chilean investigation, and the limits of Chile's claim were announced on 9 November 1940:

> All lands, islets, reefs of rock, glaciers (pack-ice), already known or to

> be discovered, and their respective territorial waters, in the sector between longitudes 53° and 90° west, constitute the Chilean Antarctic or Chilean Antarctic Territory. (Bush 1982, Vol. 2, p. 311)

The western limit of 90° west corresponds to no Chilean territory. The island called Juan Fernandez lies about 80° west and Easter Island is in the vicinity of 110° west. Possibly 90° west was selected because it was as close as Chile could claim to Peter I Island, which had already been claimed by Norway and which lies 10 nautical miles west of meridian 90° west. The eastern limit of Chile's claim does not correspond with any Chilean possession outside Antarctica, but it lies 25 nautical miles east of Clarence Island, which is the easternmost feature of the South Shetland Islands. This group has been claimed by Chile.

Argentina responded quickly to this Chilean claim on 12 November 1940. Among the points made was the fact that Argentina could have justly issued a declaration of the same class: '. . . did it not think that because such a declaration would be unilateral it would not have improved such rights and titles in any way' (Bush 1982, Vol. 1, p. 609). However, this lofty view was speedily abandoned and Argentina's claim in Antarctica was made clear in a map produced by the Instituto Geografico Militar (Bush 1982, Vol. 1, p. 610). This map showed the limits of Argentina's claim to be meridians 25° and 75° west and parallel 60° south. The Chilean authorities must have been surprised by the western boundary since in 1906 Argentina had asserted that Chile was not entitled to claim any territory in Antarctica east of meridian 67° west, which passes 'through Cape Horn'.

In a note to the British authorities dated 15 February 1943 Argentina announced a new western boundary along meridian 68° 34′ west. This meridian was first defined as a boundary dividing Tierra del Fuego in the agreement between Argentina and Chile on 23 July 1881. This claim was reiterated in a further note to the British representatives on 3 June 1946:

> To this is added its [Argentina's] indisputable right to lands situated south of the 60th parallel between the meridians 25° and 68°34′ west longitude. (United States Naval War College 1950, p. 222)

Five months later a new western boundary was published on a map produced by the Instituto Geografico Militar. This map showed the boundary along meridian 74° west, which lies just west of the mountain Cerro Bertrand, the most westerly point of Argentina. On 28 February 1957 Argentina reestablished the national territory of Tierra del Fuego, Antarctica, and islands of the South Atlantic, and defined the Antarctic sector in the following terms:

> . . . the Argentine Antarctic Sector contained within the meridians 25° west and 74° west and the parallel 60° south. (Bush 1982, Vol. 2, p. 26)

The Argentinian claim has not been changed since this declaration.

Although Chile did not set a northern limit to its claim to territory in Antarctica in the original declaration, its views on that matter have become clear. On 11 June 1961 Chile defined the boundaries of the *departamentos, comunas subdelegaciones*, and districts. In the Departamento de Tierra Magallanes the Comuna-Subdelegaciones de la Antarctica lay south of the Bellingshausen Sea and Drake's Passage. It was divided into two districts. Piloto Pardo is bounded by Drake's Passage on the north, Bransfield Strait on the south, and meridians 53° and 64° west; Tierra de O'Higgins occupies the remainder of the sector. On some Chilean maps the location of the northern boundary of districts through Drake's Passage and the Bellingshausen Sea coincides with parallel 60° south.

International boundaries and the Antarctic Treaty

The Antarctic Treaty was drafted in December 1959 following the successful period of international co-operation during the International Geophysical Year. By the middle of 1961 the 12 countries named in the preamble to the treaty had ratified it.

These states pledged themselves to preserve Antarctica for peaceful purposes, to foster unhindered scientific activity in the continent, and to preserve and conserve the living resources of Antarctica. The provisions of the short treaty apply to the area south of 60° south, including all ice shelves, but the rights of states in respect of the high seas south of 60° south are not affected.

In effect, these 12 states appointed themselves trustees of Antarctica for the international community, and they specifically stated in the treaty their conviction that it furthered the purposes and principles embodied in the Charter of the United Nations. Articles 9 and 13 made arrangements for other countries to accede to the treaty and this has been done by a number of countries, including Brazil and India in 1983.

By any standards the Antarctic Treaty has been an excellent example of a device to avoid international discord and promote international co-operation. If the treaty continues indefinitely into the future, there is no reason why that desirable situation should not continue, and in such a case the claimed international boundaries will continue to exist without significance because the claimant states will not attempt to assert the sovereignty which would normally be attached to territorial claims.

With the conclusion of the Law of the Sea Conference in December 1982 the thoughts of some statesmen turned to another area which might be considered to be part of the common heritage of mankind. Malaysia has led an attack on the nature of the Antarctic Treaty, arguing that it is a cosy club of wealthy, developed countries designed to preserve for those members the wealth that the continent will undoubtedly yield eventually. It is uncertain whether the introduction of the question of Antarctica into the proceedings of the United Nations by Malaysia and other states will eventually result in a new international treaty dealing with the continent. However, some of the difficulties of negotiating a new treaty to replace the Antarctic Treaty can be identified.

First, there will be a debate on whether economic exploitation should be permitted on the continent and in its surrounding seas. Some countries with strong conservation lobbies might oppose any economic use of Antarctica. Some countries with distant fishing fleets might support fishing in the seas around Antarctica yet oppose drilling for oil on the continental shelf. Some poor countries might argue in favour of economic exploitation, providing any profits derived from such activities were distributed in a way to reduce their poverty. Countries which produce oil for the world's market might argue in favour of restrictions on production of petroleum from Antarctica, in the same manner that countries producing nickel were able to restrict production levels of manganese nodules from the deep seabed in the 1982 Convention on the Law of the Sea.

If it was decided that some economic exploitation was appropriate, there would then be debate over whether that exploitation should be undertaken by individual countries or some international organization on behalf of the United Nations.

If national exploitation was permitted, there would be considerable argument about the royalties which should be due to the international authority. Countries with the techniques to exploit the resources of Antarctica would want those payments kept to a minimum, whereas those without the necessary skills would favour charging the highest possible taxes.

There would be a very difficult debate over the composition of any international executive to supervise the economic exploitation of Antarctica and over voting procedures within that body.

It appears to be very unlikely that any new treaty negotiated within the United Nations would provide for existing sovereign claims to be maintained. It is therefore possible that if a new treaty is created some of the countries currently claiming territory in Antarctica will refuse to join, and some of them might assert the sovereignty which has been dormant under the terms of the Antarctic Treaty. This is not the place to examine the political problems that might arise from such an assertion. Instead, attention will be focused only on the technical boundary problems that would attend the enforcement of sovereign rights in Antarctica.

In view of the fact that there is unlikely to be any reason to police the land boundaries of Antarctica at present, even if sovereign rights were asserted, it appears that the technical boundary problems will lie entirely in the sea. The first boundary which any claimant state had to establish would be the baseline from which the territorial sea and the exclusive economic zone were measured. No recent convention has faced the problem of drawing baselines along ice-bound coasts. It could be argued that the normal baseline should be located along that line, furthest from the South Pole, where solid land is known to exist at sea level. That rule would be applied whether the land was exposed or covered with ice. Such a rule could only be applied with difficulty in the latter case and another solution would have to be found.

It would not be difficult to develop a case that in some circumstances a straight baseline should be drawn along the outer edge of some ice features. Indeed, that is an old idea which was recognized by the Committee on British Policy in Antarctica in 1926:

> The Committee have also considered the memorandum on the limit of Territorial Waters off Ice-Bound Coasts, circulated to the Imperial Conference. They recognise the difficulty of measuring the three-mile limit of territorial waters from any base other than the coastline of terra firma, but they think an exception might be made in the case of ice-barriers, which are to all intents and purposes a permanent extension of the land proper. (Bush 1982, Vol. 2, p. 103)

Unfortunately, even the location of ice barriers can vary by some kilometres during a season, especially if they are the source of immense tabular icebergs.

If it were only a question of delimiting territorial seas, this would not be a serious problem, because it is unthinkable that national coasts in Antarctica would be the target for spies, illegal immigrants, or smugglers. But the position of the baseline also determines the location of the exclusive economic zone. To use the marine resources of such a zone it is essential that there should be considerable areas of sea free from ice during the summer. Because the extent and duration of ice-cover vary from year to year it would be a difficult task to select a baseline which would guarantee reasonable fishing areas every year (Prescott 1984, pp. 95–6).

Although there are no unique problems about drawing lateral maritime boundaries between the adjacent coasts of neighbouring states in Antarctica, it might take longer than usual to fix either the general direction of the coast from which a generalized median line was created or the detailed coastline from which a precise line of equidistance was constructed.

Any assertion of sovereign rights by Chile, Argentina, or Britain would require resolution of the dispute over ownership of the Antarctic Peninsula. If the serious problems associated with settlement of the Beagle Channel dispute are any guide, the disagreement over the Antarctic Peninsula would be protracted.

Finally, it is necessary to mention that even if all claimant states abandoned their claims to parts of the continent, which was then placed under international administration, it would still be necessary for those countries that own sub-Antarctic islands to determine maritime boundaries around those islands in the direction of Antarctica.

Britain, Norway, South Africa, France, and Australia own islands on or near the Antarctic Convergence, where cold surface water flowing away from the continent meets warmer sub-Antarctic water flowing southwards. The islands of Norway, South Africa, and France do not share a common continental margin with Antarctica, and claims to an economic zone 200 nautical miles from these islands would not extend south of parallel 60° south. It is only necessary, therefore, to consider the South Sandwich and South Georgia Islands of Britain and the Heard and McDonald Islands of Australia.

Southern Thule in the South Sandwich Group lies only 40 nautical miles north of parallel 60° south, so any claim to a full exclusive economic zone from this island will extend into the area covered by the present Antarctic Treaty. Both the South Sandwich Islands and the South Georgia Islands lie on a sinuous portion of the continental margin linking South America and the Antarctic Peninsula. This means that it would be necessary to fix a seabed boundary somewhere along this Scotia Ridge.

The Heard and McDonald Islands lie on the Kerguelen-Gaussberg Plateau. This aseismic ridge, which is distinguished from the surrounding seabed by morphological and geophysical characteristics, extends for about 700 nautical miles south–south-east to parallel 64° south. The apparent limit of Australia's claim to this continental margin, according to the 1982 Convention, lies in parallel 64°48′ south – very much closer to Antarctica than to Heard Island.

Conclusions

Antarctica is the only continent which contains unclaimed territory; that is, the sector of Ellsworth and Byrd Lands between 90° and 150° west. So long as the Antarctic Treaty continues in its present form, that sector will remain unclaimed and the international boundaries proclaimed by Chile, Argentina, Britain, Norway, Australia, France, and New Zealand will be of no relevance to activities carried out by scientists and administrators on the continent.

In applying the provision of the Antarctic Treaty that relates to the conservation of living resources, the members of the treaty have already created areas defined by specific boundaries. Some of these boundaries delimit sites of special significance, such as breeding grounds for penguins, and others regulate the hunting of seals and fishing. Attempts were also being made in 1986 to fix the limits within which any minerals régime would apply. Although there is no proposal to mine on the continent or the continental shelf, the member states believe that it would be wise to have a minerals régime in place *before* any such proposals are made.

If the Antarctic Treaty collapsed, which is unlikely, or if an attempt was made to replace it with a new treaty which abolished sovereign claims, which has already started, some of the claimant states might decide to assert their sovereignty over sectors of the continent. In such a case the boundary problems which would arise would be entirely concerned with maritime limits, apart from the territorial dispute over the Antarctic Peninsula which might involve Chile, Argentina, and Britain.

References

Bush, W. M. 1982. *Antarctica and international law* (3 volumes). London: Oceana.

Gould, L. M. 1941. The charting of the South Shetlands 1819–28. *Mariner's Mirror* **27**, 206–42.

Great Britain 1948. *Statutory rules and orders*, Vol. 7. London: Statutory Publications Office.

Great Britain 1962. *Statutory instruments*. Part 1, 1 January – 30 April. London: Statutory Publications Office.

Hinks, A. R. 1939. On some misrepresentations of Antarctic history. *Geographical Journal* **94**, 309–30.

Hinks, A. R. 1940. The log of the *Hero*. *Geographical Journal* **96**, 413–30.

Hinks, A. R. 1941. Antarctica rediscovered: A reply. *Geographical Review* **31**, 491–8.

Hobbs, W. H. 1939. The discoveries of Antarctica within the American sector as revealed by maps and documents. *Transactions of the American Philosophical Society* **31**, 7–21.

Hobbs, W. H. 1941. Early maps of Antarctica: true and false. *Papers of the Michigan Academy of Science, Arts and Letters* **26**, 401–5.

International Court of Justice 1956. *Antarctic cases: UK v. Argentina; UK v. Chile*. The Hague: International Court of Justice.

Lovering, J. F. & J. R. V. Prescott 1979. *Last of lands: Antarctica*. Melbourne: Melbourne University Press.

Martin, L. 1938. An American discovered Antarctica. In *Proceedings of the 15th International Geographical Congress 1938, Amsterdam*, Vol. 2, 215–18. Leiden: International Geographical Congress.

Martin, L. 1940. Antarctica discovered by a Connecticut Yankee: Captain Nathaniel Brown Palmer. *Geographical Review* **30**, 529–52.

New Zealand Gazette 1923. 2815 p. 1. Wellington: Government Printer.

O'Connell, D. P. & A. Riordan 1971. *Opinions on imperial constitutional law*. Sydney: Law Book.

Polar Record 1948. Antarctic claims – recent diplomatic exchanges between Great Britain, Argentina and Chile **5**, 228–44.

Prescott, J. R. V. 1984. Boundaries in Antarctica. In *Australia's Antarctic policy options*, S. Harris (ed.), 83–112. Canberra: Centre for Resource and Environment Studies.

United States Department of State 1929. *Foreign relations of the United States*, Vol. 3. Washington, DC: Government Printer.

United States Naval War College 1950. *International law documents 1948–9*. Washington, DC: Government Printer.

Index

References to individual boundary segments will be found under the name of the country which comes first alphabetically. For example, the 'Benin–Nigeria boundary' will not appear under 'Nigeria–Benin boundary'. Modern names have been used in the index so 'Benin–Nigeria boundary' is used even though the reference is to the period when Benin was the French colony of Dahomey. References to figures are shown in italic type, e.g. *10.1*.

For Product Safety Concerns and Information please contact our EU
representative GPSR@taylorandfrancis.com
Taylor & Francis Verlag GmbH, Kaufingerstraße 24, 80331 München, Germany

www.ingramcontent.com/pod-product-compliance
Ingram Content Group UK Ltd.
Pitfield, Milton Keynes, MK11 3LW, UK
UKHW022010180926
475701UK00003B/296

* 9 7 8 1 1 3 8 8 1 4 2 0 2 *